Hilbert-Type Integral Inequalities

Bicheng Yang

Hilbert-Type Integral Inequalities

Contents

Preface

One Hundred years ago, in 1908, H. Wely published the well known Hilbert's inequality. In 1925, G. H. Hardy gave a best extension of it by introducing one pair of conjugate exponents (p, q), named in Hardy-Hilbert's inequality. The Hilbert-type inequalities are a more wide class of analysis inequalities which are with the bilinear kernels, including Hardy-Hilbert's inequality as the particular case. These inequalities are important in analysis and its applications. By making a great effort of mathematicians in the world at about one hundred years, the theory of Hilbert-type inequalities has now come into being. This book is a monograph about the theory of Hilbert-type integral inequalities with the homogeneous kernels of real number-degree and its applications. Using the methods of Real Analysis, Functional Analysis and Operator Theory, and following the way of weight functions and real analysis, the author introduces an independent parameter and two pairs of conjugate exponents to establish a number of Hilbert-type integral inequalities with the homogeneous kernels of real number-degree and the best constant factors, including some multiple integral inequalities and multivariable integral inequalities. The equivalent forms and the reverses with the best constant factors are also considered. As applications, the author also considers some Hilbert-type integral inequalities with the non-homogeneous kernels, the Hardy-type integral inequalities as the particular kernels and a large number of particular examples.

For reading and understanding this book, readers should hold the basic knowledge of real analysis and functional analysis. This book is suited to the people who are interested in the fields of analysis inequalities and real analysis. The author expects this book can help many readers to make good progresses in research for Hilbert-type integral inequalities and their applications.

Bicheng Yang

Guangdong Education Institute

Hilbert-Type Integral Inequalities

1. Introduction

Bicheng Yang

Department of Mathematics, Guangdong Education Institute, Guangzhou, Guangdong 510303, P. R. China; E-mail: bcyang@pub.guangzhou.gd.cn

Abstract: In this chapter, we introduce some evolvements for the theory and methods of Hilbert-type inequalities, including Hilbert's inequality. We must emphasize some excellent works on Hilbert-type inequalities and Hilbert-type operators with multi-parameters in recent years, which have made more developments in this Context. This chapter will enhance the understanding of the readers of the content of the following several chapters.

1.1. HILBERT'S INEQUALITIES AND HILBERT'S OPERATOR

1.1.1. RESEARCH BACKGROUND OF HILBERT'S INEQUALITIES AND HILBERT'S OPERATOR

In 1908, H. Weyl published the following well known Hilbert's inequality (Weyl ID 1908) [1]: If $\{a_n\}$ and $\{b_n\}$ are real sequences, satisfying $0<\sum_{n=1}^{\infty} a_n^2<\infty$ and $0<\sum_{n=1}^{\infty} a_n^2<\infty$, gives

$$\sum_{n=1}^{\infty}\sum_{m=1}^{\infty}\frac{1}{m+n}a_m b_n<\pi\Big(\sum_{n=1}^{\infty}a_n^2\sum_{n=1}^{\infty}b_n^2\Big)^{\frac{1}{2}}, \quad (1.1.1)$$

where the constant factor π is the best possible. We named (1.1.1) as Hilbert's inequality. The best possible property of the constant factor π was proved by Schur in 1911 (Schur JM 1991) [2]. He also gave the following integral analogue of (1.1.1) at the same time: If $f(x)$ and $g(x)$ are measurable functions,

$$0<\int_0^{\infty} f^2(x)dx<\infty$$

and $0<\int_0^{\infty} g^2(x)dx<\infty$, then we get

$$\int_0^{\infty}\int_0^{\infty}\frac{1}{x+y}f(x)g(y)dxdy$$
$$<\pi\Big(\int_0^{\infty} f^2(x)dx\int_0^{\infty} g^2(x)dx\Big)^{\frac{1}{2}}, \quad (1.1.2)$$

where the constant factor π is still the best possible. We call (1.1.2) as Hilbert's integral inequality. Inequalities (1.1.1) and (1.1.2) are important in analysis and applications. We can see a number of improvements and extensions in vast mathematics literature, especially in the books (Hardy CMP 1934) [3], (Mitrinovic KAP 1991) [4], (Kuang SSTP 2004) [5], (Hu WUP 2007) [6].

We may express inequality (1.1.1) by using the form of operator as follows: If l^2 is a space of real sequences, and $T:l^2\to l^2$ is a linear operator for any $a=\{a_m\}_{m=1}^{\infty}\in l^2$, there exists a $c=\{c_n\}_{n=1}^{\infty}\in l^2$, satisfying

$$c_n:=(Ta)(n)=\sum_{m=1}^{\infty}\frac{a_m}{m+n},\ n\in\mathbf{N} \quad (1.1.3)$$

($\mathbf{N}$ is the set of positive integers). Hence, for any $b=\{b_n\}_{n=1}^{\infty}\in l^2$, we may indicate the inner product of Ta and b as:

$$(Ta,b)=(c,b)=\sum_{n=1}^{\infty}\Big(\sum_{m=1}^{\infty}\frac{a_m}{m+n}\Big)b_n$$
$$=\sum_{n=1}^{\infty}\sum_{m=1}^{\infty}\frac{1}{m+n}a_m b_n. \quad (1.1.4)$$

Expressing the norm of a as $\|a\|_2=(\sum_{n=1}^{\infty}a_n^2)^{\frac{1}{2}}$, in view of (1.1.4), inequality (1.1.1) may be rewritten as

$$(Ta,b)<\pi\|a\|_2\|b\|_2, \quad (1.1.5)$$

where $\|a\|_2,\|b\|_2>0$. We may prove that T is a bounded operator and obtain the norm as $\|T\|=\pi$ (Wilhelm AJM 1950) [7]. We call T Hilbert's operator with the kernel $\frac{1}{m+n}$. For $\|a\|_2>0$, the equivalent form of (1.1.5) is given as $\|Ta\|_2<\pi\|a\|_2$, e.t.

$$\sum_{n=1}^{\infty}\Big(\sum_{m=1}^{\infty}\frac{a_m}{m+n}\Big)^2<\pi^2\sum_{n=1}^{\infty}a_n^2, \quad (1.1.6)$$

where the constant factor π^2 is still the best possible. Obviously, inequalities (1.1.6) and (1.1.1) are equivalent (Hardy CUP 1934) [3].

Similarly, if $L^2(0,\infty)$ is a real function space, we may define Hilbert's integral operator as $\widetilde{T}:L^2(0,\infty)\to L^2(0,\infty)$: For any $f\in L^2(0,\infty)$, there exists a $h=\tilde{T}f\in L^2(0,\infty)$, satisfying

$$(\tilde{T}f)(y) = h(y) := \int_0^\infty \frac{f(x)}{x+y}dx, y \in (0,\infty).$$
(1.1.7)

Hence for any $g \in L^2(0,\infty)$, we may still indicate the inner product of $\tilde{T}f$ and g as follows:

$$(\tilde{T}f, g) = \int_0^\infty (\int_0^\infty \frac{1}{x+y} f(x)dx) g(y)dy = \int_0^\infty \int_0^\infty \frac{1}{x+y} f(x)g(y)dxdy. \quad (1.1.8)$$

Setting the norm of f as $\| f \|_2 = (\int_0^\infty f^2(x)dx)^{\frac{1}{2}}$, if $\| f \|_2, \| g \|_2 > 0$, then (1.1.2) may be rewritten as

$$(\tilde{T}f, g) < \pi \| f \|_2 \| g \|_2 . \quad (1.1.9)$$

We have $\| \tilde{T} \| = \pi$ (Carleman U 1923) [8] and inequality $\| \tilde{T}f \|_2 < \pi \| f \|_2$, which may be rewritten to the equivalent form of (1.1.2) as (Hardy CUP 1934) [3]:

$$\int_0^\infty (\int_0^\infty \frac{f(x)}{x+y}dx)^2 dy < \pi^2 \int_0^\infty f^2(x)dx, \quad (1.1.10)$$

where the constant factor π^2 is still the best possible. It is obvious that inequality (1.1.10) is the integral analogue of (1.1.6).

1.1.2. THE MORE ACCURATE HILBERT'S INEQUALITY

If we set the subscripts m, n of the double series from 0 to infinity, then we may rewrite inequality (1.1.1) equivalently in the following form:

$$\sum_{n=0}^\infty \sum_{m=0}^\infty \frac{a_m b_n}{m+n+2} < \pi (\sum_{n=0}^\infty a_n^2 \sum_{n=0}^\infty b_n^2)^{\frac{1}{2}}, \quad (1.1.11)$$

where the constant factor π is still the best possible. Obviously, we may raise the following question: is there a positive constant $\alpha(< 2)$, that makes inequality (1.1.11) still valid as we replace 2 by α in the kernel $\frac{1}{m+n+2}$? The answer is positive. That is the following more accurate Hilbert's inequality (for short, Hilbert's inequality) (Hardy CUP 1934) [3]:

$$\sum_{n=0}^\infty \sum_{m=0}^\infty \frac{a_m b_n}{m+n+1} < \pi (\sum_{n=0}^\infty a_n^2 \sum_{n=0}^\infty b_n^2)^{\frac{1}{2}}, \quad (1.1.12)$$

where the constant factor π is the best possible. Since for $a_m, b_n \geq 0, \alpha \geq 1$, we find

$$\sum_{n=0}^\infty \sum_{m=0}^\infty \frac{a_m b_n}{m+n+\alpha} \leq \sum_{n=0}^\infty \sum_{m=0}^\infty \frac{a_m b_n}{m+n+1},$$

then by (1.1.12) and for $\alpha \geq 1$, we have

$$\sum_{n=0}^\infty \sum_{m=0}^\infty \frac{a_m b_n}{m+n+\alpha} < \pi (\sum_{n=0}^\infty a_n^2 \sum_{n=0}^\infty b_n^2)^{\frac{1}{2}}. \quad (1.1.13)$$

For $1 \leq \alpha < 2$, inequality (1.1.13) is a refinement of (1.1.11), which is equivalently a refinement of (1.1.1). Obviously, we have a refinement of (1.1.6), which is equivalent to (1.1.13) as follows:

$$\sum_{n=0}^\infty (\sum_{m=0}^\infty \frac{a_m}{m+n+\alpha})^2 < \pi^2 \sum_{n=0}^\infty a_n^2 (1 \leq \alpha < 2). \quad (1.1.14)$$

For $0 < \alpha < 1$, in 1936, Ingham gave (Ingham JLMC 1936)[9]: If $\alpha \geq \frac{1}{2}$, then we have

$$\sum_{n=0}^\infty \sum_{m=0}^\infty \frac{a_m a_n}{m+n+\alpha} \leq \pi \sum_{n=0}^\infty a_n^2; \quad (1.1.15)$$

if $0 < \alpha < \frac{1}{2}$, then we get

$$\sum_{n=0}^\infty \sum_{m=0}^\infty \frac{a_m a_n}{m+n+\alpha} \leq \frac{\pi}{\sin(\alpha\pi)} \sum_{n=0}^\infty a_n^2. \quad (1.1.16)$$

It is interesting that if we set $x = X + \frac{\alpha}{2}, y = Y + \frac{\alpha}{2}$, $F(X) = f(X + \frac{\alpha}{2})$ and $G(Y) = g(Y + \frac{\alpha}{2})$ ($\alpha \in \mathbf{R}$, **R** is the set of real numbers) in (1.1.2), then we obtain

$$\int_{-\frac{\alpha}{2}}^\infty \int_{-\frac{\alpha}{2}}^\infty \frac{1}{X+Y+\alpha} F(X)G(Y)dXdY < \pi (\int_{-\frac{\alpha}{2}}^\infty F^2(X)dX \int_{-\frac{\alpha}{2}}^\infty G^2(X)dX)^{\frac{1}{2}}. \quad (1.1.17)$$

It is said that for $\alpha \geq \frac{1}{2}$, inequality (1.1.17) is an integral analogue of (1.1.5) (for $G = F$) and for $0 < \alpha < \frac{1}{2}$, inequality (1.1.17) is not an integral analogue of (1.1.16), since two constant factors are different.

By using the improved Euler-Maclaurin summation formula and introducing one parameter, a few authors gave some more accurate Hilbert-type inequalities as (1.1.13) (see (Yang MIA 2004) [10], (Yang JXNU 2005)[11], (Yang CM 2005)[12], (Yang AMS 2006) [13], (Yang JM 2007)[14], (Yang AMS 2007)[15], (Yang IMF 2007)[16], (Zhong JZU 2008)[17]).

1.1.3. HILBERT'S INEQUALITY WITH ONE PAIR OF CONJUGATE EXPONENTS

In 1925, by introducing one pair of conjugate exponents $(p,q)(\frac{1}{p}+\frac{1}{q}=1)$, Hardy and Riesz gave an extension of (1.1.1) as follows (Hardy PLMS 1925) [18]: If $p > 1$, $a_n, b_n \geq 0$, such that $0 < \sum_{n=1}^\infty a_n^p < \infty$ and $0 < \sum_{n=1}^\infty b_n^q < \infty$, then we obtain

$$\sum_{n=1}^\infty \sum_{m=1}^\infty \frac{a_m b_n}{m+n} < \frac{\pi}{\sin(\frac{\pi}{p})} (\sum_{n=1}^\infty a_n^p)^{\frac{1}{p}} (\sum_{n=1}^\infty b_n^q)^{\frac{1}{q}}, \quad (1.1.18)$$

where the constant factor $\frac{\pi}{\sin(\pi/p)}$ is the best possible. The equivalent form of (1.1.18) is as follows:

$$\sum_{n=1}^{\infty}\Big(\sum_{m=1}^{\infty}\frac{a_m}{m+n}\Big)^p < \Big[\frac{\pi}{\sin(\pi/p)}\Big]^p\sum_{n=1}^{\infty}a_n^p, \quad (1.1.19)$$

where the constant factor $[\frac{\pi}{\sin(\pi/p)}]^p$ is still the best possible. In the same way, inequalities (1.1.12) and (1.1.14) (for $\alpha=1$) may be extended to the following equivalent forms (Hardy CUP 1934) [3]:

$$\sum_{n=0}^{\infty}\sum_{m=0}^{\infty}\frac{a_m b_n}{m+n+1} < \frac{\pi}{\sin(\frac{\pi}{p})}\Big(\sum_{n=0}^{\infty}a_n^p\Big)^{\frac{1}{p}}\Big(\sum_{n=0}^{\infty}b_n^q\Big)^{\frac{1}{q}}, \quad (1.1.20)$$

$$\sum_{n=0}^{\infty}\Big(\sum_{m=0}^{\infty}\frac{a_m}{m+n+1}\Big)^p < \Big[\frac{\pi}{\sin(\pi/p)}\Big]^p\sum_{n=0}^{\infty}a_n^p, \quad (1.1.21)$$

where the constant factors $\frac{\pi}{\sin(\pi/p)}$ and $[\frac{\pi}{\sin(\pi/p)}]^p$ are the best possible. And the equivalent integral analogues of (1.1.18) and (1.1.19) are given as:

$$\int_0^{\infty}\int_0^{\infty}\frac{1}{x+y}f(x)g(y)dxdy$$
$$< \frac{\pi}{\sin(\frac{\pi}{p})}\Big(\int_0^{\infty}f^p(x)dx\Big)^{\frac{1}{p}}\Big(\int_0^{\infty}g^q(x)dx\Big)^{\frac{1}{q}}, \quad (1.1.22)$$

$$\int_0^{\infty}\Big(\int_0^{\infty}\frac{1}{x+y}f(x)dx\Big)^p dy < \Big[\frac{\pi}{\sin(\frac{\pi}{p})}\Big]^p\int_0^{\infty}f^p(x)dx. \quad (1.1.23)$$

We call (1.1.18) and (1.1.20) as Hardy-Hilbert's inequality and call (1.1.22) as Hardy-Hilbert's integral inequality.

Inequality (1.1.20) may be expressed in the form of operator as follows: If l^p is a space of real sequences, $T_p : l^p \to l^p$ is a linear operator, such that for any non-negative sequence $a=\{a_m\}_{m=1}^{\infty}\in l^p$, there exists $T_p a = c = \{c_n\}_{n=1}^{\infty}\in l^p$, satisfying

$$c_n := (T_p a)(n) = \sum_{m=0}^{\infty}\frac{a_m}{m+n+1},\ n\in \mathbf{N}_0 \quad (1.1.24)$$

($\mathbf{N}_0 = \mathbf{N}\bigcup\{0\}$). And for any non-negative sequence $b=\{b_n\}_{n=1}^{\infty}\in l^q$, we can indicate the formal inner product of $T_p a$ and b as:

$$(T_p a, b) = \sum_{n=0}^{\infty}\Big(\sum_{m=0}^{\infty}\frac{a_m}{m+n+1}\Big)b_n$$
$$= \sum_{n=0}^{\infty}\sum_{m=0}^{\infty}\frac{1}{m+n+1}a_m b_n. \quad (1.1.25)$$

Setting the norm of a as $\| a \|_p = (\sum_{n=0}^{\infty}a_n^p)^{\frac{1}{p}}$, then inequality (1.1.20) may be rewritten as

$$(T_p a, b) < \frac{\pi}{\sin(\pi/p)}\| a \|_p \| b \|_q, \quad (1.1.26)$$

where $\| a \|_p, \| b \|_q > 0$. We call T_p Hardy-Hilbert's operator with the kerne l $\frac{1}{m+n+1}$.

Similarly, if $L^p(0,\infty)$ is a real function space, we may define the following Hardy-Hilbert's integral operator $\tilde{T}_p : L^p(0,\infty) \to L^p(0,\infty)$ as: for any $f(\geq 0)$ $\in L^p(0,\infty)$, there exists a $h=\tilde{T}f\in L^p(0,\infty)$, satisfying

$$(\tilde{T}_p f)(y) = h(y) := \int_0^{\infty}\frac{1}{x+y}f(x)dx,$$
$$y\in(0,\infty). \quad (1.1.27)$$

And for any $g(\geq 0)\in L^q(0,\infty)$, we can indicate the formal inner product of $\tilde{T}_p f$ and g as:

$$(\tilde{T}_p f, g) = \int_0^{\infty}\int_0^{\infty}\frac{1}{x+y}f(x)g(y)dxdy. \quad (1.1.28)$$

Setting the norm of f as $\| f \|_p = (\int_0^{\infty}f^p(x)dx)^{\frac{1}{p}}$, then inequality (1.1.22) may be rewritten as

$$(\tilde{T}_p f, g) < \frac{\pi}{\sin(\pi/p)}\| f \|_p \| g \|_q. \quad (1.1.29)$$

If (p,q) is not one pair of conjugate exponents, we get the following result (Hardy CUP 1934) [3]: If $p>1, q>1, \frac{1}{p}+\frac{1}{q}\geq 1, 0<\lambda = 2-\frac{1}{p}+\frac{1}{q}\leq 1$, then

$$\sum_{n=1}^{\infty}\sum_{m=1}^{\infty}\frac{a_m b_n}{(m+n)^{\lambda}} \leq K\Big(\sum_{n=1}^{\infty}a_n^p\Big)^{\frac{1}{p}}\Big(\sum_{n=1}^{\infty}b_n^q\Big)^{\frac{1}{q}}, \quad (1.1.30)$$

where $K=K(p,q)$ relates to p,q; only for $\frac{1}{p}+\frac{1}{q}=1, \lambda = 2-\frac{1}{p}+\frac{1}{q}=1$, the constant factor K is the best possible. The integral analogues of (1.1.30) are given as follows:

$$\int_0^{\infty}\int_0^{\infty}\frac{1}{(x+y)^{\lambda}}f(x)g(y)dxdy$$
$$\leq K\Big(\int_0^{\infty}f^p(x)dx\Big)^{\frac{1}{p}}\Big(\int_0^{\infty}g^q(x)dx\Big)^{\frac{1}{q}}. \quad (1.1.31)$$

We also find an extension of (1.1.31) as (Mitrinovic CAP 1991) [4]: If $p>1, q>1, \frac{1}{p}+\frac{1}{q}>1$, $0<\lambda = 2-\frac{1}{p}+\frac{1}{q}<1$, then we get

$$\int_{-\infty}^{\infty}\int_{-\infty}^{\infty}\frac{f(x)g(y)}{|x+y|^{\lambda}}dxdy \leq k(p,q)$$
$$\times\Big(\int_{-\infty}^{\infty}f^p(x)dx\Big)^{\frac{1}{p}}\Big(\int_{-\infty}^{\infty}g^q(x)dx\Big)^{\frac{1}{q}}. \quad (1.1.32)$$

For $f(x), g(x) = 0, x\in(-\infty, 0]$, inequality (1.1.32) reduces to (1.1.31). Levin also studied the expression forms of the constant factors in (1.1.30) and (1.1.31) (Levin JIMS 1937) [19], but he could not prove their best possible property. In 1951, Bonsall considered the case of (1.1.31) in the general kernel (Bonsall JMOS 1951) [20].

1.1.4. A HILBERT-TYPE INEQUALITY WITH THE HOMOMGENEOUS KERNEL OF -1-DEGREE AND SOME PARTICULAR CASES

If $\alpha \in \mathbf{R}$, the function $k(x,y)$ is measurable in $(0,\infty)\times(0,\infty)$, satisfying for any $x,y,u>0$, $k(ux,uy)=u^{\alpha}k(x,y)$, naming $k(x,y)$ as the homogeneous function of α -degree.

Supposing that $(p,q)(\frac{1}{p}+\frac{1}{q}=1)$ is one pair of conjugate exponents with $p>1$, $k(x,y)\ge 0$ is a homogeneous function of -1 -degree in $(0,\infty)\times(0,\infty)$. If $k=\int_0^\infty k(u,1)u^{-1/p}du$ is finite, then we have $k=\int_0^\infty k(1,v)v^{-\frac{1}{q}}dv$ and the following equivalent inequalities (Hardy CUP 1934) [3]:

$$\int_0^\infty\int_0^\infty k(x,y)f(x)g(y)dxdy \le k(\int_0^\infty f^p(x)dx)^{\frac{1}{p}}(\int_0^\infty g^q(x)dx)^{\frac{1}{q}}, \quad (1.1.33)$$

$$\int_0^\infty(\int_0^\infty k(x,y)f(x)dx)^p dy \le k^p\int_0^\infty f^p(x)dx, \quad (1.1.34)$$

where the constant factors k and k^p are the best possible .

If both $k(u,1)u^{-1/p}$ and $k(1,u)u^{-1/q}$ are decreasing functions in $(0,\infty)$, then we have the following equivalent inequalities:

$$\sum_{n=1}^\infty\sum_{m=1}^\infty k(m,n)a_m b_n \le k(\sum_{n=1}^\infty a_n^p)^{\frac{1}{p}}(\sum_{n=1}^\infty b_n^q)^{\frac{1}{q}}, \quad (1.1.35)$$

$$\sum_{n=1}^\infty(\sum_{m=1}^\infty k(m,n)a_m)^p \le k^p\sum_{n=1}^\infty a_n^p. \quad (1.1.36)$$

For $0<p<1$, if $k=\int_0^\infty k(u,1)u^{-1/p}du$ is finite, then we have the reverses of (1.1.33) and (1.1.34) (Note: we have not seen any proof of (1.1.33) - (1.1.36) and reverse particular cases in (Hardy CUP 1934) [3]).

We name $k(x,y)$ the kernel of inequalities (1.1.33) - (1.1.36). If all the integrals and series in the right hand side of inequalities (1.1.33)-(1.1.36) are positive, we still obtain the following particular cases (Hardy CUP 1934) [3]: (1) For $k(x,y)=\frac{1}{x+y}$ in (1.1.33) - (1.1.36), they deduce to (1.1.22), (1.1.23), (1.1.18) and (1.1.19); (2) for $k(x,y)=\frac{1}{\max\{x,y\}}$ in (1.1.33)-(1.1.36), they deduce to the following two pairs of equivalent forms:

$$\int_0^\infty\int_0^\infty \frac{1}{\max\{x,y\}}f(x)g(y)dxdy < pq(\int_0^\infty f^p(x)dx)^{\frac{1}{p}}(\int_0^\infty g^q(x)dx)^{\frac{1}{q}}, \quad (1.1.37)$$

$$\int_0^\infty(\int_0^\infty \frac{f(x)}{\max\{x,y\}}dx)^p dy < (pq)^p\int_0^\infty f^p(x)dx; \quad (1.1.38)$$

$$\sum_{n=1}^\infty\sum_{m=1}^\infty \frac{a_m b_n}{\max\{m,n\}} < pq(\sum_{n=1}^\infty a_n^p)^{\frac{1}{p}}(\sum_{n=1}^\infty b_n^q)^{\frac{1}{q}}, \quad (1.1.39)$$

$$\sum_{n=1}^\infty(\sum_{m=1}^\infty \frac{a_m}{\max\{m,n\}})^p < (pq)^p\sum_{n=1}^\infty a_n^p; \quad (1.1.40)$$

(3) for $k(x,y)=\frac{\ln(x/y)}{x-y}$ in (1.1.33)-(1.1.36), they deduce to the following two pairs of equivalent forms:

$$\int_0^\infty\int_0^\infty \frac{\ln(x/y)}{x-y}f(x)g(y)dxdy < [\frac{\pi}{\sin(\frac{\pi}{p})}]^2(\int_0^\infty f^p(x)dx)^{\frac{1}{p}}(\int_0^\infty g^q(x)dx)^{\frac{1}{q}}, \quad (1.1.41)$$

$$\int_0^\infty[\int_0^\infty \frac{\ln(x/y)}{x-y}f(x)dx]^p dy < [\frac{\pi}{\sin(\frac{\pi}{p})}]^{2p}\int_0^\infty f^p(x)dx; \quad (1.1.42)$$

$$\sum_{n=1}^\infty\sum_{m=1}^\infty \frac{\ln(m/n)a_m b_n}{m-n} < [\frac{\pi}{\sin(\frac{\pi}{p})}]^2(\sum_{n=1}^\infty a_n^p)^{\frac{1}{p}}(\sum_{n=1}^\infty b_n^q)^{\frac{1}{q}}, \quad (1.1.43)$$

$$\sum_{n=1}^\infty[\sum_{m=1}^\infty \frac{\ln(m/n)a_m}{m-n}]^p < [\frac{\pi}{\sin(\frac{\pi}{p})}]^{2p}\sum_{n=1}^\infty a_n^p. \quad (1.1.44)$$

Note. The constant factors in the above inequalities are all the best possible. We name (1.1.39) and (1.1.43) as Hardy–Littlewood–Polya's inequalities, or for short, H-L-P inequalities, and call (1.1.37) and (1.1.41) H-L-P integral inequalities. We find that the kernels in the above inequalities are all decreasing, but this is not necessary. For example, we find the following two pairs of equivalent forms with the non-decreasing kernel (Yang JGEI 2006) [21], (Yang JJU 2007) [22]:

$$\int_0^\infty\int_0^\infty \frac{|\ln(x/y)|}{\max\{x,y\}}f(x)g(y)dxdy < (p^2+q^2)(\int_0^\infty f^p(x)dx)^{\frac{1}{p}}(\int_0^\infty g^q(x)dx)^{\frac{1}{q}}, \quad (1.1.45)$$

$$\int_0^\infty [\int_0^\infty \frac{|\ln(x/y)|}{\max\{x,y\}} f(x)dx]^p dy$$
$$< (p^2+q^2)^p \int_0^\infty f^p(x)dx; \quad (1.1.46)$$
$$\sum_{n=1}^\infty \sum_{m=1}^\infty \frac{|\ln(m/n)|}{\max\{m,n\}} a_m b_n$$
$$< (p^2+q^2)(\sum_{n=1}^\infty a_n^p)^{\frac{1}{p}}(\sum_{n=1}^\infty b_n^q)^{\frac{1}{q}}, \quad (1.1.47)$$
$$\sum_{n=1}^\infty (\sum_{m=1}^\infty \frac{|\ln(m/n)|}{\max\{m,n\}} a_m)^p < (p^2+q^2)^p \sum_{n=1}^\infty a_n^p, \quad (1.1.48)$$

where the constant factors p^2+q^2 and $(p^2+q^2)^p$ are the best possible. We call (1.1.47) Hilbert-Yang's inequality, or H-Y inequality, and call (1.1.45) as H-Y integral inequality.

1.1.5. TWO MULIPLE INEQUALITIES WITH THE HOMOGENEOUS KERNELS OF $-n+1$-DEGREE

Suppose $n \in \mathbf{N}\setminus\{1\}$, n numbers $p,q,\cdots,r$ satisfying $p,q,\cdots,r>1, \frac{1}{p}+\frac{1}{q}+\cdots+\frac{1}{r}=1$, $k(x,y,\cdots,z)\ge 0$ is a homogeneous function of $-n+1$-degree. If

$$k=\int_0^\infty \int_0^\infty \cdots \int_0^\infty k(1,y,\cdots,z)$$
$$\times y^{-\frac{1}{q}} \cdots z^{-\frac{1}{r}} dy \cdots dz \quad (1.1.49)$$

is a finite number, then we have the following multiple integral inequality (Hardy CUP 1934) [3]:

$$\int_0^\infty \int_0^\infty \cdots \int_0^\infty k(x,y,\cdots,z)$$
$$\times f(x)g(y)\cdots h(z)dxdy\cdots dz$$
$$\le k(\int_0^\infty f^p(x)dx)^{\frac{1}{p}}$$
$$\times(\int_0^\infty g^q(y)dy)^{\frac{1}{q}} \cdots (\int_0^\infty h^r(z)dz)^{\frac{1}{r}}. \quad (1.1.50)$$

If $k(1,y,\cdots,z)x^0 y^{-\frac{1}{q}}\cdots z^{-\frac{1}{r}}$,

$$k(x,1,\cdots,z)x^{-\frac{1}{p}}y^0\cdots z^{-\frac{1}{r}},\cdots,$$

$k(x,y,\cdots,1)x^{-\frac{1}{p}}y^{-\frac{1}{q}}\cdots z^0$ are all deceasing with respect to any single variable, then we have

$$\sum_{s=1}^\infty \cdots \sum_{n=1}^\infty \sum_{m=1}^\infty k(m,n,\cdots,s)a_m b_n \cdots c_s$$
$$\le k(\sum_{m=1}^\infty a_m^p)^{\frac{1}{p}}(\sum_{n=1}^\infty b_n^q)^{\frac{1}{q}} \cdots (\sum_{s=1}^\infty c_s^r)^{\frac{1}{r}}. \quad (1.1.51)$$

For $n=2$, inequalities (1.1.50) and (1.1.51) reduce respectively to (1.1.33) and (1.1.35).

1.2. MODERN RESEARCH FOR HILBERT'S INEQUALITY

1.2.1. MODERN RESEARCH FOR HILBERT'S INTEGRAL INEQUALITY

(1) In 1979, based on an improvement of Hӧlder's inequality, Hu gave a refinement of (1.1.2) as follows (Hu JJTC 1979) [23]:

$$\int_0^\infty \int_0^\infty \frac{1}{x+y} f(x)f(y)dxdy$$
$$< \pi[(\int_0^\infty f^2(x)dx)^2 - \frac{1}{4}(\int_0^\infty f^2(x)\cos\sqrt{x}dx)^2]^{\frac{1}{2}}. \quad (1.2.1)$$

Since then, he published many interesting research results (Hu WUP 2007) [6].

(2) In 1998, B. G. Pachpatte gave an inequality similar to (1.1.2) as (Pachpatte JMAA 1998) [24]:

$$\int_0^a \int_0^b \frac{1}{x+y} f(x)g(y)dxdy < \frac{1}{2}\sqrt{ab}$$
$$\times[\int_0^a (a-x)f'^2(x)dx \int_0^b (b-x)g'^2(x)dx]^{\frac{1}{2}} \quad (1.2.2)$$

($a,b>0$). Some improvements and extensions were made by (Zhao JMAA 2001)[25] (Lu TJM 2003)[26] (He JIPAM)[27]. We can see the other works of Pachpatte in (Pachpatte EBV 2005) [28].

(3) In 1998, by introducing a few independent parameters $\lambda \in (0,1]$ and $0<a<b<\infty$, Yang gave an extension of (1.1.2) as (Yang JMAA 1998) [29]:

$$\int_a^b \int_a^b \frac{f(x)g(y)}{(x+y)^\lambda} dxdy < B(\frac{\lambda}{2},\frac{\lambda}{2})[1-(\frac{a}{b})^{\frac{\lambda}{4}}]$$
$$\times(\int_a^b x^{1-\lambda} f^2(x)dx \int_a^b x^{1-\lambda} g^2(x)dx)^{\frac{1}{2}}, \quad (1.2.3)$$

where $B(u,v)$ is the Beta function. In 1999, Kuang gave another extension of (1.1.2) as (Kuang JMAA 1999)[30]: For $\lambda \in (\frac{1}{2},1]$,

$$\int_0^\infty \int_0^\infty \frac{f(x)g(y)}{x^\lambda + y^\lambda} dxdy < \frac{\pi}{\lambda \sin(\frac{\pi}{2\lambda})}$$
$$\times(\int_0^\infty x^{1-\lambda} f^2(x)dx \int_0^\infty x^{1-\lambda} g^2(x)dx)^{\frac{1}{2}}. \quad (1.2.4)$$

We can see the other works of Kuang in (Kuang SSTP 2004) [5], (Kuang JBUU 2005) [31].

(4) In 1999, by using the methods of algebra and analysis, Gao gave an improvement of (1.1.2) as (Gao JMAA 1999) [32]:

$$\int_0^\infty\int_0^\infty \frac{f(x)g(y)}{x+y}dxdy < \pi\sqrt{1-R}$$
$$\times(\int_0^\infty f^2(x)dx\int_0^\infty g^2(x)dx)^{\frac{1}{2}}, \quad (1.2.5)$$

where $R=\frac{1}{\pi}(\frac{u}{\|g\|}-\frac{v}{\|f\|})^2, u=\sqrt{\frac{2}{\pi}}(g,e)$, $v=\sqrt{2\pi}(f,e^{-x}), e(y)=\int_0^\infty \frac{e^x}{x+y}dx$. We can see the other works of Gao in (Gao JMRE 2005) [33].

(5) In 2002, by using the operator theory, K. Zhang gave an improvement of (1.1.2) as follows (Zhang JMAA 2002) [34]:

$$\int_0^\infty\int_0^\infty \frac{1}{x+y}f(x)g(y)dxdy$$
$$\le \frac{\pi}{\sqrt{2}}[\int_0^\infty f^2(x)dx\int_0^\infty g^2(x)dx$$
$$+(\int_0^\infty f(x)g(x)dx)^2]^{\frac{1}{2}}. \quad (1.2.6)$$

1.2.2. ON THE WAY OF WEIGHT COEFFICIENT FOR GIVING A STREGTHENED VERSION OF HILBERT'S INEQUALITY

In 1991, for giving an improvement of (1.1.1), Hsu raised the way of weight coefficient as follows (Hsu JMRE 1991) [35]: First, using Cauchy's inequality in the left hand side of (1.1.1), it follows

$$\sum_{n=1}^\infty\sum_{m=1}^\infty \frac{a_m b_n}{m+n} = \sum_{n=1}^\infty\sum_{m=1}^\infty [\frac{a_m}{(m+n)^{1/2}}(\frac{m}{n})^{\frac{1}{4}}][\frac{b_n}{(m+n)^{1/2}}(\frac{n}{m})^{\frac{1}{4}}]$$
$$\le \{\sum_{m=1}^\infty[\sum_{n=1}^\infty \frac{1}{m+n}(\frac{m}{n})^{\frac{1}{2}}]a_m^2\sum_{n=1}^\infty[\sum_{m=1}^\infty \frac{1}{m+n}(\frac{n}{m})^{\frac{1}{2}}]b_n^2\}^{\frac{1}{2}}. \quad (1.2.7)$$

We then define the weight coefficient

$$\omega(n):=\sum_{m=1}^\infty \frac{1}{m+n}(\frac{n}{m})^{\frac{1}{2}}, n\in \mathbf{N}, \quad (1.2.8)$$

and rewrite (1.2.7) as the following inequality :

$$\sum_{n=1}^\infty\sum_{m=1}^\infty \frac{a_m b_n}{m+n} \le \{\sum_{n=1}^\infty \omega(n)a_n^2\sum_{n=1}^\infty \omega(n)b_n^2\}^{\frac{1}{2}}. \quad (1.2.9)$$

Afterwards, setting

$$\omega(n)=\pi-\frac{\theta(n)}{n^{1/2}} \ (n\in \mathbf{N}), \quad (1.2.10)$$

where, $\theta(n):=(\pi-\omega(n))n^{1/2}$, and estimating the series of $\theta(n)$, it follows

$$\theta(n)=[\pi-\sum_{m=1}^\infty \frac{1}{m+n}(\frac{n}{m})^{1/2}]n^{1/2}$$
$$>\theta:=1.1213^+. \quad (1.2.11)$$

Hence by (1.2.10), it yields

$$\omega(n)<\pi-\frac{\theta}{n^{1/2}}, n\in \mathbf{N}, \theta=1.1213^+. \quad (1.2.12)$$

And by (1.2.9), a strengthened version of (1.1.1) is given as follows:

$$\sum_{n=1}^\infty\sum_{m=1}^\infty \frac{a_m b_n}{m+n} \le \{\sum_{n=1}^\infty[\pi-\frac{\theta}{n^{1/2}}]a_n^2\sum_{n=1}^\infty[\pi-\frac{\theta}{n^{1/2}}]b_n^2\}^{\frac{1}{2}}. \quad (1.2.13)$$

In this paper, Hsu raised an open problem to obtain the best value of θ in (1.2.13). In 1992, Gao obtained the best value of $\theta_0=1.281669^+$ (Gao HMA 1992) [36].

In the same year, by using the above way, a strengthened version of (1.1.8) was given as follows (Xu CQJM 1991) [37]:

$$\sum_{n=1}^\infty\sum_{m=1}^\infty \frac{a_m b_n}{m+n} < \{\sum_{n=1}^\infty[\frac{\pi}{\sin(\frac{\pi}{p})}-\frac{p-1}{n^{1/p}+n^{-1/q}}]a_n^p\}^{\frac{1}{p}}$$
$$\times\{\sum_{n=1}^\infty[\frac{\pi}{\sin(\frac{\pi}{p})}-\frac{q-1}{n^{1/q}+n^{-1/p}}]b_n^q\}^{\frac{1}{q}}. \quad (1.2.14)$$

In1997, by using the way of weight coefficient and the improved Euler-Maclaurin's summation formula, Yang and Gao gave a new strengthened version of (1.1.18) as (Yang AM 1997) [38]:

$$\sum_{n=1}^\infty\sum_{m=1}^\infty \frac{a_m b_n}{m+n} < \{\sum_{n=1}^\infty[\frac{\pi}{\sin(\frac{\pi}{p})}-\frac{1-\gamma}{n^{1/q}}]b_n^q\}^{\frac{1}{q}}$$
$$\times\{\sum_{n=1}^\infty[\frac{\pi}{\sin(\frac{\pi}{p})}-\frac{1-\gamma}{n^{1/q}}]b_n^q\}^{\frac{1}{q}}, \quad (1.2.15)$$

where $1-\gamma=0.42278433^+$ (γ is Euler constant) . We can see similar works in (Gao PAMS 1998) [39].

In 1998, Yang and Debnath gave another strengthened version of (1.1.18), which is an improvement of (1.2.14) (Yang IJMMS 1998) [40]. We can see some strengthened versions of (1.1.12) and (1.1.20) in (Yang HJ 1997) [41], (Yang AMS 1999) [42], (Yang PJMS 2003) [43].

1.2.3. HILBERT'S INEQUALITY WITH INDEPENDENT PARAMETERS

In 1998, by using the optimized weight coefficient and introducing an independent parameter $0<\lambda\le 1$, Yang gave an extension of (1.1.2) as (Yang JMAA 1998) [29]: If

$$0<\int_0^\infty x^{1-\lambda}f^2(x)dx<\infty$$

and $0<\int_0^\infty x^{1-\lambda}g^2(x)dx<\infty$, then

$$\int_0^\infty\int_0^\infty \frac{1}{(x+y)^\lambda}f(x)g(y)dxdy < B(\frac{\lambda}{2},\frac{\lambda}{2})$$

$$\times(\int_0^\infty x^{1-\lambda} f^2(x)dx \int_0^\infty x^{1-\lambda} g^2(x)dx)^{\frac{1}{2}}, \quad (1.2.16)$$

where the constant factor $B(\frac{\lambda}{2},\frac{\lambda}{2})$ is the best possible. The proof about the best possible property of the constant factor was given by (Yang CQJM 1998) [45], and the expressions of the Beta function $B(u,v)$ are given as follows (Wang SP 1979) [46]:

$$B(u,v) = \int_0^\infty \frac{t^{u-1}}{(1+t)^{u+v}}dt = \int_0^1 (1-t)^{u-1}t^{v-1}dt$$
$$= \int_1^\infty \frac{(t-1)^{u-1}}{t^{u+v}}dt \ (u,v>0). \quad (1.2.17)$$

Some extensions of (1.1.18), (1.1.20) and (1.1.22) were obtained by (Yang CAM 2000) [47], (Yang JMAA 2002) [48], (Yang JMAA 1999)[49] as follows: If $\lambda > 2-\min\{p,q\}$, then

$$\int_0^\infty \int_0^\infty \frac{f(x)g(y)}{(x+y)^\lambda}dxdy < B(\frac{p+\lambda-2}{p},\frac{q+\lambda-2}{q})$$
$$\times(\int_0^\infty x^{1-\lambda} f^p(x)dx)^{\frac{1}{p}}(\int_0^\infty x^{1-\lambda} g^q(x)dx)^{\frac{1}{q}}; \quad (1.2.18)$$

if $2-\min\{p,q\} < \lambda \le 2$, then

$$\sum_{n=1}^\infty \sum_{m=1}^\infty \frac{a_m b_n}{(m+n)^\lambda} < B(\frac{p+\lambda-2}{p},\frac{q+\lambda-2}{q})$$
$$\times(\sum_{n=1}^\infty n^{1-\lambda} a_n^p)^{\frac{1}{p}}(\sum_{n=1}^\infty n^{1-\lambda} b_n^q)^{\frac{1}{q}}, \quad (1.2.19)$$

$$\sum_{n=0}^\infty \sum_{m=0}^\infty \frac{a_m b_n}{(m+n+1)^\lambda} < B(\frac{p+\lambda-2}{p},\frac{q+\lambda-2}{q})$$
$$\times\{\sum_{n=0}^\infty (n+\tfrac{1}{2})^{1-\lambda} a_n^p\}^{\frac{1}{p}}\{\sum_{n=0}^\infty (n+\tfrac{1}{2})^{1-\lambda} b_n^q\}^{\frac{1}{p}}, \quad (1.2.20)$$

where the constant factor $B(\frac{p+\lambda-2}{p},\frac{q+\lambda-2}{q})$ is the best possible (assuming that the right hand side of the above inequalities are all positive numbers). Yang also proved that (1.2.19) is valid for $p=q=2$ and $0<\lambda\le 4$ (Yang JNUMB 2001) [50].

Yang also gave other best extensions of (1.1.18) and (1.1.20) as (Yang CAM 2002)[51] (Yang AM 2006) [52]: If $0<\lambda\le\min\{p,q\}$, then

$$\sum_{n=1}^\infty \sum_{m=1}^\infty \frac{a_m b_n}{m^\lambda+n^\lambda} < \frac{\pi}{\lambda\sin(\pi/p)}$$
$$\times\{\sum_{n=1}^\infty n^{(p-1)(1-\lambda)} a_n^p\}^{\frac{1}{p}}\{\sum_{n=1}^\infty n^{(q-1)(1-\lambda)} b_n^q\}^{\frac{1}{q}}; \quad (1.2.21)$$

if $0<\lambda\le 1$, then

$$\sum_{n=0}^\infty \sum_{m=0}^\infty \frac{a_m b_n}{(m+\frac{1}{2})^\lambda+(n+\frac{1}{2})^\lambda} < \frac{\pi}{\lambda\sin(\frac{\pi}{p})}$$
$$\times\{\sum_{n=0}^\infty (n+\tfrac{1}{2})^{p-1-\lambda} a_n^p\}^{\frac{1}{p}}\{\sum_{n=0}^\infty (n+\tfrac{1}{2})^{q-1-\lambda} b_n^q\}^{\frac{1}{q}}. \quad (1.2.22)$$

In 2004, Yang discovered the following dual form of (1.1.18) (Yang ANH 2004) [53]:

$$\sum_{n=1}^\infty \sum_{m=1}^\infty \frac{a_m b_n}{m+n} < \frac{\pi}{\sin(\frac{\pi}{p})}(\sum_{n=1}^\infty n^{p-2} a_n^p)^{\frac{1}{p}}(\sum_{n=1}^\infty n^{p-2} b_n^q)^{\frac{1}{q}}. \quad (1.2.23)$$

Inequality (1.2.23) is similar to (1.1.18) but different, and for $p=q=2$, both of them reduce to (1.1.1). For $\lambda=1$, inequality (1.2.22) also reduces to the dual form of (1.1.20) as follows:

$$\sum_{n=0}^\infty \sum_{m=0}^\infty \frac{a_m b_n}{m+n+1} < \frac{\pi}{\sin(\frac{\pi}{p})}$$
$$\times\{\sum_{n=0}^\infty (n+\tfrac{1}{2})^{p-2} a_n^p\}^{\frac{1}{p}}\{\sum_{n=0}^\infty (n+\tfrac{1}{2})^{q-2} b_n^q\}^{\frac{1}{q}}. \quad (1.2.24)$$

We can see some best extensions of the H-L-P inequalities such as (1.1.37)-(1.1.48) in (Yang MIA 2003)[54] , (Yang JJU 2004)[55], (Yang CJEM 2004) [56], (Yang JMRE 2005)[57] , (Yang SJM 2005)[58] , (Yang BBMS 2006)[59] , (Wang AJMAA 2006)[60], by introducing some independent parameters.

In 2001, by introducing some independent parameters, Hong gave a multiple integral inequality, which is an extension of (1.2.18) (Hong AMS 2001) [61]. He and Gao gave the similar work for particular conjugate exponents (He JSU 2002) [62]. For making an improvement of their works, Yang gave the following multiple integral inequality, which is a best extension of (1.2.18) (Yang CAM 2003) [63]: If $n\in \mathbf{N}\setminus\{1\}$, $p_i>1$, $\sum_{i=1}^n \frac{1}{p_i}=1$, $\lambda > n-\min_{1\le i\le n}\{p_i\}$, $f_i(t)\ge 0$ and $0<\int_0^\infty t^{n-1-\lambda} f_i^{p_i}(t)dt<\infty$ $(i=1,2,\cdots,n)$, then we have

$$\int_0^\infty \cdots \int_0^\infty \frac{\prod_{i=1}^n f_i(x_i)}{(\sum_{i=1}^n x_i)^\lambda}dx_1\cdots dx_n$$
$$< \frac{1}{\Gamma(\lambda)}\prod_{i=1}^n \Gamma(\frac{p_i+\lambda-n}{p_i})\{\int_0^\infty t^{n-1-\lambda} f_i^{p_i}(t)dt\}^{\frac{1}{p_i}}, \quad (1.2.25)$$

where the constant factor $\frac{1}{\Gamma(\lambda)}\prod_{i=1}^n \Gamma(\frac{p_i+\lambda-n}{p_i})$ is the best possible. In particular, for $\lambda=n-1$, it follows

$$\int_0^\infty \cdots \int_0^\infty \frac{\prod_{i=1}^n f_i(x_i)}{(\sum_{i=1}^n x_i)^{n-1}} dx_1 \cdots dx_n$$

$$< \frac{1}{(n-2)!} \prod_{i=1}^n \Gamma(\frac{p_i - 1}{p_i}) \{\int_0^\infty f_i^{p_i}(t) dt\}^{\frac{1}{p_i}}. \quad (1.2.26)$$

In 2003, Yang and Rassias introduced the way of weight coefficient and considered its applications to Hilbert-type inequalities. They summarized how to use the way of weight coefficient to obtain some new improvements and generalizations of the Hilbert-type inequalities (Yang MIA 2003) [64]. Since then, a number of authors discussed this problem (see (Yang JGEI 2005) [65]-(Chen AMH 2007) [85]). But how to give a best extension of inequalities (1.2.23) and (1.1.18), this was solved in 2004 by introducing two pairs of conjugate exponents.

1.2.4. HILBERT-TYPE INEQUALITIES WITH MULTI-PARAMETERS

In 2004, by introducing an independent parameter $\lambda > 0$ and two pairs of conjugate exponents (p,q) and (r,s) $(\frac{1}{p}+\frac{1}{q}=1, \frac{1}{r}+\frac{1}{s}=1)$, Yang gave an extension of (1.1.2) as (Yang AJMAA 2004) [86]: If $p, r > 1$ and the integrals of the right hand side are positive constants, then we obtain

$$\int_0^\infty \int_0^\infty \frac{1}{x^\lambda + y^\lambda} f(x) g(y) dx dy < \frac{\pi}{\lambda \sin(\frac{\pi}{r})}$$

$$\times \{\int_0^\infty x^{p(1-\frac{\lambda}{r})-1} f^p(x) dx\}^{\frac{1}{p}} \{\int_0^\infty x^{q(1-\frac{\lambda}{s})-1} g^q(x) dx\}^{\frac{1}{q}}, \quad (1.2.27)$$

where the constant factor $\frac{\pi}{\lambda \sin(\pi/r)}$ is the best possible. For $\lambda = 1, r = q, s = p$, inequality (1.2.27) reduces to (1.1.22); for $\lambda = 1, r = p, s = q$, (1.2.27) reduces to the following dual form of (1.1.22):

$$\int_0^\infty \int_0^\infty \frac{1}{x+y} f(x) g(y) dx dy < \frac{\pi}{\sin(\frac{\pi}{p})}$$

$$\times \{\int_0^\infty x^{p-2} f^p(x) dx\}^{\frac{1}{p}} \{\int_0^\infty x^{q-2} g^q(x) dx\}^{\frac{1}{q}}. \quad (1.2.28)$$

In 2005, by introducing an independent parameter $\lambda > 0$ and two pairs of generalized conjugate exponents $(p_1, \cdots, p_n)$, $(r_1, \cdots, r_n)$ $(p_i, r_i > 1$, $\sum_{i=1}^n \frac{1}{p_i} = 1$, $\sum_{i=1}^n \frac{1}{r_i} = 1)$, Yang et al. gave a multiple inequality as follows (Yang MIA 2005)[87]:

$$\int_0^\infty \cdots \int_0^\infty \frac{\prod_{i=1}^n f_i(x_i)}{(\sum_{i=1}^n x_i)^\lambda} dx_1 \cdots dx_n$$

$$< \frac{1}{\Gamma(\lambda)} \prod_{i=1}^n \Gamma(\frac{\lambda}{r_i}) \{\int_0^\infty t^{p_i(1-\frac{\lambda}{r_i})-1} f_i^{p_i}(t) dt\}^{\frac{1}{p_i}}, \quad (1.2.29)$$

where the constant factor $\frac{1}{\Gamma(\lambda)} \prod_{i=1}^n \Gamma(\frac{\lambda}{r_i})$ is the best possible. For $r_i = \frac{p_i \lambda}{p_i - \lambda + n}$ $(i = 1, 2, \cdots, n)$, inequality (1.2.29) reduces to (1.2.25); for $n = 2$, inequality (1.2.29) reduces to the following:

$$\int_0^\infty \int_0^\infty \frac{1}{(x+y)^\lambda} f(x) g(y) dx dy < B(\frac{\lambda}{r}, \frac{\lambda}{s})$$

$$\times \{\int_0^\infty x^{p(1-\frac{\lambda}{r})-1} f^p(x) dx\}^{\frac{1}{p}} \{\int_0^\infty x^{q(1-\frac{\lambda}{s})-1} g^q(x) dx\}^{\frac{1}{q}}. \quad (1.2.30)$$

It is obvious that inequality (1.2.30) is another best extension of (1.1.22).

In 2006, Hong gave multivariable integral inequalities as (Hong JIA 2006) [88]: If

$$R_+^n = \{(x_1, x_2, \cdots, x_n); x_i > 0, i = 1, 2, \cdots, n\},$$

$$\|x\|_\alpha = \{\sum_{i=1}^n x_i^\alpha\}^{\frac{1}{\alpha}}, \alpha, \lambda, \beta > 0, f, g \ge 0,$$

$$0 < \int_{R_+^n} \|x\|_\alpha^{p(n-\frac{\lambda\beta}{r})-n} f^p(x) dx$$

and $0 < \int_{R_+^n} \|x\|_\alpha^{q(n-\frac{\lambda\beta}{s})-n} g^q(x) dx$, then we get

$$\int\int_{R_+^n \times R_+^n} \frac{f(x) g(y)}{(\|x\|_\alpha^\beta + \|y\|_\alpha^\beta)^\lambda} dx dy < \frac{\Gamma^n(\frac{1}{\alpha})}{\beta \alpha^{n-1} \Gamma(\frac{n}{\alpha})} B(\frac{\lambda}{r}, \frac{\lambda}{s})$$

$$\times \{\int_0^\infty \|x\|_\alpha^{p(n-\frac{\lambda\beta}{r})-n} f^p(x) dx\}^{\frac{1}{p}}$$

$$\times \{\int_0^\infty \|x\|_\alpha^{q(n-\frac{\lambda\beta}{s})-n} g^q(x) dx\}^{\frac{1}{q}}, \quad (1.2.31)$$

where the constant factor $\frac{\Gamma^n(1/\alpha)}{\beta \alpha^{n-1} \Gamma(n/\alpha)} B(\frac{\lambda}{r}, \frac{\lambda}{s})$ is the best possible. In particular, for $n = 1$, (1.2.31) reduces to Hong's work in (Hong JIPAM 2005) [89]; for $n = 1$, $\beta = 1$, (1.2.31) reduces to (1.2.30). In 2007, Zhong and Yang generalized Hong's work to the general kernel and also proposed the revision (Zhong JIA 2007) [90].

We can see another Hilbert's inequality with two parameters as follows (Yang JIPAM 2005) [91]:

$$\sum_{n=1}^\infty \sum_{m=1}^\infty \frac{a_m b_n}{(m^\alpha + n^\alpha)^\lambda} < \frac{1}{\alpha} B(\frac{\lambda}{r}, \frac{\lambda}{s})$$

$$\times \{\sum_{n=1}^\infty n^{p(1-\frac{\alpha\lambda}{r})-1} a_n^p\}^{\frac{1}{p}} \{\sum_{n=1}^\infty n^{q(1-\frac{\alpha\lambda}{s})-1} b_n^q\}^{\frac{1}{q}}, \quad (1.2.32)$$

where $\alpha, \lambda > 0, \alpha\lambda \le \min\{r, s\}$. For $\alpha = 1$, it follows $0 < \lambda \le \min\{r, s\}$ and

$$\sum_{n=1}^\infty \sum_{m=1}^\infty \frac{a_m b_n}{(m+n)^\lambda} < B(\frac{\lambda}{r}, \frac{\lambda}{s})$$

$$\times\{\sum_{n=1}^{\infty} n^{p(1-\frac{\lambda}{r})-1} a_n^p\}^{\frac{1}{p}} \{\sum_{n=1}^{\infty} n^{q(1-\frac{\lambda}{s})-1} b_n^q\}^{\frac{1}{q}} . \quad (1.2.33)$$

For $\lambda = 1, r = q$, inequality (1.2.33) reduces to (1.1.18), and for $\lambda = 1, r = p$, (1.2.33) reduces to (1.2.23). Also we can see the reverse form as follows (Yang JSCNU 2005) [92]:

$$\sum_{n=0}^{\infty}\sum_{m=0}^{\infty} \frac{a_m b_n}{(m+n+1)^2}$$
$$> 2\{\sum_{n=0}^{\infty}[1-\frac{1}{4(n+1)^2}]\frac{a_n^p}{2n+1}\}^{\frac{1}{p}}\{\sum_{n=0}^{\infty}\frac{b_n^q}{2n+1}\}^{\frac{1}{q}}, \quad (1.2.34)$$

where $0 < p < 1, \frac{1}{p}+\frac{1}{q}=1$. The other results on the reverse Hilbert-type inequalities are found in (Yang JJU 2004)[93], (Yang PAM 2006)[94], (Yang IMF 2006)[95], (Yang MPT 2006)[96], (Yang IJPAM 2006)[97], (Xi JIA 2007)[98], (Yang AMS 2006)[99].

In 2006, Xin gave a best extension of H-L-P integral inequality (1.1.41) as (Xin JIPAM 2006) [100]:

$$\int_0^\infty\int_0^\infty \frac{\ln(x/y)}{x^\lambda - y^\lambda} f(x)g(y)dxdy < [\frac{\pi}{\lambda\sin(\frac{\pi}{r})}]^2$$
$$\times\{\int_0^\infty x^{p(1-\frac{\lambda}{r})-1} f^p(x)dx\}^{\frac{1}{p}}\{\int_0^\infty x^{q(1-\frac{\lambda}{s})-1} g^q(x)dx\}^{\frac{1}{q}} ; \quad (1.2.35)$$

Zhong et.al gave an extension of another H-L-P integral inequality (1.1.37) as (Zhong JJU 2007)[101]:

$$\int_0^\infty\int_0^\infty \frac{1}{\max\{x^\lambda, y^\lambda\}} f(x)g(y)dxdy < \frac{rs}{\lambda}$$
$$\times\{\int_0^\infty x^{p(1-\frac{\lambda}{r})-1} f^p(x)dx\}^{\frac{1}{p}}\{\int_0^\infty x^{q(1-\frac{\lambda}{s})-1} g^q(x)dx\}^{\frac{1}{q}} ; \quad (1.2.36)$$

Zhong et al. also gave the reverse form of (1.2.36) (Zhong PAM 2008)[102]. For some particular kernel and parameters, Yang gave (Yang JSU 2007) [103]:

$$\sum_{n=1}^{\infty}\sum_{m=1}^{\infty} \frac{a_m b_n}{(\sqrt{m}+\sqrt{n})\sqrt{\max\{m,n\}}}$$
$$< 4\ln 2(\sum_{n=1}^{\infty} n^{\frac{p}{2}-1} a_n^p)^{\frac{1}{p}}(\sum_{n=1}^{\infty} n^{\frac{q}{2}-1} b_n^q)^{\frac{1}{q}} ; \quad (1.2.37)$$

He also gave (Yang JXU 2006) [104]:

$$\sum_{n=1}^{\infty}\sum_{m=1}^{\infty} \frac{a_m b_n}{(m+an)^2+n^2} < (\frac{\pi}{2} - \arctan a)$$
$$\times(\sum_{n=1}^{\infty}\frac{1}{n}a_n^p)^{\frac{1}{p}}(\sum_{n=1}^{\infty}\frac{1}{n}b_n^q)^{\frac{1}{q}} \ (a \geq 0) . \quad (1.2.38)$$

By using the residue theory, Yang obtained (Yang JYU 2008) [105]

$$\int_0^\infty\int_0^\infty \frac{f(x)g(y)}{(x+ay)(x+by)(x+cy)} dxdy$$
$$< k(\int_0^\infty \frac{1}{t^{1+p/2}} f^p(t)dt)^{\frac{1}{p}}(\int_0^\infty \frac{1}{t^{1+q/2}} g^q(t)dt)^{\frac{1}{q}}, \quad (1.2.39)$$

where $k = \frac{\pi}{(\sqrt{a}+\sqrt{b})(\sqrt{b}+\sqrt{c})(\sqrt{a}+\sqrt{c})} (a,b,c>0)$. The constant factors in the above new inequalities are all the best possible. We can see other new works in (Xie JJU 2007) [106], (Xie NSJXU 2007) [107], (Xie JMAA 2007)[108], (Xie SJM 2007)[109], (Li BAMS 2007) [110], (He CMA 2008)[111], (Yang JGEI 2007)[112].

1.3. OPERATOR EXPRESSIONS AND BASIC HILBERT-TYPE INEQUALITIES

1.3.1. MORDERN RESEARCH FOR HILBERT-TYPE OPERATORS

Supposing that H is a separable Hilbert space $T: H \to H$ is a bounded self-adjoined semi -positive definite operator. In 2002, K. Zhang gave the following inequality (Zhang JMAA 2002) [32]:

$$(a,Tb)^2 \leq \frac{\|T\|^2}{2}(\|a\|^2\|b\|^2 + (a,b)^2)(a,b \in H), \quad (1.3.1)$$

where (a,b) is the inner product of a and b, and $\|a\| = \sqrt{(a,a)}$ is the norm of a. Since the Hilbert integral operator $\tilde{T}$ is defined by (1.1.7) satisfying the conditions of (1.3.1), and $\|\tilde{T}\| = \pi$, then inequality (1.1.2) may be improved as (1.2.6). Since the operator T_p defined by (1.1.24) (for $p = 2$) satisfies the conditions of (1.3.1) (Wilhelm AJM 1950) [7], we may improve (1.1.12) to the following form:

$$\sum_{n=0}^{\infty}\sum_{m=0}^{\infty} \frac{a_m b_n}{m+n+1} \leq \frac{\pi}{\sqrt{2}}[(\sum_{n=0}^{\infty} a_n^2 \sum_{n=0}^{\infty} b_n^2) + (\sum_{n=0}^{\infty} a_n b_n)^2]^{\frac{1}{2}} . \quad (1.3.2)$$

The key of applying (1.3.1) is to obtain the norm of the operator and the property of semi-definite. Now, we consider the concept and properties of Hilbert-type integral operator as follows.

Suppose $p > 1, \frac{1}{p}+\frac{1}{q}=1, L^r(0,\infty)$ $(r = p,q)$ are real normal linear spaces, $k(x,y)$ is a non-negative symmetric measurable function in $(0,\infty)\times(0,\infty)$ satisfying $k(x,y) = k(y,x)$ and

$$\int_0^\infty k(x,t)(\frac{x}{t})^{\frac{1}{r}} dt = k_0(p) \in \mathbf{R} (x > 0) . \quad (1.3.3)$$

We define an integral operator as

$$T: L^r(0,\infty) \to L^r(0,\infty) \ (r = p,q)$$

for any $f (\geq 0) \in L^p(0,\infty)$, there exists $h = Tf \in L^p(0,\infty)$, such that

$$(Tf)(y) = h(y) := \int_0^\infty k(x,y)f(x)dx (y > 0) \; ; \quad (1.3.4)$$

or for any $g(\geq 0) \in L^q(0,\infty)$, there exists $\tilde{h} = Tg \in L^q(0,\infty)$, such that

$$(Tg)(x) = \tilde{h}(x) := \int_0^\infty k(x,y)g(y)dy (x > 0). \quad (1.3.5)$$

Then we have the following expression of inner product:

$$(Tf,g) = \int_0^\infty \int_0^\infty k(x,y)f(x)g(y)dxdy = (f,Tg).$$

In 2006, Yang prove that the operator T defined by (1.3.4) or (1.3.5) are bounded and $\| T \| \leq k_0(p)$ (Yang JMAA 2006) [113]. The following are some results in this paper: If $\varepsilon > 0$ is small enough, the integral $\int_0^\infty k(x,t)(\frac{x}{t})^{\frac{1+\varepsilon}{r}} dt$ $(r = p,q; x > 0)$ convergences to a constant $k_\varepsilon(p)$ independent of x, and $k_\varepsilon(p) = k_0(p) + o(1)$ $(\varepsilon \to 0^+)$, then we have $\| T \| = k_0(p)$; if $\| T \| > 0$, and for $f,g \geq 0$, $f \in L^p(0,\infty), g \in L^q(0,\infty)$, $\| f \|_p, \| g \|_q > 0$, we have the following equivalent inequalities:

$$(Tf,g) < \| T \| \cdot \| f \|_p \| g \|_q, \quad (1.3.6)$$

$$\| Tf \|_p < \| T \| \cdot \| f \|_p. \quad (1.3.7)$$

Some particular cases are considered in this paper.

Yang also considered some properties of Hilbert-type integral operator (for $p = q = 2$) (Yang AMS 2007) [114]; when the homogeneous kernel is -1-degree, Yang considered some sufficient conditions to make $\| T \| = k_0(p) > 0$ (Yang TJM 2008) [115]. We can see some properties of the Hilbert-type operator in the disperse space in (Yang JMAA 2007) [116], (Yang BBMS 2006)[117], (Yang JIA 2007)[118], (Yang IJPAM 2008)[119]. A multiple integral operator is scored by (Arpad JIA 2006) [120]. In 2009, Yang summarized the above results in (Yang SP 2009) [121].

1.3.2. SOME BASIC HILBERT-TYPE INEQUALITIES

If the Hilbert-type integral inequality relates to a symmetric homogeneous kernel of -1-degree and the best constant factor, which is more brief and exhibits barbarism form and does not relate to any conjugate exponent (such as (1.1.2)), then we call it basic Hilbert-type integral inequality. Its series analogue (if exists) is called a basic Hilbert-type inequality. If the kernel of basic Hilbert-type (integral) inequality relates to a parameter, and the inequality can not be obtained by a simple transform to a basic Hilbert-type (integral) inequality, then we call it a basic Hilbert-type (integral) inequality with a parameter. For example, we call the following integral inequality (i.e. (1.1.2)):

$$\int_0^\infty \int_0^\infty \frac{1}{x+y} f(x)g(y)dxdy < \pi(\int_0^\infty f^2(x)dx \int_0^\infty g^2(x)dx)^{\frac{1}{2}} \quad (1.3.8)$$

and the following H-L-P integral inequalities (for $p = q = 2$ in (1.1.37) and (1.1.41))

$$\int_0^\infty \int_0^\infty \frac{1}{\max\{x,y\}} f(x)g(y)dxdy < 4(\int_0^\infty f^2(x)dx \int_0^\infty g^2(x)dx)^{\frac{1}{2}}, \quad (1.3.9)$$

$$\int_0^\infty \int_0^\infty \frac{\ln(x/y)}{x-y} f(x)g(y)dxdy < \pi^2(\int_0^\infty f^2(x)dx \int_0^\infty g^2(x)dx)^{\frac{1}{2}} \quad (1.3.10)$$

basic Hilbert-type integral inequalities.

In 2006, Yang gave the following H-Y integral inequality (Yang JGEI 2006) [21]:

$$\int_0^\infty \int_0^\infty \frac{|\ln(x/y)|}{\max\{x,y\}} f(x)g(y)dxdy < 8(\int_0^\infty f^2(x)dx \int_0^\infty g^2(x)dx)^{\frac{1}{2}}; \quad (1.3.11)$$

in 2008, Yang also gave the following H-Y integral inequalities (Yang CM 2008) [122], (Yang JGEI 2008) [123]:

$$\int_0^\infty \int_0^\infty \frac{|\ln(x/y)|}{x+y} f(x)g(y)dxdy < c_0(\int_0^\infty f^2(x)dx \int_0^\infty g^2(x)dx)^{\frac{1}{2}}, \quad (1.3.12)$$

$$\int_0^\infty \int_0^\infty \frac{\arctan\sqrt{x/y}}{x+y} f(x)g(y)dxdy < \frac{\pi^2}{4}(\int_0^\infty f^2(x)dx \int_0^\infty g^2(x)dx)^{\frac{1}{2}}, \quad (1.3.13)$$

where the constant factors $c_0 = 8\sum_{n=1}^{\infty} \frac{(-1)^n}{(2n-1)^2} = 7.3277^+$ and $\frac{\pi^2}{4}$ are the best possible. We still call (1.3.11)-(1.3.13) basic Hilbert-type integral inequalities.

In 2005, Yang gave the following H-Y integral inequality with a parameter (Yang JJU 2005) [124], (Yang JHU 2005) [125]:

$$\int_0^\infty\int_0^\infty \frac{1}{|x-y|^\lambda} f(x)g(y)dxdy < 2B(\tfrac{\lambda}{2},1-\lambda) \times(\int_0^\infty x^{1-\lambda} f^2(x)dx\int_0^\infty x^{1-\lambda} g^2(x)dx)^{\frac{1}{2}}, \quad (1.3.14)$$

where the constant factor $2B(\frac{\lambda}{2},1-\lambda)(0<\lambda<1)$ is the best possible. As in (1.2.16), i.e.

$$\int_0^\infty\int_0^\infty \frac{1}{(x+y)^\lambda} f(x)g(y)dxdy < B(\tfrac{\lambda}{2},\tfrac{\lambda}{2}) \times(\int_0^\infty x^{1-\lambda} f^2(x)dx\int_0^\infty x^{1-\lambda} g^2(x)dx)^{\frac{1}{2}} \quad (1.3.15)$$

$(\lambda>0)$, we call (1.3.14) and (1.3.15) basic Hilbert-type integral inequalities with a parameter.

It is noticed that the following inequality (for $p=r=2$ in (1.2.27))

$$\int_0^\infty\int_0^\infty \frac{1}{x^\lambda+y^\lambda} f(x)g(y)dxdy < \frac{\pi}{\lambda} \times(\int_0^\infty x^{1-\lambda} f^2(x)dx\int_0^\infty x^{1-\lambda} g^2(x)dx)^{\frac{1}{2}} \ (\lambda>0), \quad (1.3.16)$$

is not a basic Hilbert-type integral inequality with a parameter. By setting $x=X^\lambda, y=Y^\lambda$ in (1.3.8), we may get (1.3.16). Similarly, neither (1.2.35) nor (1.2.36) (for $p=r=2$) are basic Hilbert-type integral inequalities with a parameter.

Also we find the following basic Hilbert-type inequalities:

$$\sum_{n=1}^\infty\sum_{m=1}^\infty \frac{a_m b_n}{m+n} < \pi(\sum_{n=1}^\infty a_n^2 \sum_{n=1}^\infty b_n^2)^{\frac{1}{2}}, \quad (1.3.17)$$

$$\sum_{n=1}^\infty\sum_{m=1}^\infty \frac{a_m b_n}{\max\{m,n\}} < 4(\sum_{n=1}^\infty a_n^2 \sum_{n=1}^\infty b_n^2)^{\frac{1}{2}}, \quad (1.3.18)$$

$$\sum_{n=1}^\infty\sum_{m=1}^\infty \frac{\ln(m/n)}{m-n} a_m b_n < \pi^2(\sum_{n=1}^\infty a_n^2 \sum_{n=1}^\infty b_n^2)^{\frac{1}{2}}, \quad (1.3.19)$$

$$\sum_{n=1}^\infty\sum_{m=1}^\infty \frac{|\ln(m/n)|a_m b_n}{\max\{m,n\}} < 8(\sum_{n=1}^\infty a_n^2 \sum_{n=1}^\infty b_n^2)^{\frac{1}{2}}, \quad (1.3.20)$$

$$\sum_{n=1}^\infty\sum_{m=1}^\infty \frac{a_m b_n}{(m+n)^\lambda} < B(\tfrac{\lambda}{2},\tfrac{\lambda}{2}) \times(\sum_{n=1}^\infty n^{1-\lambda} a_n^2 \sum_{n=1}^\infty n^{1-\lambda} b_n^2)^{\frac{1}{2}} (0<\lambda\le 4). \quad (1.3.21)$$

Among them, inequality (1.3.21) is called Hilbert-type inequality with a parameter.

By simple operation of the kernels in basic Hilbert-type inequalities, we may get some new Hilbert-type inequalities. For example, we found (Li IJMMS 2006) [126], (Xie JJU 2007) [127]

$$\int_0^\infty\int_0^\infty \frac{f(x)g(y)}{x+y+\max\{x,y\}} dxdy < c(\int_0^\infty f^2(x)dx\int_0^\infty g^2(x)dx)^{\frac{1}{2}}, \quad (1.3.22)$$

where the constant factor $c=\sqrt{2}(\pi-\arctan\sqrt{2})$ is the best possible. Still we found the following inequality (Yang JMAA 2006) [113]:

$$\int_0^\infty\int_0^\infty \frac{|x-y|}{(\max\{x,y\})^2} f(x)g(y)dxdy < \frac{8}{3}(\int_0^\infty f^2(x)dx\int_0^\infty g^2(x)dx)^{\frac{1}{2}}, \quad (1.3.23)$$

where the constant factor $\frac{8}{3}$ is the best possible. We can obtain some new Hilbert-type inequalities in the same way (Li IJMMS 2007) [128], (Li JIA 2007) [129], (Xie KMJ 2008) [130].

1.4. REFERENCES

1. Weyl H. Singulare integral gleichungen mit besonderer berucksichtigung des fourierschen integral theorems. Gottingen : Inaugeral-Dissertation, 1908.
2. Schur I. Bernerkungen sur Theorie der beschrankten Bilinearformen mit unendlich vielen veranderlichen. Journal of Math., 1911; 140: 1-28.
3. Hardy GH, Littlewood JE, Polya G. Inequalities. Cambridge : Cambridge University Press, 1934.
4. Mitrinovic J E, Pecaric J E, Fink A M. Inequalities involving functions and their integrals and derivatives. Boston: Kluwer Acaremic Publishers, 1991.
5. Kuang JC. Applied inequalities. Jinan: Shandong Science Technic Press, 2004.
6. Hu K. Some problems in analysis inequalities. Wuhan: Wuhan University Press, 2007.
7. Wilhelm M. On the spectrum of Hilbert's matrix. Amer J. Math., 1950; 72: 699-704.
8. Carleman T. Sur les equations integrals singulieres a noyau reel et symetrique. Uppsala, 1923.
9. Ingham AE. A note on Hilbert's inequality. J. London Math. Soc., 1936; 11: 237-240.
10. Yang BC. On a new Hardy-Hilbert's type inequality. Math. Inequalities & Applications, 2004; 7(3): 355-363.
11. Yang BC. A more accurate Hardy-Hilbert's type inequality. Journal of Xinyang Normal University, 2005; 18(2): 140-142.
12. Yang BC. A more accurate Hilbert-type inequality. College Mathematics, 2005; 21(5): 99-102.
13. Yang BC. On a more accurate Hardy-Hilbert's type inequality and its applications, Acta Mathematica Sinica, 2006; 49(3): 363-368.
14. Yang BC. A more accurate Hilbert type inequality. Journal of Mathematics, 2007; 27(6): 673-678.
15. Yang BC. On an extension of Hardy-Hilbert's type inequality and a reverse. Acta Mathematica Sinica, Chinese Series, 2007; 50(4): 861-868.
16. Yang BC. On a more accurate Hilbert's type inequality. International Mathematical Forum, 2007; 2(37): 1831~1837.
17. Zhong JH, Yang BC. On an extension of a more accurate Hilbert-type inequality. Journal of Zhejiang University (Science Edition), 2008; 35(2): 121-124.
18. Hardy GH. Note on a theorem of Hilbert concerning series of positive term. Proceedings of the London Mathematical Society, 1925; 23: 45-46.
19. Levin V. Two remarks on Hilbert's double series theorem. J. Indian Math. Soc., 1937; 11: 111~115.

20. Bonsall FF. Inequalities with non-conjugate parameter. J. Math. Oxford Ser. 1951; 2(2): 135-150.
21. Yang BC. On a basic Hilbert-type inequality. Journal of Guangdong Education Institute (Natural Science), 2006; 26(3): 1-5.
22. Yang BC. A Hilbert-type inequality with two pairs of conjugate exponents. Journal of Jilin University (Science Edition), 2007; 45(4): 524-528.
23. Hu K. A few important inequalities. Journal of Jianxi Teacher's College (Natural Science), 1979; 3(1): 1-4.
24. Pachpatte BG. On some new inequalities similar to Hilbert's inequality. J. Math. Anal. Appl., 1998; 226: 166-179.
25. Zhao CJ, Debnath L. Some new type Hilbert integral inequalities. J. Math. Anal. Appl., 2001; 262: 411~418.
26. Lu ZX. Some new inverse type Hilbert-Pachpatte inequalities. Tamkang Journal of Mathematics, 2003; 34(2): 155-161.
27. He B, Li YJ. On several new inequalities close to Hilbert-Pachpatte's inequality. J. Ineq. in Pure and Applied Math., 2006; 7(4), Art.154: 1-9.
28. Pachpatte BG. Mathematical inequalities. Netherland: Elsevier B. V., 2005.
29. Yang BC. On Hilbert's integral inequality. J. Math. Anal. Appl., 1998; 220: 778-785.
30. Kuang JC. On new extension of Hilbert's integral inequality. J. Math. Anal. Appl., 1999; 235: 608-614.
31. Kuang JC. New progress in inequality study in China. Journal of Beijing Union University (Natural Science), 2005; 19(1): 29- 37.
32. Gao MZ. On the Hilbert inequality . J. for Anal. Appl., 1999; 18 (4): 1117-1122.
33. Gao MZ, Hsu LC. A survey of various refinements and generalizations of Hilbert's inequalities. J. Math. Res. Exp., 2005; 25(2): 227-243.
34. Zhang KW . A bilinear inequality. J. Math. Anal. Appl., 2002; 271: 288~296.
35. Hsu LC, Wang YJ. A refinement of Hilbert's double series theorem. J. Math. Res. Exp., 1991; 11(1): 143~144.
36. Gao MZ. A note on Hilbert double series theorem. Hunan Mathematical Annal, 1992; 12(1-2): 143-147.
37. Xu LZ, Guo YK. Note on Hardy-Riesz's extension of Hilbert's inequality. Chin. Quart. J. Math., 1991; 6(1): 75-77.
38. Yang BC, Gao MZ. On a best value of Hardy-Hilbert's inequality. Advances in Math., 1997; 26(2): 159-164.
39. Gao MZ, Yang BC. On the extended Hilbert's inequality. Proc. Amer. Math. Soc., 1998; 126(3): 751-759.
40. Yang BC, Debnath L.On new strengthened Hardy-Hilbert's inequality. Internat. J. Math. & Math. Soc., 1998; 21(2): 403-408.
41. Yang BC. A refinement of Hilbert's inequality. Huanghuai Journal, 1997; 13(2): 47-51.
42. Yang BC. On a strengthened version of the more accurate Hardy-Hilbert's inequality. Acta Mathematica Sineca, 1999; 42(6): 1103-1110.
43. Yang BC, Debnath L. A strengthened Hardy-Hilbert's inequality. Proceedings of the Jangjeon Mathematical Society, 2003; 6(2): 119-124.
44. Yang BC. On generalization of Hardy-Hilbert's integral inequalities, Acta Math. Sineca, 1998; 41(4): 839-844.
45. Yang BC. A note on Hilbert's integral inequalities. Chinese Quarterly J. Math., 1998; 13(4): 83-86.
46. Wang ZQ, Guo DR. Introduction to special functions. Beijing: Science Press, 1979.
47. Yang BC. A general Hardy-Hilbert's integral inequality with a best value. Chinese Annal of Mathematics, 2000; 21A(4): 401-408.
48. Yang BC, Debnath L. On the extended Hardy-Hilbert's inequality. J. Math. Anal. Appl., 2002; 272: 187-199.
49. Yang BC, Debnath L. On a new generalization of Hardy-Hilbert's inequality. J. Math. Anal. Appl., 1999; 233: 484-497.
50. Yang BC. On a genealization of Hilbert's double series theorem. Journal of Nanjing University-Mathematical Biquarterly, 2001; 18(1): 145-151.
51. Yang BC. On a general Hardy-Hilbert's inequality. Chinese Annal of Mathematics, 2002; 23A(2): 247-254.
52. Yang BC. A dual Hardy-Hilbert's inequality and generalizations. Advances in Math., 2006; 35(1): 102-108.
53. Yang BC. On new extensions of Hilbert's inequality. Acta Math. Hungar., 2004; 104(4): 291-299.
54. Yang BC. On a new inequality similar to Hardy-Hilbert's inequality. Math. Ineq. Appl., 2003; 6(1): 37-44.
55. Yang BC. Best generalization of Hilbert's type of inequality. Journal of Jilin Univ. (Science Edition), 2004; 42(1): 30-34.
56. Yang BC. On a generalization of the Hilbert's type inequality and its applications. Chinese Journal of Engineering Mathematics, 2004; 21(5): 821-824.
57. Yang BC. Generalization of the Hilbert's type inequality with best constant factor and its applications. J. Math. Res. Exp., 2005; 25(2): 341-346.
58. Yang BC. On Mulholand's integral inequality. Soochow Journal of Mathematics, 2005; 31(4): 573-580.
59. Yang BC. A new Hilbert-type inequality. Bull. Belg. Math. Soc., 2006; 13: 479-487.
60. Wang WH, Yang BC. A strengthened Hardy-Hilbert's type inequality. The Australian Journal of Mathematical Analysis and Applications, 2006; 3(2), Art.17: 1-7.
61. Hong Y. All-side generalization about Hardy-Hilbert integral inequalities, Acta Mathematica Sinica, 2001; 44(4): 619-626.
62. He LP, Yu JM, Gao MZ. An extension of Hilbert's integral inequality. Journal of Shaoguan University (Natural Science) , 2002; 23(3): 25-30.
63. Yang BC. On a multiple Hardy-Hilbert's integral inequality. Chinese Annal of Mathematics, 2003, ; 24A(6): 743-750.
64. Yang BC, Rassias TM. On the way of weight coefficient and research for Hilber-type inequalities. Math. Ineq. Appl., 2003; 6(4): 625-658.
65. Yang BC. On the way of weight function and research for Hilbert's type integral inequalities. Journal of Guangdong Education Institute (Natural Science), 2005; 25(3): 1-6.
66. Sulaiman WT. On Hardy-Hilbert's integral inequality. J. Ineq. in Pure & Appl. Math., 2004; 5(2), Art.25: 1-9.
67. Brnetic I, Pecaric J. Generalization of Hilbert's integral inequality. Math. Ineq. & Appl., 2004; 7(2): 199-205.
68. Krnic M, Gao MZ, Pecaric J, Gao X M. On the best constant in Hilbert's inequality. Math. Ineq. & Appl., 2005; 8(2): 317-329.
69. Brnet I, Krnic M, Pecaric J. Multiple Hilbert and Hardy-Hilbert inequalities with non-conjugate parameters . Bull. Austral. Math. Soc., 2005; 71: 447-457.
70. Krnic M, Pecaric J. General Hilbert's and Hardy's inequalities. Math. Ineq. & Appl., 2005; 8(1): 29-51.
71. Sulaiman WT. New ideas on Hardy-Hilbert's integral inequality (I). Pan American Math. J., 2005; 15(2): 95-100.
72. Salem SR. Some new Hilbert type inequalities. Kyungpook Math. J., 2006; 46: 19-29.
73. Laith EA. On some extensions of Hardy-Hilbert's inequality and applications. Journal of Inequalities and Applications, volume 2008; Article ID 546828, 14 pages. Doi: 10.1155/ 2008/546829.
74. Jia WJ, Gao MZ, Debnath L. Some new improvement of the Hardy -Hilbert inequality with applications. International Journal of Pure and Applied Math., 2004; 11(1): 21-28.
75. Lu ZX. On new generalizations of Hilbert's inequalities. Tamkang Journal of Mathematics, 2004; 35(1): 77-86.
76. Xie H, Lu Z. Discrete Hardy-Hilbert's inequalities in $\mathbb{R}^n$. Northeast. Math., 2005; 21(1): 87-94.
77. Gao MZ. A new Hardy-Hilbert's type inequality for double series and its applications. The Australian Journal of Mathematical Analysis and Appl., 2005; 3(1), Art.13: 1-10.
78. He LP, Gao MZ, Jia WJ. On a new strengthened Hardy-Hilbert's inequality. J. Math. Res. Exp., 2006; 26(2): 276-282.

79. He LP, Jia WJ, Gao MZ. A Hardy-Hilbert's type inequality with gamma function and its applications. Integral Transforms and Special functions, 2006; 17(5): 355-363.
80. Jia WJ, Gao MZ, Gao XM. On an extension of the Hardy-Hilbert theorem. Studia Scientiarum Mathematicarum Hungarica, 2005; 42(1): 21-35.
81. Gao MZ, Jia WJ, Gao XM, On an improvement of Hardy-Hilbert's inequality. J. Math., 2006; 26(6): 647-651
82. Sun BJ. Best generalization of a Hilbert type inequality. J. Ineq. in Pure & Applied Math., 2006; 7(3), Art.113: 1-7.
83. Wang WH, Xin DM. On a new strengthened version of a Hardy-Hilbert type inequality and applications. J. Ineq. in Pure & Applied Math., 2006; 7(5), Art.180: 1-7.
84. Xu JS. Hardy-Hilbert's inequalities with two parameters. Advances in Mathematics, 2007; 36(2): 189-198.
85. Chen ZQ, Xu JS. New extensions of Hilbert's inequality with multiple parameters. Acta Math. Hungar., 2007; 117(4): 383-400.
86. Yang BC. On an extension of Hilbert's integral inequality with some parameters. The Australian Journal of Math. Analysis and Applications, 2004; 1(1), Art.11: 1-8.
87. Yang BC, Brnetc I, Krnic M, Pecaric J. Generalization of Hilbert and Hardy-Hilbert integral inequalities. Math. Ineq. & Appl., 2005; 8(2): 259-272.
88. Hong Y. On multiple Hardy-Hilbert integral inequalities with some parameters. Journal of Inequalities and Applications, Vol. 2006; Art.ID94960: 1-11.
89. Hong Y. On Hardy-Hilbert integral inequalities with some parameters. J. Ineq. in Pure & Applied Math., 2005; 6(4), Art. 92: 1-10.
90. Zhong WY, Yang BC. On Multiple's Hardy-Hilbert integral inequality with kernel. Journal of Inequalities and Applications, Vol.2007; Art.ID 27962, 17 pages, doi: 10.1155/ 2007/27962.
91. Yang BC. On best extensions of Hardy-Hilbert's inequality with two parameters. J. Ineq. in Pure & Applied Math., 2005; 6(3), Art. 81: 1-15.
92. Yang BC. A reverse of the Hardy-Hilbert's type inequality. Journal of Southwest China Normal University (Natural Science), 2005; 30(6): 1012-1015.
93. Yang BC. A reverse Hardy-Hilbert's integral inequality. Journal of Jilin University, 2004; 42(4): 489-493.
94. Yang BC. On a reverse of Hardy-Hilbert's integral inequality. Pure and Appl.Math., 2006; 22(3): 312-317.
95. Yang BC. On an extended Hardy-Hilbert's inequality and some reversed form. International Mathematical Forum, 2006; 1(39): 1905~1912.
96. Yang BC. A reverse of the Hardy-Hilbert's inequality. Math. in Practice and Theory, 2006; 36(11): 207-212.
97. Yang BC. On a reverse of a Hardy-Hilbert type inequality. J. Ineq. in Pure & Applied Math., 2006; 7(3), Art.115: 1-7.
98. Xi GW. A reverse Hardy-Hilbert-type inequality. Journal of Inequalities and Appl., Vol.2007; Art.ID79758: 1-7.
99. Yang BC. On a relation to Hardy-Hilbert's inequality and Mulholland's inequality. Acta Mathematica Sinica, 2006; 49 (3): 559-566.
100. Xin DM. Best generalization of Hardy-Hilbert's inequality with multi-parameters. J. Ineq. in Pure and Applied Math., 2006; 7(4), Art.153: 1-8.
101. Zhong WI, Yang BC. A best extension of Hilbert inequality involving several parameters. Journal of Jinan University (Natural Science), 2007; 28(1): 20-23.
102. Zhong WI, Yang BC. A reverse Hilbert's type integral inequality with some parameters and the equivalent forms. Pure and Applied Mathematics, 2008; 24(2): 401-407.
103. Yang BC. A new Hilbert-type inequality. Journal of Shanghai Univ. (Natural Science), 2007; 13(3): 274-278.
104. Yang BC. A bilinear inequality with the kernel of -2-order homogeneors. Journal of Xiamen University (Natural Science), 2006; 45(6): 752-755.
105. Yang BC. A Hilbert-type integral inequality with the kernel of -3-order homogeneous. Journal of Yunnam University, 2008; 30(4): 325-330.
106. Xie ZT. A new Hilbert-type inequality with the kernel of 3 μ -homogeneous. Journal of Jilin University (Science Edition), 2007; 45(3): 369-373.
107. Xie ZT, Zheng Z. A Hilbert-type inequality with parameters. J. Xiangtan Univ. (Natural Science), 2007;29(3): 24-28.
108. Xie ZT, Zheng Z. A Hilbert-type integral inequality whose kernel is a homogeneous form of degree -3 . J. Math. Anal. Appl., 2007; 339: 324-331.
109. Xie ZT, Zheng Z. A new Hilbert-type integral inequality and Its reverse. Soochow Journal of Math., 2007; 33(4): 751-759.
110. Li YJ, He B. On inequalities of Hilbert's type. Bull. Austral. Math. Soc., 2007; 76: 1-13.
111. He B, Qian Y, Li YJ. On analogues of the Hilbert's inequality. Comm. in Math. Anal., 2008; 4(2): 47-53.
112. Yang BC. On a Hilbert-type inequality with the homogeneous kernel of -3-order. Journal of Guangdong Education Institute (Natural Science), 2007; 27(5): 1-5.
113. Yang BC. On the norm of an integral operator and applications. J. Math. Anal. Appl., 2006; 321: 182-192.
114. Yang BC. On the norm of a self-adjoint operator and a new bilinear integral inequality. Acta Mathematica Sinica, English Series, 2007; 23(7): 1311-1316.
115. Yang BC. On the norm of a certain self-adjoint integral operator and applications to bilinear integral inequalities. Taiwan Journal of Mathematics, 2008; 12(2): 315-324.
116. Yang BC. On the norm of a Hilbert's type linear operator and applications. J. Math. Anal. Appl., 2007; 325: 529-541.
117. Yang BC. On the norm of a self-adjoint operator and applications to Hilbert's type inequalities. Bull. Belg. Math. Soc., 2006; 13: 577-584.
118. Yang BC. On a Hilbert-type operator with a symmetric homogeneous kernel of -1-order and applications. Journal of Inequalities and Applications, Volume 2007; Article ID 47812, 9 pages, doi: 10.1155/2007/47812.
119. Yang BC. On the norm of a linear operator and its applications. Indian Journal of Pure and Applied Mathematics, 2008; 39(3): 237-250.
120. Arpad B, Choonghong O. Best constants for certain multilinear integral operator. Journal of Inequalities and Applications, Vol.2006; Art. ID28582: 1-12.
121. Yang BC. On the norm of operator and Hilbert-type inequalities. Science Press, 2009.
122. Yang BC. On a basic Hilbert-type integral inequality and extensions. College Mathematics, 2008; 24(1): 87-92.
123. Yang BC. A basic Hilbert-type integral inequality with the homogeneous kernel of -1-degree and extensions. Journal of Guangdong Education Institute (Natural Science), 2008; 28(3): 1-10.
124. Yang BC. A new Hilbert-type integral inequality and its generalization. Journal of Jilin University (Science Edition), 2005; 43(5): 580-584.
125. Yang BC, Liang HW. A new Hilbert-type integral inequality with a parameter. Journal of Henan University (Natural Science), 2005; 35(4): 4-8.
126. Li YJ, He B. A new Hilbert-type integral inequality and the equivalent form. Internat. J. Math. & Math. Soc., Vol. 2006; Art.ID45378: 1~6.
127. Xie CH. A best extension of a new Hilbert-type inequality. Journal of Jinan Univ. (Natural Science), 2007; 28(1): 24-27.
128. Li YJ, Qian Y, He B. On further analogs of Hilbert's inequality. Internat. J. Math. & Math. Soc., Vol. 2007; Art. ID76329: 1~6.
129. Li YJ, Wang Z, He B. Hilbert's type linear operator and some extensions of Hilbert's inequality. J. Ineq & Appl., Vol.2007; Art. ID82138: 1~10.
130. Xie ZT, Yang BC. A new Hilbert-type integral inequality with some parameters and its reverse . Kyungpook Math. J., 2008; 48: 93-100.

2. Hilbert-Type Integral Inequalities with Multi-Parameters

Bicheng Yang

Department of Mathematics, Guangdong Institute of Education, Guangzhou, Guangdong 510303, P. R. China; E-mail: bcyang@pub.guangzhou.gd.cn

Abstract: In this chapter, we build some Hilbert-type integral inequalities with two pairs of conjugate exponents and an independent parameter, and consider their operator characterizations. The equivalent forms and the reverses are also given. Some sufficient conditions of bounded operator are established and a number of particular inequalities are deduced. The proofs of some basic theorems are used the limit theorems in the theory of Lebesgue integrals and the technique of real analysis.

2.1. HILBERT-TYPE INTEGRAL INEQUALITIES WITH NO CONJUGATE EXPONENT

2.1.1. SOME BASIC RESULTS

As we know in 1.1.4, for $\lambda \in \mathbf{R}$, $k_\lambda(x,y)$ is a homogenous function of $-\lambda$ -degree, if $k_\lambda(ux,uy) = u^{-\lambda} k_\lambda(x,y)$, for any $u,x,y>0$.

Theorem 2.1.1 Suppose that $k_1(x,y)(\ge 0)$ is a homogeneous function of -1-degree and

$$k := \int_0^\infty k_1(1,u) u^{-\frac{1}{2}} du$$

is a finite number. If $f, g \in L^2(0,\infty)$, then

$$I := \int_0^\infty \int_0^\infty k_1(x,y) f(x) g(y) dxdy$$

$$\le k\left(\int_0^\infty f^2(x)dx \int_0^\infty g^2(x)dx\right)^{\frac{1}{2}}, \qquad (2.1.1)$$

where the constant factor k is the best possible.

Proof Setting $v = \frac{1}{u}$, we find

$$\tilde{k} := \int_0^\infty k_1(u,1) u^{-\frac{1}{2}} du = \int_0^\infty k_1(1,v) v^{-\frac{1}{2}} dv = k. \qquad (2.1.2)$$

By using Cauchy's inequality with weight (Kuang SSTP 2004)[1], it follows

$$I = \int_0^\infty \int_0^\infty k_1(x,y)\left[\left(\tfrac{x}{y}\right)^{\frac{1}{4}} f(x)\right]\left[\left(\tfrac{y}{x}\right)^{\frac{1}{4}} g(y)\right] dxdy$$

$$\le \left[\int_0^\infty \int_0^\infty k_1(x,y)\left(\tfrac{x}{y}\right)^{\frac{1}{2}} f^2(x) dxdy \times \int_0^\infty \int_0^\infty k_1(x,y)\left(\tfrac{y}{x}\right)^{\frac{1}{2}} g^2(y) dxdy\right]^{\frac{1}{2}}$$

$$= \left\{\int_0^\infty \left[\int_0^\infty k_1(x,y)\left(\tfrac{x}{y}\right)^{\frac{1}{2}} dy\right] f^2(x) dx \times \int_0^\infty \left[\int_0^\infty k_1(x,y)\left(\tfrac{y}{x}\right)^{\frac{1}{2}} dx\right] g^2(y) dy\right\}^{\frac{1}{2}}$$

$$= \left(\int_0^\infty \omega(x) f^2(x) dx \int_0^\infty \varpi(y) g^2(y) dy\right)^{\frac{1}{2}}, \quad (2.1.3)$$

where define the weight functions $\omega(x)$ and $\varpi(y)$ as

$$\omega(x) := \int_0^\infty k_1(x,y)\left(\tfrac{x}{y}\right)^{\frac{1}{2}} dy, x \in (0,\infty), \quad (2.1.4)$$

$$\varpi(y) := \int_0^\infty k_1(x,y)\left(\tfrac{y}{x}\right)^{\frac{1}{2}} dx, y \in (0,\infty). \quad (2.1.5)$$

For fixed $x>0$, setting $u = \frac{y}{x}$, since $k_1(x,y)$ is a homogeneous function of -1-degree, we obtain

$$\omega(x) = \int_0^\infty k_1(1,u) u^{-\frac{1}{2}} du = k; \qquad (2.1.6)$$

for fixed $y>0$, setting $u = \frac{x}{y}$, then by (2.1.2), it follows

$$\varpi(y) = \int_0^\infty k_1(u,1) u^{-\frac{1}{2}} du = \tilde{k} = k. \qquad (2.1.7)$$

Substitution (2.1.6) -(2.1.7) to (2.1.3), we have (2.1.1).

For $n \in \mathbf{N} \setminus \{1\}$, setting $f_n(x)$ as:

$$f_n(x) = 0, x \in (0,1); f_n(x) = x^{(-1-\frac{1}{n})/2}, x \in [1,\infty),$$

if there exists a non-negative number $k' \le k$, such that (2.1.1) is still valid as we replace k by k', then in particular, we find

$$I_n := \int_0^\infty \int_0^\infty k_1(x,y) f_n(x) f_n(y) dxdy$$

$$\le k' \int_0^\infty f_n^2(x) dx = nk'. \qquad (2.1.8)$$

By Fubini theorem (Kuang HEP 1996) [2], we find

$$k' \ge \frac{I_n}{n} = \frac{1}{n}\int_1^\infty x^{(-1-\frac{1}{n})/2}\left[\int_1^\infty k_1(x,y) y^{(-1-\frac{1}{n})/2} dy\right] dx$$

$$\overset{u=x/y}{=} \frac{1}{n}\int_1^\infty x^{-1-\frac{1}{n}}\left[\int_0^x k_1(u,1) u^{(-1+\frac{1}{n})/2} du\right] dx$$

$$= \frac{1}{n}\int_1^\infty x^{-1-\frac{1}{n}}\left[\int_0^1 k_1(u,1) u^{(-1+\frac{1}{n})/2} du\right] dx$$

$$+\int_1^\infty x^{-1-\frac{1}{n}}[\int_1^x k_1(u,1)u^{(-1+\frac{1}{n})/2}du]dx$$
$$=\int_0^1 k_1(u,1)u^{(-1+\frac{1}{n})/2}du$$
$$+\frac{1}{n}\int_1^\infty k_1(u,1)u^{(-1+\frac{1}{n})/2}(\int_u^\infty x^{-1-\frac{1}{n}}dx)du$$
$$=\int_0^1 k_1(u,1)u^{(-1+\frac{1}{n})/2}du$$
$$+\int_1^\infty k_1(u,1)u^{-(1+\frac{1}{n})/2}du. \qquad (2.1.9)$$

By using Fatou lemma (Kuang HEP 1996) [2], (2.1.2) and (2.1.9), it follows

$$k=\int_0^\infty k_1(u,1)u^{-\frac{1}{2}}du$$
$$=\int_0^1 k_1(u,1)u^{-\frac{1}{2}}du+\int_1^\infty k_1(u,1)u^{-\frac{1}{2}}du$$
$$=\int_0^1 \lim_{n\to\infty} k_1(u,1)u^{(-1+\frac{1}{n})/2}du$$
$$+\int_1^\infty \lim_{n\to\infty} k_1(u,1)u^{-(1+\frac{1}{n})/2}du$$
$$\le \underline{\lim}_{n\to\infty}[\int_0^1 k_1(u,1)u^{(-1+\frac{1}{n})/2}du$$
$$+\int_1^\infty k_1(u,1)u^{-(1+\frac{1}{n})/2}du]\le k'.$$

Therefore, $k'=k$ is the best value of (2.1.1).

Theorem 2.1.2 As the assumptions of Theorem 2.1.1, if $k>0$ and $0<\int_0^\infty f^2(x)dx<\infty$, then

$$J:=\int_0^\infty(\int_0^\infty k_1(x,y)f(x)dx)^2dy$$
$$<k^2\int_0^\infty f^2(x)dx; \qquad (2.1.10)$$

if we also have $0<\int_0^\infty g^2(x)dx<\infty$, then (2.1.1) takes the strict sign-inequality, i.e.

$$I=\int_0^\infty\int_0^\infty k_1(x,y)f(x)g(y)dxdy$$
$$<k(\int_0^\infty f^2(x)dx\int_0^\infty g^2(x)dx)^{\frac{1}{2}}, \qquad (2.1.11)$$

where the constant factors k^2 and k are all the best possible, and (2.1.11) is equivalent to (2.1.10).

Proof By Cauchy's inequality with weight (Kuang SSTP 2004) [1], (2.1.5) and (2.1.7), for $y\in(0,\infty)$, we obtain

$$(\int_0^\infty k_1(x,y)f(x)dx)^2$$
$$=\{\int_0^\infty k_1(x,y)[(\tfrac{x}{y})^{\frac{1}{4}}f(x)][(\tfrac{y}{x})^{\frac{1}{4}}]dx\}^2$$
$$\le[\int_0^\infty k_1(x,y)(\tfrac{x}{y})^{\frac{1}{2}}f^2(x)dx]$$
$$\times[\int_0^\infty k_1(x,y)(\tfrac{y}{x})^{\frac{1}{2}}dx]$$
$$=k\int_0^\infty k_1(x,y)(\tfrac{x}{y})^{\frac{1}{2}}f^2(x)dx. \qquad (2.1.12)$$

Then by Fubini theorem (Kuang HEP 1996) [2], (2.1.4) and (2.1.6), it follows

$$J\le k\int_0^\infty[\int_0^\infty k_1(x,y)(\tfrac{x}{y})^{\frac{1}{2}}f^2(x)dx]dy$$
$$=k\int_0^\infty[\int_0^\infty k_1(x,y)(\tfrac{x}{y})^{\frac{1}{2}}dy]f^2(x)dx$$
$$=k^2\int_0^\infty f^2(x)dx. \qquad (2.1.13)$$

We conform that (2.1.12) takes the strict sign-inequality, otherwise, since $k>0$, there exists $y\in(0,\infty)$, such that $k_1(x,y)$ is not zero a.e. in $(0,\infty)$ and there exist constants A and B, satisfying they are not all zero and

$$A(\tfrac{x}{y})^{\frac{1}{2}}f^2(x)=B(\tfrac{y}{x})^{\frac{1}{2}} \quad \text{a.e. in}(0,\infty)$$

(Kuang SSTP 2004) [1], i.e.

$$Axf^2(x)=By \quad \text{a.e. in}(0,\infty).$$

Since $A\neq 0$ (otherwise, $A=B=0$), then

$$f^2(x)=\tfrac{By}{Ax} \quad \text{a.e. in}(0,\infty),$$

which contradicts the fact that $0<\int_0^\infty f^2(x)dx<\infty$. Hence both (2.1.12) and (2.1.13) take the strict sign-inequalities, and we have (2.1.10).

By Cauchy's inequality (Kuang SSTP 2004) [1], we obtain

$$I=\int_0^\infty(\int_0^\infty k_1(x,y)f(x)dx)(g(y))dy$$
$$\le J^{1/2}(\int_0^\infty g^2(y)dy)^{\frac{1}{2}}. \qquad (2.1.14)$$

Then by (2.1.10), since $0<\int_0^\infty g^2(y)dy<\infty$, we have (2.1.11).

On the other hand, suppose (2.1.11) is valid. Setting

$$g(y):=\int_0^\infty k_1(x,y)f(x)dx \ (y>0),$$

it is obvious that $\int_0^\infty g^2(y)dy>0$. By (2.1.13) and $\int_0^\infty f^2(x)dx<\infty$, we conform that

$$\int_0^\infty g^2(y)dy=J<\infty.$$

Then by (2.1.11), we find

$$0<\int_0^\infty g^2(y)dy=J=I$$

$$< k(\int_0^\infty f^2(x)dx\int_0^\infty g^2(y)dy)^{\frac{1}{2}} < \infty,$$

$$J^{\frac{1}{2}} = \{\int_0^\infty g^2(y)dy\}^{\frac{1}{2}} < k(\int_0^\infty f^2(x)dx)^{\frac{1}{2}}.$$

Hence we have (2.1.10), which is equivalent to (2.1.11).

By Theorem 2.1.1, it follows that the constant factor k in (2.1.11) is the best possible. We conform that the constant k^2 in (2.1.10) is the best possible, otherwise we can get a contradiction by (2.1.14) that the constant k in (2.1.11) is not the best possible. □

Theorem 2.1.3 Suppose that for $\lambda \in \mathbf{R}$, $k_\lambda(x,y)$ (≥ 0) is a homogeneous function of $-\lambda$-degree in $(0,\infty)\times(0,\infty)$, such that

$$0 < k_\lambda := \int_0^\infty k_\lambda(1,u)u^{\frac{\lambda}{2}-1}du < \infty.$$

If $0 < \int_0^\infty x^{1-\lambda} f^2(x)dx < \infty$ and

$$0 < \int_0^\infty x^{1-\lambda} g^2(x)dx < \infty,$$

then we have the following equivalent inequalities:

$$\int_0^\infty\int_0^\infty k_\lambda(x,y)f(x)g(y)dxdy$$

$$< k_\lambda(\int_0^\infty x^{1-\lambda}f^2(x)dx\int_0^\infty x^{1-\lambda}g^2(x)dx)^{\frac{1}{2}}, \tag{2.1.15}$$

$$\int_0^\infty y^{\lambda-1}(\int_0^\infty k_\lambda(x,y)f(x)dx)^2dy$$

$$< k_\lambda^2\int_0^\infty x^{1-\lambda}f^2(x)dx, \tag{2.1.16}$$

where the constant factors k_λ and k_λ^2 are the best possible.

Proof It is obvious that $k_1(x,y) := k_\lambda(x,y)(xy)^{\frac{\lambda-1}{2}}$ is a homogeneous function of -1-degree in $(0,\infty)\times(0,\infty)$. Since

$$k = \int_0^\infty k_1(1,u)u^{-\frac{1}{2}}du$$

$$= \int_0^\infty k_\lambda(1,u)u^{\frac{\lambda}{2}-1}du = k_\lambda \in (0,\infty), \tag{2.1.17}$$

setting $F(x) := x^{\frac{1-\lambda}{2}}f(x)$, $G(x) := x^{\frac{1-\lambda}{2}}g(x)$, then by Theorem 2.1.2, we obtain the following equivalent inequalities:

$$\int_0^\infty\int_0^\infty k_1(x,y)F(x)G(y)dxdy$$

$$< k_\lambda(\int_0^\infty F^2(x)dx\int_0^\infty G^2(x)dx)^{\frac{1}{2}}, \tag{2.1.18}$$

$$\int_0^\infty(\int_0^\infty k_1(x,y)F(x)dx)^2dy$$

$$< k_\lambda^2\int_0^\infty F^2(x)dx. \tag{2.1.19}$$

Substitution the expressions of F, G and $k_1(x,y)$ to (2.1.18) and (2.1.19), by simplification, we obtain (2.1.15) and (2.1.16).

It is obvious that inequalities (2.1.15) and (2.1.16) are respectively equivalent to (2.1.11) and (2.1.10). Since inequalities (2.1.11) and (2.1.10) are equivalent, then inequalities (2.1.15) and (2.1.16) are also equivalent.

If the constant factor k_λ in (2.1.15) is not the best possible for a $\lambda \in \mathbf{R}$, then there exists a positive number $k_\lambda' < k_\lambda$, such that (2.1.15) is still valid as we replace k_λ by k_λ'. It follows that (2.1.18) is still valid as we replace k_λ by k_λ', which is a contradiction. Hence the constant factor k_λ in (2.1.15) is the best possible. By the same way, we can prove that the constant factor k_λ^2 in (2.1.16) is the best possible.□

2.1.2. SOME PARTICULAR HILBERT-TYPE INTEGRAL INEQUALITIES WITH NO CONJUGATE EXPONENT

In the following examples, the conditions

$$0 < \int_0^\infty x^{1-\lambda}f^2(x)dx < \infty,$$

$0 < \int_0^\infty x^{1-\lambda}g^2(x)dx < \infty$ and the conclusions that the constant factors are the best possible are omitted.

Example 2.1.4 If $\lambda, \alpha > 0, k_\lambda(x,y) = \frac{1}{(x^\alpha+y^\alpha)^{\lambda/\alpha}}$, setting $v = u^\alpha$, then we obtain

$$k_\lambda = \int_0^\infty k_\lambda(1,u)u^{\frac{\lambda}{2}-1}du = \int_0^\infty \frac{u^{\frac{\lambda}{2}-1}}{(1+u^\alpha)^{\lambda/\alpha}}du$$

$$= \frac{1}{\alpha}\int_0^\infty \frac{v^{\frac{\lambda}{2\alpha}-1}}{(1+v)^{\lambda/\alpha}}dv = \frac{1}{\alpha}B(\frac{\lambda}{2\alpha},\frac{\lambda}{2\alpha}) > 0.$$

Due to Theorem 2.1.3, we have the following equivalent inequalities:

$$\int_0^\infty\int_0^\infty \frac{f(x)g(y)}{(x^\alpha+y^\alpha)^{\lambda/\alpha}}dxdy < \frac{1}{\alpha}B(\frac{\lambda}{2\alpha},\frac{\lambda}{2\alpha})$$

$$\times(\int_0^\infty x^{1-\lambda}f^2(x)dx\int_0^\infty x^{1-\lambda}g^2(x)dx)^{\frac{1}{2}}, \tag{2.1.20}$$

$$\int_0^\infty y^{\lambda-1}[\int_0^\infty \frac{f(x)}{(x^\alpha+y^\alpha)^{\lambda/\alpha}}dx]^2dy$$

$$< [\frac{1}{\alpha}B(\frac{\lambda}{2\alpha},\frac{\lambda}{2\alpha})]^2\int_0^\infty x^{1-\lambda}f^2(x)dx. \tag{2.1.21}$$

In particular, (1) for $\alpha = 1, \lambda > 0$, they deduce to the following basic Hilbert-type integral inequality with a parameter and the equivalent form (Yang CQJM 1998) [3]:

$$\int_0^\infty \int_0^\infty \frac{1}{(x+y)^\lambda} f(x)g(y)dxdy < B(\tfrac{\lambda}{2},\tfrac{\lambda}{2})$$
$$\times(\int_0^\infty x^{1-\lambda} f^2(x)dx \int_0^\infty x^{1-\lambda} g^2(x)dx)^{\frac{1}{2}}, \quad (2.1.22)$$

$$\int_0^\infty y^{\lambda-1}[\int_0^\infty \frac{f(x)}{(x+y)^\lambda}dx]^2 dy$$
$$< [B(\tfrac{\lambda}{2},\tfrac{\lambda}{2})]^2 \int_0^\infty x^{1-\lambda} f^2(x)dx; \quad (2.1.23)$$

(2) for $\alpha = \lambda > 0$, they deduce to the following equivalent inequalities:

$$\int_0^\infty \int_0^\infty \frac{1}{x^\lambda + y^\lambda} f(x)g(y)dxdy$$
$$< \tfrac{\pi}{\lambda}(\int_0^\infty x^{1-\lambda} f^2(x)dx \int_0^\infty x^{1-\lambda} g^2(x)dx)^{\frac{1}{2}}, \quad (2.1.24)$$

$$\int_0^\infty y^{\lambda-1}[\int_0^\infty \frac{1}{x^\lambda + y^\lambda} f(x)dx]^2 dy$$
$$< (\tfrac{\pi}{\lambda})^2 \int_0^\infty x^{1-\lambda} f^2(x)dx; \quad (2.1.25)$$

(3) for $\lambda = \alpha = 1$, they deduce to the following basic Hilbert-type integral inequality and the equivalent form (Hardy CUP 1934) [4]:

$$\int_0^\infty \int_0^\infty \frac{1}{x+y} f(x)g(y)dxdy$$
$$< \pi(\int_0^\infty f^2(x)dx \int_0^\infty g^2(x)dx)^{\frac{1}{2}}, \quad (2.1.26)$$

$$\int_0^\infty (\int_0^\infty \frac{f(x)}{x+y}dx)^2 dy < \pi^2 \int_0^\infty f^2(x)dx. \quad (2.1.27)$$

Example 2.1.5 If $\lambda > 0, k_\lambda(x,y) = \frac{1}{(\max\{x,y\})^\lambda}$, then we obtain

$$k_\lambda = \int_0^\infty k_\lambda(1,u)u^{\frac{\lambda}{2}-1}du = \int_0^\infty \frac{u^{\frac{\lambda}{2}-1}}{(\max\{1,u\})^\lambda}du$$
$$= \int_0^1 u^{\frac{\lambda}{2}-1}du + \int_1^\infty \frac{1}{u^\lambda}u^{\frac{\lambda}{2}-1}du = \frac{4}{\lambda} > 0.$$

Due to Theorem 2.1.3, we have the following equivalent inequalities (Yang CJEM 2004) [5]:

$$\int_0^\infty \int_0^\infty \frac{1}{(\max\{x,y\})^\lambda} f(x)g(y)dxdy$$
$$< \tfrac{4}{\lambda}(\int_0^\infty x^{1-\lambda} f^2(x)dx \int_0^\infty x^{1-\lambda} g^2(x)dx)^{\frac{1}{2}}, (2.1.28)$$

$$\int_0^\infty y^{\lambda-1}[\int_0^\infty \frac{1}{(\max\{x,y\})^\lambda} f(x)dx]^2 dy$$
$$< (\tfrac{4}{\lambda})^2 \int_0^\infty x^{1-\lambda} f^2(x)dx. \quad (2.1.29)$$

In particular, for $\lambda = 1$, we have the following basic Hilbert-type integral inequality and the equivalent form (Hardy CUP 1934) [4]:

$$\int_0^\infty \int_0^\infty \frac{1}{\max\{x,y\}} f(x)g(y)dxdy$$
$$< 4(\int_0^\infty f^2(x)dx \int_0^\infty g^2(x)dx)^{\frac{1}{2}}, \quad (2.1.30)$$

$$\int_0^\infty (\int_0^\infty \frac{f(x)dx}{\max\{x,y\}})^2 dy < 16 \int_0^\infty f^2(x)dx. \quad (2.1.31)$$

Example 2.1.6 If $\lambda > 0, k_\lambda(x,y) = \frac{\ln(x/y)}{x^\lambda - y^\lambda}$, setting $v = u^\lambda$, since $\int_0^\infty \frac{\ln v}{v-1} v^{-\frac{1}{2}}dv = \pi^2$ (Hardy CUP 1934) [4], we obtain

$$k_\lambda = \int_0^\infty k_\lambda(1,u)u^{\frac{\lambda}{2}-1}du = \int_0^\infty \frac{\ln u}{u^\lambda - 1}u^{\frac{\lambda}{2}-1}du$$
$$= \frac{1}{\lambda^2}\int_0^\infty \frac{\ln v}{v-1}v^{-\frac{1}{2}}dv = (\tfrac{\pi}{\lambda})^2 > 0.$$

Due to Theorem 2.1.3, we have the following equivalent inequalities (Yang JM 2007) [6]:

$$\int_0^\infty \int_0^\infty \frac{\ln(x/y)}{x^\lambda - y^\lambda} f(x)g(y)dxdy$$
$$< (\tfrac{\pi}{\lambda})^2(\int_0^\infty x^{1-\lambda} f^2(x)dx \int_0^\infty x^{1-\lambda} g^2(x)dx)^{\frac{1}{2}}, \quad (2.1.32)$$

$$\int_0^\infty y^{\lambda-1}[\int_0^\infty \frac{\ln(x/y)}{x^\lambda - y^\lambda} f(x)dx]^2 dy$$
$$< (\tfrac{\pi}{\lambda})^4 \int_0^\infty x^{1-\lambda} f^2(x)dx. \quad (2.1.33)$$

In particular, for $\lambda = 1$, we have the following basic Hilbert-type integral inequality and the equivalent form (Hardy CUP 1934) [4]:

$$\int_0^\infty \int_0^\infty \frac{\ln(x/y)}{x-y} f(x)g(y)dxdy$$
$$< \pi^2(\int_0^\infty f^2(x)dx \int_0^\infty g^2(x)dx)^{\frac{1}{2}}, \quad (2.1.34)$$

$$\int_0^\infty [\int_0^\infty \frac{\ln(x/y)}{x-y} f(x)dx]^2 dy < \pi^4 \int_0^\infty f^2(x)dx. \quad (2.1.35)$$

Example 2.1.7 If $\lambda > 0, k_\lambda(x,y) = \frac{|\ln(x/y)|}{(\max\{x,y\})^\lambda}$, then we obtain

$$k_\lambda = \int_0^\infty k_\lambda(1,u)u^{\frac{\lambda}{2}-1}du = \int_0^\infty \frac{|\ln u| u^{\frac{\lambda}{2}-1}}{(\max\{u,1\})^\lambda}du$$
$$= \int_0^1 (-\ln u)u^{\frac{\lambda}{2}-1}du + \int_1^\infty \frac{\ln u}{u^\lambda}u^{\frac{\lambda}{2}-1}du = \frac{8}{\lambda^2} > 0.$$

Due to Theorem 2.1.3, we have the following equivalent inequalities (Yang JSCNU 2007) [7]:

$$\int_0^\infty \int_0^\infty \frac{|\ln(x/y)|}{(\max\{x,y\})^\lambda} f(x)g(y)dxdy$$
$$< \frac{8}{\lambda^2}(\int_0^\infty x^{1-\lambda} f^2(x)dx \int_0^\infty x^{1-\lambda} g^2(x)dx)^{\frac{1}{2}}, \quad (2.1.36)$$

$$\int_0^\infty y^{\lambda-1}[\int_0^\infty \frac{|\ln(x/y)|}{(\max\{x,y\})^\lambda} f(x)dx]^2 dy$$

$$< (\tfrac{8}{\lambda^2})^2 \int_0^\infty x^{1-\lambda} f^2(x)dx. \quad (2.1.37)$$

In particular, for $\lambda = 1$, we have the following basic Hilbert-type integral inequality and the equivalent form:

$$\int_0^\infty \int_0^\infty \frac{|\ln(x/y)|}{\max\{x,y\}} f(x)g(y)dxdy < 8(\int_0^\infty f^2(x)dx \int_0^\infty g^2(x)dx)^{\frac{1}{2}}, \quad (2.1.38)$$

$$\int_0^\infty [\int_0^\infty \frac{|\ln(x/y)|}{\max\{x,y\}} f(x)dx]^2 dy < 64\int_0^\infty f^2(x)dx. \quad (2.1.39)$$

Example 2.1.8 If $\lambda > 0, k_\lambda(x,y) = \frac{|\ln(x/y)|}{x^\lambda + y^\lambda}$, then we obtain

$$k_\lambda = \int_0^\infty k_\lambda(1,u)u^{\frac{\lambda}{2}-1}du = \int_0^\infty \frac{|\ln u|}{u^\lambda+1} u^{\frac{\lambda}{2}-1}du$$

$$= \int_0^1 \frac{(-\ln u)}{u^\lambda+1} u^{\frac{\lambda}{2}-1}du + \int_1^\infty \frac{\ln u}{u^\lambda+1} u^{\frac{\lambda}{2}-1}du$$

$$= \frac{2}{\lambda^2}\int_0^1 \frac{(-\ln v)}{v+1} v^{\frac{-1}{2}}du$$

$$= \frac{2}{\lambda^2}\int_0^1 \sum_{k=0}^\infty (-1)^k(-\ln v)v^{k-\frac{1}{2}}dv$$

$$= \frac{2}{\lambda^2}\sum_{k=0}^\infty (-1)^k \int_0^1 (-\ln v)v^{k-\frac{1}{2}}dv$$

$$= \frac{2}{\lambda^2}\sum_{k=0}^\infty \frac{(-1)^k}{(k+1/2)} \int_0^1 (-\ln v)dv^{k+\frac{1}{2}}$$

$$= \frac{8}{\lambda^2}\sum_{k=0}^\infty \frac{(-1)^k}{(2k+1)^2} = \frac{c_0}{\lambda^2},$$

where $c_0 = 8\sum_{k=0}^\infty \frac{(-1)^k}{(2k+1)^2} = 0.7377^+$. By Theorem 2.1.3, we have the following equivalent inequalities:

$$\int_0^\infty \int_0^\infty \frac{|\ln(x/y)|}{x^\lambda+y^\lambda} f(x)g(y)dxdy < \frac{c_0}{\lambda^2}(\int_0^\infty x^{1-\lambda} f^2(x)dx \int_0^\infty x^{1-\lambda} g^2(x)dx)^{\frac{1}{2}}, \quad (2.1.40)$$

$$\int_0^\infty y^{\lambda-1}[\int_0^\infty \frac{|\ln(x/y)|}{x^\lambda+y^\lambda} f(x)dx]^2 dy < (\frac{c_0}{\lambda^2})^2 \int_0^\infty x^{1-\lambda} f^2(x)dx. \quad (2.1.41)$$

In particular, for $\lambda = 1$, we have the following basic Hilbert-type integral inequality and the equivalent form (Yang CM 2008) [8]:

$$\int_0^\infty \int_0^\infty \frac{|\ln(x/y)|}{x+y} f(x)g(y)dxdy < c_0(\int_0^\infty f^2(x)dx \int_0^\infty g^2(x)dx)^{\frac{1}{2}}, \quad (2.1.42)$$

$$\int_0^\infty [\int_0^\infty \frac{|\ln(x/y)|}{x+y} f(x)dx]^2 dy < c_0^2 \int_0^\infty f^2(x)dx. \quad (2.1.43)$$

Example 2.1.9 If $\lambda > 0, \alpha > -1$,

$$k_\lambda(x,y) = \frac{\arctan^\alpha (x/y)^{\lambda/2}}{x^\lambda+y^\lambda},$$

setting $u = v^{-2/\lambda}$, then we obtain

$$k_\lambda = \int_0^\infty k_\lambda(1,u)u^{\frac{\lambda}{2}-1}du = \int_0^\infty \frac{\arctan^\alpha (1/u)^{\lambda/2}}{u^\lambda+1} u^{\frac{\lambda}{2}-1}du$$

$$= \frac{2}{\lambda}\int_0^\infty \frac{\arctan^\alpha v}{v^2+1}dv = \frac{2}{\lambda(\alpha+1)}(\frac{\pi}{2})^{\alpha+1}.$$

Due to Theorem 2.1.3, we have the following equivalent inequalities:

$$\int_0^\infty \int_0^\infty \frac{\arctan^\alpha (x/y)^{\lambda/2}}{x^\lambda+y^\lambda} f(x)g(y)dxdy < \frac{2(\pi/2)^{\alpha+1}}{\lambda(\alpha+1)}(\int_0^\infty x^{1-\lambda} f^2(x)dx \int_0^\infty x^{1-\lambda} g^2(x)dx)^{\frac{1}{2}}, \quad (2.1.44)$$

$$\int_0^\infty y^{\lambda-1}[\int_0^\infty \frac{\arctan^\alpha (x/y)^{\lambda/2}}{x^\lambda+y^\lambda} f(x)dx]^2 dy < [\frac{2}{\lambda(\alpha+1)}(\frac{\pi}{2})^{\alpha+1}]^2 \int_0^\infty x^{1-\lambda} f^2(x)dx. \quad (2.1.45)$$

In particular, for $\lambda = \alpha = 1$, we have the following basic Hilbert-type integral inequality and the equivalent form (Yang JGEI 2008) [9]:

$$\int_0^\infty \int_0^\infty \frac{\arctan\sqrt{x/y}}{x+y} f(x)g(y)dxdy < \frac{\pi^2}{4}(\int_0^\infty f^2(x)dx \int_0^\infty g^2(x)dx)^{\frac{1}{2}}, \quad (2.1.46)$$

$$\int_0^\infty [\int_0^\infty \frac{\arctan\sqrt{x/y}}{x+y} f(x)dx]^2 dy < \frac{\pi^4}{16}\int_0^\infty f^2(x)dx. \quad (2.1.47)$$

Example 2.1.10 If $0 < \lambda < \alpha, k_\lambda(x,y) = \frac{1}{|x^\alpha - y^\alpha|^{\lambda/\alpha}}$, setting $v = u^\lambda$, then by (1.2.17), we obtain

$$k_\lambda = \int_0^\infty k_\lambda(1,u)u^{\frac{\lambda}{2}-1}du = \int_0^\infty \frac{u^{\frac{\lambda}{2}-1}}{|1-u^\alpha|^{\lambda/\alpha}}du$$

$$= \frac{1}{\alpha}[\int_0^1 (1-v)^{(1-\frac{\lambda}{\alpha})-1} v^{\frac{\lambda}{2\alpha}-1}dv + \int_1^\infty \frac{(v-1)^{(1-\lambda/\alpha)-1}}{v^{1-\lambda/(2\alpha)}}dv]$$

$$= \frac{2}{\alpha}B(1-\frac{\lambda}{\alpha}, \frac{\lambda}{2\alpha}) > 0.$$

Due to Theorem 2.1.3, we have the following equivalent inequalities:

$$\int_0^\infty \int_0^\infty \frac{f(x)g(y)}{|x^\alpha - y^\alpha|^{\lambda/\alpha}} dxdy < \frac{2}{\alpha}B(1-\frac{\lambda}{\alpha}, \frac{\lambda}{2\alpha}) \times(\int_0^\infty x^{1-\lambda} f^2(x)dx \int_0^\infty x^{1-\lambda} g^2(x)dx)^{\frac{1}{2}}, \quad (2.1.48)$$

$$\int_0^\infty y^{\lambda-1}\left(\int_0^\infty \frac{f(x)}{|x^\alpha-y^\alpha|^{\lambda/\alpha}}dx\right)^2 dy$$
$$<[\tfrac{2}{\alpha}B(1-\tfrac{\lambda}{\alpha},\tfrac{\lambda}{2\alpha})]^2\int_0^\infty x^{1-\lambda}f^2(x)dx. \quad (2.1.49)$$

In particular, for $\alpha=1$, $0<\lambda<1$, we have the following basic Hilbert-type integral inequality with a parameter and the equivalent form (Yang JJU 2005) [10]:

$$\int_0^\infty\int_0^\infty \frac{f(x)g(y)}{|x-y|^\lambda}dxdy<2B(1-\lambda,\tfrac{\lambda}{2})$$
$$\times\left(\int_0^\infty x^{1-\lambda}f^2(x)dx\int_0^\infty x^{1-\lambda}g^2(x)dx\right)^{\frac{1}{2}}; \quad (2.1.50)$$
$$\int_0^\infty y^{\lambda-1}\left(\int_0^\infty \frac{f(x)}{|x-y|^\lambda}dx\right)^2 dy$$
$$<[2B(1-\lambda,\tfrac{\lambda}{2})]^2\int_0^\infty x^{1-\lambda}f^2(x)dx. \quad (2.1.51)$$

Example 2.1.11 If $\alpha>0, b>-1$,

$$k_{2\alpha}(x,y)=\frac{1}{x^{2\alpha}+2bx^\alpha y^\alpha+y^{2\alpha}},$$

setting $v=u^{-\lambda}$, then we obtain

$$k_{2\alpha}=\int_0^\infty k_{2\alpha}(1,u)u^{-1+(2\alpha)/2}du$$
$$=\int_0^\infty \frac{u^{-1+\alpha}}{1+2bu^\alpha+u^{2\alpha}}du=\frac{1}{\alpha}\int_0^\infty \frac{dv}{v^2+2bv^2+1}$$
$$=\begin{cases}\frac{1}{\alpha\sqrt{b^2-1}}\ln(b+\sqrt{b^2-1}),\ b>1\\ \frac{1}{\alpha},\quad b=1\\ \frac{1}{\alpha\sqrt{1-b^2}}(\frac{\pi}{2}-\arctan\frac{b}{\sqrt{1-b^2}}),\ -1<b<1.\end{cases} \quad (2.1.52)$$

Due to Theorem 2.1.3 (for $\lambda=2\alpha$), we have the following equivalent inequalities:

$$\int_0^\infty\int_0^\infty \frac{f(x)g(y)}{x^{2\alpha}+2bx^\alpha y^\alpha+y^{2\alpha}}dxdy<k_{2\alpha}$$
$$\times\left(\int_0^\infty x^{1-2\alpha}f^2(x)dx\int_0^\infty x^{1-2\alpha}g^2(x)dx\right)^{\frac{1}{2}}, \quad (2.1.53)$$
$$\int_0^\infty y^{2\alpha-1}\left(\int_0^\infty \frac{f(x)dx}{x^{2\alpha}+2bx^\alpha y^\alpha+y^{2\alpha}}\right)^2 dy$$
$$<k_{2\alpha}^2\int_0^\infty x^{1-2\alpha}f^2(x)dx. \quad (2.1.54)$$

In particular, (1) for $b=\frac{1}{2}$, we have

$$\int_0^\infty\int_0^\infty \frac{f(x)g(y)dxdy}{x^{2\alpha}+x^\alpha y^\alpha+y^{2\alpha}}<\frac{2\pi}{3\sqrt{3}\alpha}$$
$$\times\left(\int_0^\infty x^{1-2\alpha}f^2(x)dx\int_0^\infty x^{1-2\alpha}g^2(x)dx\right)^{\frac{1}{2}}, \quad (2.1.55)$$
$$\int_0^\infty y^{2\alpha-1}\left(\int_0^\infty \frac{f(x)}{x^{2\alpha}+x^\alpha y^\alpha+y^{2\alpha}}dx\right)^2 dy$$
$$<\left(\frac{2\pi}{3\sqrt{3}\alpha}\right)^2\int_0^\infty x^{1-2\alpha}f^2(x)dx; \quad (2.1.56)$$

(2) for $\alpha=1$, we have the following basic Hilbert-type integral inequality with a parameter and the equivalent form :

$$\int_0^\infty\int_0^\infty \frac{f(x)g(y)}{x^2+2bxy+y^2}dxdy$$
$$<k_2\left(\int_0^\infty \tfrac{1}{x}f^2(x)dx\int_0^\infty \tfrac{1}{x}g^2(x)dx\right)^{\frac{1}{2}}, \quad (2.1.57)$$
$$\int_0^\infty y\left(\int_0^\infty \frac{f(x)dx}{x^2+2bxy+y^2}\right)^2 dy<k_2^2\int_0^\infty \tfrac{1}{x}f^2(x)dx. \quad (2.1.58)$$

Example 2.1.12 If $\alpha>0, k_0(x,y)=\frac{(\min\{x,y\})^\alpha}{x^\alpha+y^\alpha}$, then we obtain

$$k_0=\int_0^\infty k_0(1,u)u^{-1}du$$
$$=\int_0^\infty \frac{(\min\{1,u\})^\alpha}{1+u^\alpha}u^{-1}du=2\int_0^1 \frac{u^{\alpha-1}}{1+u^\alpha}du$$
$$=\frac{2}{\alpha}\int_0^1 \frac{1}{1+u^\alpha}du^\alpha=\frac{2}{\alpha}\ln(u^\alpha+1)\big|_0^1=\frac{2\ln 2}{\alpha}.$$

Due to Theorem 2.1.3 (for $\lambda=0$), we have the following equivalent inequalities:

$$\int_0^\infty\int_0^\infty \frac{(\min\{x,y\})^\alpha}{x^\alpha+y^\alpha}f(x)g(y)dxdy$$
$$<\frac{2\ln 2}{\alpha}\left(\int_0^\infty xf^2(x)dx\int_0^\infty xg^2(x)dx\right)^{\frac{1}{2}}, \quad (2.1.59)$$
$$\int_0^\infty \frac{1}{y}\left[\int_0^\infty \frac{(\min\{x,y\})^\alpha}{x^\alpha+y^\alpha}f(x)dx\right]^2 dy$$
$$<\left(\frac{2\ln 2}{\alpha}\right)^2\int_0^\infty xf^2(x)dx. \quad (2.1.60)$$

Example 2.1.13 If $\alpha>0, k_0(x,y)=\left(\frac{\min\{x,y\}}{\max\{x,y\}}\right)^\alpha$, then we obtain

$$k_0=\int_0^\infty k_0(1,u)u^{-1}du=\int_0^\infty \left(\frac{\min\{1,u\}}{\max\{1,u\}}\right)^\alpha u^{-1}du$$
$$=2\int_0^1 u^{\alpha-1}du=\frac{2}{\alpha}.$$

Due to Theorem 3.1.3 (for $\lambda=0$), we have the following equivalent inequalities:

$$\int_0^\infty\int_0^\infty \left(\frac{\min\{x,y\}}{\max\{x,y\}}\right)^\alpha f(x)g(y)dxdy$$
$$<\frac{2}{\alpha}\left(\int_0^\infty xf^2(x)dx\int_0^\infty xg^2(x)dx\right)^{\frac{1}{2}}, \quad (2.1.61)$$
$$\int_0^\infty \frac{1}{y}\left[\int_0^\infty \left(\frac{\min\{x,y\}}{\max\{x,y\}}\right)^\alpha f(x)dx\right]^2 dy$$
$$<\left(\frac{2}{\alpha}\right)^2\int_0^\infty xf^2(x)dx. \quad (2.1.62)$$

Example 2.1.14 If $0<\alpha<1$, $k_0(x,y)=\left(\frac{\min\{x,y\}}{|x-y|}\right)^\alpha$,

then by (1.2.17), we obtain

$$k_0 = \int_0^\infty k_0(1,u)u^{-1}du = \int_0^\infty (\tfrac{\min\{1,u\}}{|1-u|})^\alpha u^{-1}du$$
$$= 2\int_0^1 (1-u)^{(1-\alpha)-1}u^{\alpha-1}du = 2B(1-\alpha,\alpha).$$

Due to Theorem 2.1.3 (for $\lambda = 0$), we have the following equivalent inequalities:

$$\int_0^\infty \int_0^\infty (\tfrac{\min\{x,y\}}{|x-y|})^\alpha f(x)g(y)dxdy$$
$$< 2B(1-\alpha,\alpha)(\int_0^\infty xf^2(x)dx\int_0^\infty xg^2(x)dx)^{\frac{1}{2}}, \quad (2.1.63)$$

$$\int_0^\infty \tfrac{1}{y}[\int_0^\infty (\tfrac{\min\{x,y\}}{|x-y|})^\alpha f(x)dx]^2 dy$$
$$< 4[B(1-\alpha,\alpha)]^2 \int_0^\infty xf^2(x)dx. \quad (2.1.64)$$

2.1.3. OPERATOR CHARACTERIZATIONS OF HILBERT-TYPE INTEGRAL INEQUALITIES WITH NO CONJUGATE EXPONENTS

Suppose that $L^2(0,\infty)$ is a real function space, $k_1(x,y)(\geq 0)$ is a homogeneous function of -1-degree, such that $k := \int_0^\infty k_1(1,u)u^{-\frac{1}{2}}du$ is a finite number. If $f \in L^2(0,\infty)$, define a linear operator $T: L^2(0,\infty) \to L^2(0,\infty)$ as

$$(Tf)(y) = \int_0^\infty k_1(x,y)f(x)dx, \quad y \in (0,\infty). \quad (2.1.65)$$

By (2.1.14), we find $Tf \in L^2(0,\infty)$ and

$$\| Tf \|_2 = [\int_0^\infty (\int_0^\infty k_1(x,y)f(x)dx)^2 dy]^{\frac{1}{2}}$$
$$\leq k(\int_0^\infty f^2(x)dx)^{\frac{1}{2}} = k\| f \|_2 \quad .$$

(2.1.66)

In view of the norm of T as (Taylor HWS 1980) [11]

$$\| T \| := \sup_{(f\neq\theta)} \{\tfrac{\|Tf\|}{\|f\|}; f \in L^2(0,\infty)\},$$

It follows $\| T \| \leq k$. Due to (2.1.60), the constant factor k is the best possible, then in virtue of $\| Tf \|_2 \leq \| T \| \cdot \| f \|_2$, it follows $k \leq \| T \|$, and then $\| T \| = k$. If $g \in L^2(0,\infty)$, then we can indicate the inner product of Tf and g as

$$(Tf,g) = \int_0^\infty (\int_0^\infty k_1(x,y)f(x)dx)g(y)dy$$
$$= \int_0^\infty \int_0^\infty k_1(x,y)f(x)g(y)dxdy. \quad (2.1.67)$$

Hence we may express (2.1.1) by the operator and the norm as

$$(Tf,g) \leq \| T \| \cdot \| f \|_2 \cdot \| g \|_2 .$$

If $k = \| T \| > 0$ and $\| f \|_2 > 0, \| g \|_2 > 0$, then we may rewrite (2.1.11) and (2.1.12) to the following equivalent forms:

$$\| Tf \|_2 < \| T \| \cdot \| f \|_2, \quad (2.1.68)$$
$$(Tf,g) < \| T \| \cdot \| f \|_2 \cdot \| g \|_2, \quad (2.1.69)$$

where the constant factor $\| T \| = \int_0^\infty k(1,u)u^{-\frac{1}{2}}du$ is the best possible.

Setting $\omega(x) = x^{1-\lambda}$ $(x \in (0,\infty))$, and

$$L^2_\omega(0,\infty)$$
$$:= \{f; \| f \|_{2,\omega} = \{\int_0^\infty \omega(x)f^2(x)dx\}^{\frac{1}{2}} < \infty\},$$

then Theorem 2.1.3 my be rewritten to

Theorem 2.1.15 Suppose that $\lambda \in \mathbf{R}, k_\lambda(x,y)(\geq 0)$ is a homogeneous function of $-\lambda$-degree in $(0,\infty)\times(0,\infty)$, such that

$$0 < k_\lambda = \int_0^\infty k_\lambda(1,u)u^{\frac{\lambda}{2}-1}du < \infty.$$

If $f \in L^2_\omega(0,\infty)$, define the operator

$$T_\lambda : L^2_\omega(0,\infty) \to L^2_{\omega^{-1}}(0,\infty) \text{ as:}$$

$$(T_\lambda f)(y) = \int_0^\infty k_\lambda(x,y)f(x)dx, y \in (0,\infty). \quad (2.1.70)$$

Then for $g \in L^2_\omega(0,\infty)$, satisfying $\| g \|_{2,\omega} > 0$ and $\| f \|_{2,\omega} > 0$, we have the following equivalent inequalities:

$$(T_\lambda f,g) < \| T_\lambda \| \cdot \| f \|_{2,\omega} \cdot \| g \|_{2,\omega}, \quad (2.1.71)$$
$$\| T_\lambda f \|_{2,\omega^{-1}} < \| T_\lambda \| \cdot \| f \|_{2,\omega}, \quad (2.1.72)$$

where the constant factor

$$\| T_\lambda \| = k_\lambda = \int_0^\infty k_\lambda(1,u)u^{\frac{\lambda}{2}-1}du$$

is the best possible.

2.1.4. SOME HILBERT-TYPE INTEGRAL INEQUALITIES WITH VARIABLES AS PARAMETERS BUT WITH NO CONJUGATE EXPONETS

Theorem 2.1.16 Suppose that $\lambda \in \mathbf{R}$, $k_\lambda(x,y)$ (≥ 0) is a homogeneous function of $-\lambda$-degree in $(0,\infty)\times(0,\infty)$, such that

$$0 < k_\lambda = \int_0^\infty k_\lambda(1,u)u^{\frac{\lambda}{2}-1}du < \infty,$$

$u(x)$ is a strict increasing derivable function and $u(a^+)=0, u(b^-)=\infty$. Setting

$$\tilde{k}_\lambda(x,y):=k_\lambda(u(x),u(y)),$$

if $0<\int_a^b \frac{u^{1-\lambda}(x)}{u'(x)}f^2(x)dx<\infty$ and

$$0<\int_a^b \frac{u^{1-\lambda}(x)}{u'(x)}g^2(x)dx<\infty,$$

then we have

$$\int_a^b\int_a^b \tilde{k}_\lambda(x,y)f(x)g(y)dxdy < k_\lambda\left(\int_a^b \frac{u^{1-\lambda}(x)}{u'(x)}f^2(x)dx\int_a^b \frac{u^{1-\lambda}(x)}{u'(x)}g^2(x)dx\right)^{\frac{1}{2}}, \tag{2.1.73}$$

$$\int_a^b \frac{u'(y)}{u^{1-\lambda}(y)}\left(\int_a^b \tilde{k}_\lambda(x,y)f(x)dx\right)^2dy < k_\lambda^2\int_a^b \frac{u^{1-\lambda}(x)}{u'(x)}f^2(x)dx, \tag{2.1.74}$$

where the constant factors k_λ and $k_\lambda{}^2$ are all the best possible, and inequalities (2.1.73) and (2.1.74) are equivalent.

Proof Replacing x and y respectively by $u(x)$ and $u(y)$ in (2.1.15) and (2.1.16), by simplifications, then replacing $u'(x)f(u(x))$ and $u'(y)g(u(y))$ respectively by $f(x)$ and $g(y)$, we find (2.1.73) and (2.1.74). It is obvious that inequalities (2.1.15) and (2.1.73) are equivalent; so are (2.1.16) and (2.1.74). Hence inequalities (2.1.73) and (2.1.74) are equivalent. We can prove that the constant factors in (2.1.73) and (2.1.74) are the best possible by the equivalent relationship of them. □

Example 2.1.17 If $u(x)=x^\alpha(\alpha>0; x>0)$, by (2.1.73) and (2.1.74), we have

$$\int_0^\infty\int_0^\infty k_\lambda(x^\alpha,y^\alpha)f(x)g(y)dxdy < \frac{k_\lambda}{\alpha}\left(\int_0^\infty x^{1-\alpha\lambda}f^2(x)dx\int_0^\infty x^{1-\alpha\lambda}g^2(x)dx\right)^{\frac{1}{2}}, \tag{2.1.75}$$

$$\int_0^\infty y^{\alpha\lambda-1}\left(\int_0^\infty k_\lambda(x^\alpha,y^\alpha)f(x)dx\right)^2dy < \left(\frac{k_\lambda}{\alpha}\right)^2\int_0^\infty x^{1-\alpha\lambda}f^2(x)dx. \tag{2.1.76}$$

Example 2.1.18 If $u(x)=\ln x \ (x\in(1,\infty))$, by (2.1.73) and (2.1.74), we have

$$\int_1^\infty\int_1^\infty k_\lambda(\ln x,\ln y)f(x)g(y)dxdy < k_\lambda\left[\int_1^\infty x(\ln x)^{1-\lambda}f^2(x)dx\int_1^\infty x(\ln x)^{1-\lambda}g^2(x)dx\right]^{\frac{1}{2}}, \tag{2.1.77}$$

$$\int_1^\infty \frac{(\ln y)^{\lambda-1}}{y}\left(\int_1^\infty k_\lambda(\ln x,\ln y)f(x)dx\right)^2dy < k_\lambda^2\int_1^\infty x(\ln x)^{1-\lambda}f^2(x)dx. \tag{3.1.78}$$

In particular, (1) setting $k_\lambda(x,y)=\frac{1}{(x+y)^\lambda}$, we obtain the following equivalent inequalities (Yang JIPAM 2005) [12]:

$$\int_1^\infty\int_1^\infty \frac{1}{(\ln xy)^\lambda}f(x)g(y)dxdy < B\left(\frac{\lambda}{2},\frac{\lambda}{2}\right) \times\left[\int_1^\infty x(\ln x)^{1-\lambda}f^2(x)dx\int_1^\infty x(\ln x)^{1-\lambda}g^2(x)dx\right]^{\frac{1}{2}}, \tag{2.1.79}$$

$$\int_1^\infty \frac{(\ln y)^{\lambda-1}}{y}\left[\int_1^\infty \frac{f(x)}{(\ln xy)^\lambda}dx\right]^2dy < \left(B\left(\frac{\lambda}{2},\frac{\lambda}{2}\right)\right)^2\int_1^\infty x(\ln x)^{1-\lambda}f^2(x)dx. \tag{2.1.80}$$

(2) for $\lambda=1$ in (2.1.79), replacing $f(x)$ and $g(y)$ respectively by $\frac{1}{x}f(x)$ and $\frac{1}{y}g(y)$, we find the following Mulholland's integral inequality (Yang MIA 2005) [13]:

$$\int_1^\infty\int_1^\infty \frac{1}{xy\ln xy}f(x)g(y)dxdy < \pi\left(\int_1^\infty \frac{1}{x}f^2(x)dx\int_1^\infty \frac{1}{x}g^2(x)dx\right)^{\frac{1}{2}}. \tag{2.1.81}$$

Example 2.1.19 If $u(x)=e^x(x\in(-\infty,\infty))$, by (2.1.73) and (2.1.74), we have

$$\int_{-\infty}^\infty\int_{-\infty}^\infty k_\lambda(e^x,e^y)f(x)g(y)dxdy < k_\lambda\left(\int_{-\infty}^\infty e^{-\lambda x}f^2(x)dx\int_{-\infty}^\infty e^{-\lambda x}g^2(x)dx\right)^{\frac{1}{2}}, \tag{2.1.82}$$

$$\int_{-\infty}^\infty e^{\lambda y}\left(\int_{-\infty}^\infty k_\lambda(e^x,e^y)f(x)dx\right)^2dy < k_\lambda^2\int_{-\infty}^\infty e^{-\lambda x}f^2(x)dx. \tag{2.1.83}$$

Example 2.1.20 If $u(x)=\tan x \ (x\in(0,\frac{\pi}{2}))$, by (2.1.73) and (2.1.74), we have

$$\int_0^{\frac{\pi}{2}}\int_0^{\frac{\pi}{2}} k_\lambda(\tan x,\tan y)f(x)g(y)dxdy < k_\lambda\left(\int_0^{\frac{\pi}{2}} \frac{\tan^{1-\lambda}x}{\sec^2 x}f^2(x)dx\int_0^{\frac{\pi}{2}} \frac{\tan^{1-\lambda}x}{\sec^2 x}g^2(x)dx\right)^{\frac{1}{2}}, \tag{2.1.84}$$

$$\int_0^{\frac{\pi}{2}} \frac{\sec^2 y}{\tan^{1-\lambda}y}\left(\int_0^{\frac{\pi}{2}} k_\lambda(\tan x,\tan y)f(x)dx\right)^2dy < k_\lambda^2\int_0^{\frac{\pi}{2}} \frac{\tan^{1-\lambda}x}{\sec^2 x}f^2(x)dx. \tag{2.1.85}$$

Example 2.1.21 If $u(x)=\sec x-1 \ (x\in(0,\frac{\pi}{2}))$, by (2.1.73) and (2.1.74), we have

$$\int_0^{\frac{\pi}{2}}\int_0^{\frac{\pi}{2}} k_\lambda(\sec x-1,\sec y-1)f(x)g(y)dxdy$$
$$<k_\lambda[\int_0^{\frac{\pi}{2}}\frac{(\sec x-1)^{1-\lambda}}{\sec x\tan x}f^2(x)dx\int_0^{\frac{\pi}{2}}\frac{(\sec x-1)^{1-\lambda}}{\sec x\tan x}g^2(x)dx]^{\frac{1}{2}}, \quad (2.1.86)$$

$$\int_0^{\frac{\pi}{2}}\frac{\sec y\tan y}{(\sec y-1)^{1-\lambda}}(\int_0^{\frac{\pi}{2}} k_\lambda(\sec x-1,\sec y-1)f(x)dx)^2dy$$
$$<k_\lambda^2\int_0^{\frac{\pi}{2}}\frac{(\sec x-1)^{1-\lambda}}{\sec x\tan x}f^2(x)dx. \quad (2.1.87)$$

2.2. HILBERT-TYPE INTEGRAL INEQUALITIES WITH MULTI-PARAMETERS

In this book, for short, we write $p>1(0<p<1)$ to express the condition $\frac{1}{p}+\frac{1}{q}=1$ with $p>1,q>1$ ($\frac{1}{p}+\frac{1}{q}=1$ with $0<p<1,q<0$) , and write $r,s>1$ $(r,s\neq 0)$ to express the condition $\frac{1}{r}+\frac{1}{s}=1$ with $r>1,s>1$ ($\frac{1}{r}+\frac{1}{s}=1$ with $r,s\neq 0;\ r=1,\frac{1}{s}:=0; s=1,\frac{1}{r}:=0)$.

2.2.1. HILBERT-TYPE INTEGRAL INEQUALITIES WITH MULTI-PARAMETERS AND THE OPERATOR CHARACTERIZATIONS

Theorem 2.2.1 Suppose that (p,q) and (r,s) are two pairs of conjugate exponents with $p>1,r,s\neq 0$, $\lambda\in \mathbf{R}$, $k_\lambda(x,y)(\geq 0)$ is a homogeneous function of $-\lambda$ -degree in $(0,\infty)\times(0,\infty)$, and $k_\lambda(r):=\int_0^\infty k_\lambda(u,1)u^{\frac{\lambda}{r}-1}du$ is a finite number. If $f(x),g(x)$ are non-negative measurable function in $(0,\infty)$, then we have the following equivalent inequalities:

$$I_\lambda:=\int_0^\infty\int_0^\infty k_\lambda(x,y)f(x)g(y)dxdy\leq k_\lambda(r)$$
$$\times\{\int_0^\infty x^{p(1-\frac{\lambda}{r})-1}f^p(x)dx\}^{\frac{1}{p}}\{\int_0^\infty x^{q(1-\frac{\lambda}{s})-1}g^q(x)dx\}^{\frac{1}{p}} \quad (2.2.1)$$

$$J_\lambda:=\int_0^\infty y^{\frac{p\lambda}{s}-1}(\int_0^\infty k_\lambda(x,y)f(x)dx)^pdy$$
$$\leq k_\lambda^p(r)\int_0^\infty x^{p(1-\frac{\lambda}{r})-1}f^p(x)dx. \quad (2.2.2)$$

Proof Setting $u=\frac{1}{v}$, we find

$$\tilde{k}_\lambda(s):=\int_0^\infty k_\lambda(1,u)u^{\frac{\lambda}{s}-1}du$$
$$=\int_0^\infty k_\lambda(v,1)v^{\frac{\lambda}{r}-1}dv=k_\lambda(r). \quad (2.2.3)$$

Define the weight functions $\varpi_\lambda(s,x)$ and $\omega_\lambda(r,y)$ as:

$$\varpi_\lambda(s,x):=x^{\frac{\lambda}{r}}\int_0^\infty k_\lambda(x,y)y^{\frac{\lambda}{s}-1}dy, x\in(0,\infty), \quad (2.2.4)$$

$$\omega_\lambda(r,y):=y^{\frac{\lambda}{s}}\int_0^\infty k_\lambda(x,y)x^{\frac{\lambda}{r}-1}dx, y\in(0,\infty). \quad (2.2.5)$$

Setting $u:=\frac{y}{x}$, it follows

$$\varpi_\lambda(s,x)=\int_0^\infty k_\lambda(1,u)u^{\frac{\lambda}{s}-1}du=\tilde{k}_\lambda(s);$$

setting $u=\frac{x}{y}$, by (3.2.3), we still have

$$\omega_\lambda(r,y)=\int_0^\infty k_\lambda(u,1)u^{\frac{\lambda}{r}-1}du=k_\lambda(r)$$
$$=\tilde{k}_\lambda(s)\ =\varpi_\lambda(s,x). \quad (2.2.6)$$

By H$\ddot{o}$ lder's inequality with weight and (2.2.5), in view of Fubini theorem, we find

$$(\int_0^\infty k_\lambda(x,y)f(x)dx)^p$$
$$=\{\int_0^\infty k_\lambda(x,y)[\frac{x^{(1-\frac{\lambda}{r})/q}}{y^{(1-\frac{\lambda}{s})/p}}f(x)][\frac{y^{(1-\frac{\lambda}{s})/p}}{x^{(1-\frac{\lambda}{r})/q}}]dx\}^p$$
$$\leq\int_0^\infty k_\lambda(x,y)\frac{x^{(1-\frac{\lambda}{r})(p-1)}}{y^{1-\frac{\lambda}{s}}}f^p(x)dx$$
$$\times[\int_0^\infty k_\lambda(x,y)\frac{y^{(1-\frac{\lambda}{s})(q-1)}}{x^{1-\frac{\lambda}{r}}}dx]^{p-1}$$
$$=k_\lambda^{p-1}(r)y^{1-\frac{p\lambda}{s}}\int_0^\infty k_\lambda(x,y)\frac{x^{(1-\frac{\lambda}{r})(p-1)}}{y^{1-\frac{\lambda}{s}}}f^p(x)dx, \quad (2.2.7)$$

$$J_\lambda\leq k_\lambda^{p-1}(r)$$
$$\times\int_0^\infty[\int_0^\infty k_\lambda(x,y)\frac{x^{(1-\frac{\lambda}{r})(p-1)}}{y^{1-\frac{\lambda}{s}}}f^p(x)dx]dy$$
$$=k_\lambda^{p-1}(r)\int_0^\infty[\int_0^\infty k_\lambda(x,y)\frac{x^{(1-\frac{\lambda}{r})(p-1)}}{y^{1-\frac{\lambda}{s}}}dy]f^p(x)dx. \quad (2.2.8)$$

Hence by (2.2.4) and (2.2.6), we have (2.2.2).

By H$\ddot{o}$ lder's inequality, we find

$$I_\lambda=\int_0^\infty[y^{\frac{\lambda}{s}-\frac{1}{p}}\int_0^\infty k_\lambda(x,y)f(x)dx]$$
$$\times[y^{(1-\frac{\lambda}{s})-\frac{1}{q}}g(y)]dy$$
$$\leq J_\lambda^{\frac{1}{p}}\{\int_0^\infty y^{q(1-\frac{\lambda}{s})-1}g^q(y)dy\}^{\frac{1}{q}}. \quad (2.2.9)$$

Then by (2.2.2), we have (2.2.1).

On the other hand, suppose (2.2.1) is valid. Setting

$$g(y):=y^{\frac{p\lambda}{s}-1}(\int_0^\infty k_\lambda(x,y)f(x)dx)^{p-1}(y>0),$$

if $J_\lambda=\infty$, then by (2.2.8), it follows (2.2.2) keeps the form of equality; if $J_\lambda=0$, then (2.2.2) is naturally valid; if $0<J_\lambda<\infty$, then by (2.2.1), we find

$$0<\int_0^\infty y^{q(1-\frac{\lambda}{s})-1}g^q(y)dy=J_\lambda=I_\lambda$$
$$\le k_\lambda(r)\{\int_0^\infty x^{p(1-\frac{\lambda}{r})-1}f^p(x)dx\}^{\frac{1}{p}}$$
$$\times\{\int_0^\infty y^{q(1-\frac{\lambda}{s})-1}g^q(y)dy\}^{\frac{1}{q}}<\infty, \quad (2.2.10)$$
$$\int_0^\infty y^{q(1-\frac{\lambda}{s})-1}g^q(y)dy=J_\lambda$$
$$\le k_\lambda^p(r)\int_0^\infty x^{p(1-\frac{\lambda}{r})-1}f^p(x)dx. \quad (2.2.11)$$

Hence (2.2.2) is valid, which is equivalent to (2.2.1). □

Following the conditions of Theorem 2.2.1, setting

$$\phi(x):=x^{p(1-\frac{\lambda}{r})-1},\psi(x):=x^{q(1-\frac{\lambda}{s})-1},x\in(0,\infty),$$

then $\psi^{1-p}(x)=x^{\frac{p\lambda}{s}-1}$, define the following real function spaces :

$$L_\phi^p(0,\infty):=\{f;\|f\|_{p,\phi}:=\{\int_0^\infty\phi(x)|f(x)|^p dx\}^{\frac{1}{p}}<\infty\},$$
$$L_\psi^q(0,\infty):=\{g;\|g\|_{q,\psi}:=\{\int_0^\infty\psi(x)|g(x)|^q dx\}^{\frac{1}{q}}<\infty\}$$

and Hilbert-type integral operator

$$T:L_\phi^p(0,\infty)\to L_{\psi^{1-p}}^p(0,\infty)$$

as: for any $f\in L_\phi^p(0,\infty)$, there exists

$$Tf(y)=\int_0^\infty k_\lambda(x,y)f(x)dx, y\in(0,\infty). \quad (2.2.12)$$

Hence by (2.2.2), it follows $Tf\in L_{\psi^{1-p}}^p(0,\infty)$. For $g\in L_\psi^q(0,\infty)$, define the formal inner product of Tf and g as

$$(Tf,g):=\int_0^\infty\int_0^\infty k_\lambda(x,y)f(x)g(y)dxdy. \quad (2.2.13)$$

We may rewrite (2.2.1) and (2.2.2) equivalently to the following forms:

$$(Tf,g)\le k_\lambda(r)\|f\|_{p,\phi}\|g\|_{q,\psi}, \quad (2.2.14)$$
$$\|Tf\|_{p,\psi^{1-p}}\le k_\lambda(r)\|f\|_{p,\phi}. \quad (2.2.15)$$

In view of (2.2.15), it follows that T is a bounded operator and $\|T\|\le k_\lambda(r)$.

Theorem 2.2.2 Suppose that $p>1, r,s\neq 0$, $\lambda\in\mathbf{R}$, $k_\lambda(x,y)(\ge 0)$ is a homogeneous function of $-\lambda$ -degree, such that $k_\lambda(r)=\int_0^\infty k_\lambda(u,1)u^{\frac{\lambda}{r}-1}du$ is a finite number. If $T:L_\phi^p(0,\infty)\to L_{\psi^{1-p}}^p(0,\infty)$ is an integral operator defined by (2.2.12), then we have $\|T\|=k_\lambda(r)$, and the constant factors in (2.2.14) and (2.2.15) are all the best possible.

Proof For $n\in\mathbf{N}$, setting $f_n(x)$ and $g_n(x)$ as:

$$f_n(x)=g_n(x)=0, x\in(0,1);$$
$$f_n(x)=x^{\frac{\lambda}{r}-1-\frac{1}{np}},\ g_n(x)=x^{\frac{\lambda}{s}-1-\frac{1}{nq}}, x\in[1,\infty),$$

we find

$$\|f_n\|_{p,\phi}=\{\int_1^\infty x^{p(1-\frac{\lambda}{r})-1}\cdot x^{p(\frac{\lambda}{r}-1-\frac{1}{np})}dx\}^{\frac{1}{p}}=n^{\frac{1}{p}},$$
$$\|g_n\|_{q,\psi}=\{\int_1^\infty x^{q(1-\frac{\lambda}{s})-1}\cdot x^{q(\frac{\lambda}{s}-1-\frac{1}{nq})}dx\}^{\frac{1}{q}}=n^{\frac{1}{q}}. \quad (2.2.16)$$

If there exists a real number $k\le k_\lambda(r)$, such that (2.2.14) is still valid as we replace $k_\lambda(r)$ by k. In particular, it follows

$$\frac{1}{n}(Tf_n,g_n)\le\frac{k}{n}\|f_n\|_{p,\phi}\|g_n\|_{q,\psi}=k. \quad (2.2.17)$$

Setting $u=\frac{y}{x}$ in the following, in view of Fubini theorem and (2.2.17), we obtain

$$k\ge\frac{1}{n}(Tf_n,g_n)$$
$$=\frac{1}{n}\int_0^\infty\int_0^\infty k_\lambda(x,y)f_n(x)g_n(y)dxdy$$
$$=\frac{1}{n}\int_1^\infty(\int_1^\infty k_\lambda(x,y)x^{\frac{\lambda}{r}-1-\frac{1}{np}}y^{\frac{\lambda}{s}-1-\frac{1}{nq}}dx)dy$$
$$=\frac{1}{n}\int_1^\infty y^{-1-\frac{1}{n}}(\int_0^y k_\lambda(1,u)u^{\frac{\lambda}{s}-1+\frac{1}{np}}du)dy$$
$$=\frac{1}{n}[\int_1^\infty y^{-1-\frac{1}{n}}(\int_0^1 k_\lambda(1,u)u^{\frac{\lambda}{s}-1+\frac{1}{np}}du)dy$$
$$+\int_1^\infty y^{-1-\frac{1}{n}}\int_1^y k_\lambda(1,u)u^{\frac{\lambda}{s}-1+\frac{1}{np}}dudy]$$
$$=\int_0^1 k_\lambda(1,u)u^{\frac{\lambda}{s}-1+\frac{1}{np}}du$$
$$+\frac{1}{n}\int_1^\infty k_\lambda(1,u)u^{\frac{\lambda}{s}-1+\frac{1}{np}}(\int_u^\infty y^{-1-\frac{1}{n}}dy)du$$
$$=\int_0^1 k_\lambda(1,u)u^{\frac{\lambda}{s}-1+\frac{1}{np}}du+\int_1^\infty k_\lambda(1,u)u^{\frac{\lambda}{s}-1-\frac{1}{nq}}du. \quad (2.2.18)$$

Then by (2.2.18) and Fatou lemma, we find

$$k\ge\underline{\lim}_{n\to\infty}[\int_0^1 k_\lambda(1,u)u^{\frac{\lambda}{s}-1+\frac{1}{np}}du$$
$$+\int_1^\infty k_\lambda(1,u)u^{\frac{\lambda}{s}-1-\frac{1}{nq}}du]$$

$$\geq \int_0^1 \lim_{n\to\infty} k_\lambda(1,u)u^{\frac{\lambda}{s}-1+\frac{1}{np}}du$$

$$+\int_1^\infty \lim_{n\to\infty} k_\lambda(1,u)u^{\frac{\lambda}{s}-1-\frac{1}{nq}}du$$

$$= \int_0^1 k_\lambda(1,u)u^{\frac{\lambda}{s}-1}du + \int_1^\infty k_\lambda(1,u)u^{\frac{\lambda}{s}-1}du$$

$$= \tilde{k}_\lambda(s) = k_\lambda(r).$$

Hence $k = k_\lambda(r)$ is the best value of (2.2.14). We conform that the constant factor in (2.2.15) is the best possible, otherwise, by (2.2.9), we can get a contradiction that the constant factor in (2.2.14) is not the best possible. Since the constant in (2.2.15) is the best possible, we can conclude that $\|T\| = k_\lambda(r)$.□

Theorem 2.2.3 As the assumption of Theorem 2.2.1, if $0 < k_\lambda(r) < \infty$ and

$$0 < \|f\|_{p,\phi} = \{\int_0^\infty x^{p(1-\frac{\lambda}{r})-1} f^p(x)dx\}^{\frac{1}{p}} < \infty,$$

$$0 < \|g\|_{q,\psi} = \{\int_0^\infty x^{q(1-\frac{\lambda}{s})-1} g^q(x)dx\}^{\frac{1}{q}} < \infty,$$

then we have the following equivalent inequalities :

$$\int_0^\infty \int_0^\infty k_\lambda(x,y)f(x)g(y)dxdy < k_\lambda(r)\|f\|_{p,\phi}\|g\|_{q,\psi}, \tag{2.2.19}$$

$$\int_0^\infty y^{\frac{p\lambda}{s}-1}(\int_0^\infty k_\lambda(x,y)f(x)dx)^p dy < k_\lambda^p(r)\|f\|_{p,\phi}^p, \tag{2.2.20}$$

where the constant factors

$$k_\lambda(r) = \int_0^\infty k_\lambda(u,1)u^{\frac{\lambda}{r}-1}du$$

and $k_\lambda^p(r)$ are the best possible.

Proof In view of Theorem 2.2.2, the constant factors $k_\lambda(r)$ and $k_\lambda^p(r)$ in (2.2.19) and (2.2.20) are all the best possible. If there exists a $y > 0$, such that (2.2.7) takes the form of equality, since $k_\lambda(x,y)$ is not zero *a.e.* in $x \in (0,\infty)$, there exist constants A and B, such that they are not all zero and

$$A\frac{x^{(1-\frac{\lambda}{r})(p-1)}}{y^{1-\frac{\lambda}{s}}} f^p(x) = B\frac{y^{(1-\frac{\lambda}{s})(q-1)}}{x^{1-\frac{\lambda}{r}}} \text{ a.e. in } (0,\infty).$$

i.e. $Ax^{p(1-\frac{\lambda}{r})}f^p(x) = By^{q(1-\frac{\lambda}{s})}$ a.e. in $(0,\infty)$. We conform that $A \neq 0$ (otherwise $B = A = 0$). Then $x^{p(1-\frac{\lambda}{r})-1}f^p(x) = [By^{q(1-\frac{\lambda}{s})}]/(Ax)$ a.e. in $(0,\infty)$, which contradicts the fact that $0 < \|f\|_{p,\phi} < \infty$. Hence (2.2.8) still takes the strict sign- inequality and (2.2.20) is valid. By (2.2.9) and (2.2.20), we have (2.2.19). On the other hand, suppose that (2.2.19) is valid. As in Theorem 2.2.1, still setting

$$g(y) := y^{\frac{p\lambda}{s}-1}(\int_0^\infty k_\lambda(x,y)f(x)dx)^{p-1} \ (y>0),$$

by (2.2.2) and $\|f\|_{p,\phi} < \infty$, it follows that $J_\lambda < \infty$. If $J_\lambda = 0$, then (2.2.20) is naturally valid; if $0 < \|g\|_{q,\psi} = J_\lambda^{\frac{1}{q}} < \infty$, since $0 < \|f\|_{p,\phi} < \infty$, then by (2.2.19), both (2.2.10) and (2.2.11) take the forms of strict sign-inequality. Hence (2.2.20) is valid, which is equivalent to (2.2.19).□

Remark 2.2.4 For $p = q = r = s = 2$, inequalities (2.2.19) and (2.2.20) reduce respectively to (2.1.15) and (2.1.16). Hence inequalities (2.2.19) and (2.2.20) are respectively the best extensions of (2.1.15) and (2.1.16) with two pairs of conjugate exponents.

2.2.2. SOME REVERSE HILBERT-TYPE INTEGRAL INEQUALITIES

Lemma 2.2.5 Suppose that $\lambda \in \mathbf{R}$, $k_\lambda(x,y)$ is a homogenous function of $-\lambda$ -degree in $(0,\infty)\times(0,\infty)$. If $r, s \neq 0$,

$$K_\lambda(x) := \int_0^\infty k_\lambda(u,1)u^{x-1}du$$

is finite in a neighborhood I_λ of $\frac{\lambda}{r}$, then $K_\lambda(x)$ is continuous at $\frac{\lambda}{r}$; so are

$$H_\lambda(x) := \int_0^1 k_\lambda(u,1)u^{x-1}du$$

and $J_\lambda(x) := \int_1^\infty k_\lambda(u,1)u^{x-1}du$.

Proof For any positive decreasing sequence $\{\varepsilon_n\}_{n=1}^\infty$, $\varepsilon_n \to 0(n\to\infty)$, since for $0 < u \leq 1$, $\{u^{\varepsilon_n}\}_{n=1}^\infty$ is increasing, the by Levi theorem (Kuang HEP 1996) [2], it follows

$$\lim_{n\to\infty} H_\lambda(\frac{\lambda}{r}+\varepsilon_n) = \lim_{n\to\infty}\int_0^1 k_\lambda(u,1)u^{\frac{\lambda}{r}+\varepsilon_n-1}du$$

$$= \int_0^1 \lim_{n\to\infty} k_\lambda(u,1)u^{\frac{\lambda}{r}+\varepsilon_n-1}du = H_\lambda(\frac{\lambda}{r}).$$

There exists a $n_0 \in \mathbf{N}$, such that

$$0 \leq H_\lambda(\frac{\lambda}{r}-\varepsilon_{n_0}) \leq K_\lambda(\frac{\lambda}{r}-\varepsilon_{n_0}) < \infty.$$

Since for $n \geq n_0$,

$$k_\lambda(u,1)u^{\frac{\lambda}{r}-\varepsilon_n-1} \leq k_\lambda(u,1)u^{\frac{\lambda}{r}-\varepsilon_{n_0}-1}(0<u\leq 1),$$

then by Lebesgue control convergent theorem (Kuang HEP 1996) [2], it follows

$$\lim_{n\to\infty} H_\lambda(\tfrac{\lambda}{r}-\varepsilon_n)$$
$$=\lim_{n\to\infty}\int_0^1 k_\lambda(u,1)u^{\frac{\lambda}{r}-\varepsilon_n-1}du = H_\lambda(\tfrac{\lambda}{r}).$$

Hence we conclude that $H_\lambda(x)$ is continuous at $\frac{\lambda}{r}$. By the same way, we can show that $J_\lambda(x)$ is continuous at $\frac{\lambda}{r}$. Therefore

$$K_\lambda(x) = H_\lambda(x) + I_\lambda(x)$$

is continuous at $\frac{\lambda}{r}$.

Note 2.2.6 As the assumption of Lemma 2.2.5, since

$$\tilde{K}_\lambda(x) := \int_0^\infty k_\lambda(1,u)u^{(\lambda-x)-1}du = K_\lambda(x),$$
$$\tilde{H}_\lambda(x) := \int_0^1 k_\lambda(1,u)u^{(\lambda-x)-1}du = I_\lambda(x)$$

and $\tilde{I}_\lambda(x) := \int_1^\infty k_\lambda(1,u)u^{(\lambda-x)-1}du = H(x)$, then $\tilde{K}_\lambda(x), \tilde{H}_\lambda(x)$ and $\tilde{I}_\lambda(x)$ are all continuous at $\frac{\lambda}{s}$.

Theorem 2.2.7 As the assumption of Theorem 2.2.1, for $0<p<1$, if there exists a neighborhood I_λ of $\frac{\lambda}{r}$, such that

$$0 < K_\lambda(x) = \int_0^\infty k_\lambda(u,1)u^{x-1}du < \infty, x\in I_\lambda$$

and $0<\|f\|_{p,\phi}<\infty$, $0<\|g\|_{q,\psi}<\infty$, then we have the following equivalent inequalities:

$$I_\lambda = \int_0^\infty\int_0^\infty k_\lambda(x,y)f(x)g(y)dxdy$$
$$> k_\lambda(r)\|f\|_{p,\phi}\|g\|_{q,\psi}, \qquad (2.2.21)$$
$$J_\lambda = \int_0^\infty y^{\frac{p\lambda}{s}-1}\left(\int_0^\infty k_\lambda(x,y)f(x)dx\right)^p dy$$
$$> k_\lambda^p(r)\|f\|_{p,\phi}^p, \qquad (2.2.22)$$
$$L_\lambda = \int_0^\infty x^{\frac{q\lambda}{r}-1}\left(\int_0^\infty k_\lambda(x,y)g(y)dy\right)^q dx$$
$$< k_\lambda^q(r)\|g\|_{q,\psi}^q, \qquad (2.2.23)$$

where the constant factors

$$k_\lambda(r) = \int_0^\infty k_\lambda(u,1)u^{\frac{\lambda}{r}-1}du$$

and $k_\lambda^\rho(r)(\rho = p,q)$ are the best possible.

Proof By the reverse Hölder's inequality (Kuang SSTP 2004) [1] and the same way of obtaining (2.2.2), we can find

$$J_\lambda = \int_0^\infty y^{\frac{p\lambda}{s}-1}\left(\int_0^\infty k_\lambda(x,y)f(x)dx\right)^p dy$$
$$\ge k_\lambda^p(r)\|f\|_{p,\phi}^p, \qquad (2.2.24)$$

and the reverse of (2.2.9) as

$$I_\lambda \ge J_\lambda^{\frac{1}{p}}\left\{\int_0^\infty y^{q(1-\frac{\lambda}{s})-1}g^q(y)dy\right\}^{\frac{1}{q}}. \qquad (2.2.25)$$

By the same way of Theorem 2.2.3, it follows that (2.2.24) keeps the form of strict sign-inequality and (2.2.22) is valid. Due to (2.2.25) and (3.2.22), we can obtain. (2.2.21). On the other hand, suppose that (2.2.21) is valid. Still setting

$$g(y) := y^{\frac{p\lambda}{s}-1}\left(\int_0^\infty k_\lambda(x,y)f(x)dx\right)^{p-1},$$

by (2.2.24) and $0<\|f\|_{p,\phi}<\infty$, we conform that

$$J_\lambda = \int_0^\infty y^{q(1-\frac{\lambda}{s})-1}g^q(y)dy > 0.$$

If $J_\lambda = \infty$, then (2.2.22) is naturally valid; if $0<J_\lambda<\infty$, then by (2.2.21), it follows

$$\infty > \int_0^\infty y^{q(1-\frac{\lambda}{s})-1}g^q(y)dy = J_\lambda = I_\lambda$$
$$> k_\lambda(r)\left\{\int_0^\infty x^{p(1-\frac{\lambda}{r})-1}f^p(x)dx\right\}^{\frac{1}{p}}$$
$$\times\left\{\int_0^\infty y^{q(1-\frac{\lambda}{s})-1}g^q(y)dy\right\}^{\frac{1}{q}} > 0,$$
$$\int_0^\infty y^{q(1-\frac{\lambda}{s})-1}g^q(y)dy = J_\lambda$$
$$> k_\lambda^p(r)\int_0^\infty x^{p(1-\frac{\lambda}{r})-1}f^p(x)dx.$$

Hence (2.2.22) is valid, which is equivalent to (2.2.21).

By the same way of obtaining (2.2.2) and the reverse Hölder's inequality, we obtain

$$L_\lambda^{\frac{1}{q}} = \left\{\int_0^\infty x^{\frac{q\lambda}{r}-1}\left(\int_0^\infty k_\lambda(x,y)g(y)dy\right)^q dx\right\}^{\frac{1}{q}}$$
$$\ge k_\lambda(r)\|g\|_{q,\psi}, \qquad (2.2.26)$$
$$I_\lambda = \int_0^\infty x^{\frac{1}{q}-\frac{\lambda}{r}}f(x)\left[x^{\frac{-1}{q}+\frac{\lambda}{r}}\int_0^\infty k_\lambda(x,y)g(y)dy\right]dx$$
$$\ge \|f\|_{p,\phi}L_\lambda^{\frac{1}{q}}. \qquad (2.2.27)$$

By the same way of Theorem 2.2.4, it follows that (2.2.26) keeps the form of strict sign-inequality and (2.2.23) is valid, since $q<0$. By (2.2.27) and (2.2.23), we can obtain (2.2.21). On the other hand, suppose that (2.2.21) is valid. Setting

$$f(x) = x^{\frac{q\lambda}{r}-1}\left(\int_0^\infty k_\lambda(x,y)g(y)dy\right)^{q-1},$$

by (2.2.26) and $\|g\|_{q,\psi}>0$, we conform that $L_\lambda^{\frac{1}{q}}>0$ and then $L_\lambda = \int_0^\infty x^{p(1-\frac{\lambda}{r})-1}f^p(x)dx < \infty$.

If $L_\lambda = 0$, then (2.2.23) is naturally valid; if $0<L_\lambda<\infty$, then by (2.2.21), it follows

$$\|f\|_{p,\phi}^p = L_\lambda = I_\lambda > k_\lambda(r)\|f\|_{p,\phi}\|g\|_{q,\psi},$$

$$\| f \|_{p,\phi}^{p-1} = L_\lambda^{\frac{1}{q}} > k_\lambda(r) \| g \|_{q,\psi}.$$

Hence (2.2.23) is valid, which is equivalent to (2.2.21). Therefore inequalities (2.2.21), (2.2.22) and (2.2.23) are equivalent.

For a large enough $n \in \mathbf{N}$, setting $f_n(x)$ and $g_n(x)$ as: $f_n(x) = g_n(x) = 0, x \in (0,1)$,

$$f_n(x) = x^{\frac{\lambda}{r}-1-\frac{1}{np}}, g_n(x) = x^{\frac{\lambda}{s}-1-\frac{1}{nq}}, x \in [1,\infty),$$

we find $\| f_n \|_{p,\phi} = n^{\frac{1}{p}}$, $\| g_n \|_{q,\psi} = n^{\frac{1}{q}}$. If there exists a constant $K \geq k_\lambda(r)$, such that (2.2.21) is still valid as we replace $k_\lambda(r)$ by K. In particular, by Note 2.2.6, it follows

$$\begin{aligned} K &= K \cdot \tfrac{1}{n} \| f_n \|_{p,\phi} \| g_n \|_{q,\psi} \\ &< \tfrac{1}{n} \int_0^\infty \int_0^\infty k_\lambda(x,y) f_n(x) g_n(y) dxdy \\ &\leq \tfrac{1}{n} \int_1^\infty y^{\frac{\lambda}{s}-1-\frac{1}{nq}} \Big(\int_0^\infty k_\lambda(x,y) x^{\frac{\lambda}{r}-1-\frac{1}{np}} dx \Big) dy \\ &= \tfrac{1}{n} \int_1^\infty y^{-1-\frac{1}{n}} \Big(\int_0^\infty k_\lambda(1,\mathrm{u}) \mathrm{u}^{\frac{\lambda}{r}-1-\frac{1}{np}} d\mathrm{u} \Big) dy \\ &= \int_0^\infty k_\lambda(1,u) u^{\frac{\lambda}{s}-1+\frac{1}{np}} du \\ &\leq \int_0^1 k_\lambda(1,u) u^{\frac{\lambda}{s}-1} du + \int_1^\infty k_\lambda(1,u) u^{\frac{\lambda}{s}-1+\frac{1}{np}} du \\ &= \int_0^1 k_\lambda(1,u) u^{\frac{\lambda}{s}-1} du \\ &\quad + \int_1^\infty k_\lambda(1,u) u^{\frac{\lambda}{s}-1} du + o(1) \end{aligned} \quad (2.2.28)$$

For $n \to \infty$ in (3.2.28), we obtain

$$K \leq \tilde{k}_\lambda(s) = k_\lambda(r).$$

Hence $K = k_\lambda(r)$ is the best value of (2.2.21). We conform that the constant factor in (2.2.22) (or (2.2.23)) is the best possible, otherwise, by (2.2.25) (or (2.2.27)), we can get a contradiction that the constant factor in (2.2.21) is not the best possible.□

2.2.3. SOME PARICULAR CASES

In the following examples, the conditions $f(x)$, $g(x) \geq 0, \phi(x) = x^{p(1-\frac{\lambda}{r})-1}, \psi(x) = x^{q(1-\frac{\lambda}{s})-1}$,

$$0 < \| f \|_{p,\phi} = \Big\{ \int_0^\infty x^{p(1-\frac{\lambda}{r})-1} f^p(x) dx \Big\}^{\frac{1}{p}} < \infty,$$

$$0 < \| g \|_{q,\psi} = \Big\{ \int_0^\infty x^{q(1-\frac{\lambda}{s})-1} g^q(x) dx \Big\}^{\frac{1}{q}} < \infty,$$

and the conclusions that the constant factors are the best possible are omitted.

Example 2.2.8 If $r, s > 1, \lambda, \alpha > 0$,

$$k_\lambda(x,y) = \frac{1}{(x^\alpha + y^\alpha)^{\lambda/\alpha}},$$

setting $v = u^\alpha$, we obtain

$$k_\lambda(r) = \int_0^\infty k_\lambda(u,1) u^{\frac{\lambda}{r}-1} du = \int_0^\infty \frac{u^{\frac{\lambda}{r}-1}}{(u^\alpha+1)^{\lambda/\alpha}} du$$

$$= \frac{1}{\alpha} \int_0^\infty \frac{1}{(v+1)^{\lambda/\alpha}} v^{\frac{\lambda}{r\alpha}-1} dv = \frac{1}{\alpha} B(\tfrac{\lambda}{r\alpha}, \tfrac{\lambda}{s\alpha}).$$

Then by Theorem 2.2.3 and Theorem 2.2.7, (1) for $p > 1$, we have the following equivalent inequalities:

$$\int_0^\infty \int_0^\infty \frac{1}{(x^\alpha + y^\alpha)^{\lambda/\alpha}} f(x) g(y) dxdy < \frac{1}{\alpha} B(\tfrac{\lambda}{r\alpha}, \tfrac{\lambda}{s\alpha}) \| f \|_{p,\phi} \| g \|_{q,\psi}, \quad (2.2.29)$$

$$\int_0^\infty y^{\frac{p\lambda}{s}-1} \Big[\int_0^\infty \frac{\mathrm{f(x)}}{(x^\alpha + y^\alpha)^{\lambda/\alpha}} dx \Big]^p dy < \Big[\frac{1}{\alpha} B(\tfrac{\lambda}{r\alpha}, \tfrac{\lambda}{s\alpha})\Big]^p \| f \|_{p,\phi}^p; \quad (2.2.30)$$

(2) for $0 < p < 1$, we have the equivalent reverses of (2.2.29) and (2.2.30). In particular, (i) for $\alpha = 1$, if $p > 1$, we have the following equivalent inequalities (Yang MIA 2005) [13]:

$$\int_0^\infty \int_0^\infty \frac{1}{(x+y)^\lambda} f(x) g(y) dxdy < B(\tfrac{\lambda}{r}, \tfrac{\lambda}{s}) \| f \|_{p,\phi} \| g \|_{q,\psi}, \quad (2.2.31)$$

$$\int_0^\infty y^{\frac{p\lambda}{s}-1} \Big[\int_0^\infty \frac{\mathrm{f(x)}}{(x+y)^\lambda} dx \Big]^p dy < [B(\tfrac{\lambda}{r}, \tfrac{\lambda}{s})]^p \| f \|_{p,\phi}^p; \quad (2.2.32)$$

if $0 < p < 1$, we have the reverses of (2.2.31) and (2.2.32). (ii) For $\alpha = \lambda$, if $p > 1$, we have the following equivalent inequalities (Yang AJMAA 2004) [14]:

$$\int_0^\infty \int_0^\infty \frac{1}{x^\lambda + y^\lambda} f(x) g(y) dxdy < \frac{\pi}{\lambda \sin(\pi / r)} \| f \|_{p,\phi} \| g \|_{q,\psi}, \quad (2.2.33)$$

$$\int_0^\infty y^{\frac{p\lambda}{s}-1} \Big(\int_0^\infty \frac{f(x)}{x^\lambda + y^\lambda} dx \Big)^p dy < \Big[\frac{1}{\lambda} \frac{\pi}{\sin(\pi / r)}\Big]^p \| f \|_{p,\phi}^p; \quad (2.2.34)$$

if $0 < p < 1$, we have the equivalent reverses of (2.2.33) and (2.2.34).

Remark 2.2.9 Inequalities (2.2.31) and (2.2.32) are the best extensions of (2.1.22) and (2.1.23), and for $\lambda = 1$, they deduce to the following equivalent inequalities:

$$\int_0^\infty \int_0^\infty \frac{1}{x+y} f(x) g(y) dxdy < \frac{\pi}{\sin(\pi / r)} \times \Big\{ \int_0^\infty x^{\frac{p}{s}-1} f^p(x) dx \Big\}^{\frac{1}{p}} \Big\{ \int_0^\infty x^{\frac{q}{r}-1} g^q(x) dx \Big\}^{\frac{1}{q}}, \quad (2.2.35)$$

$$\int_0^\infty y^{\frac{p}{s}-1}\left(\int_0^\infty \frac{f(x)}{x+y}dx\right)^p dy$$
$$<\left[\frac{\pi}{\sin(\pi/r)}\right]^p \int_0^\infty x^{\frac{p}{s}-1}f^p(x)dx. \quad (2.2.36)$$

For $r=\frac{p\lambda}{p+\lambda-2}$, $s=\frac{q\lambda}{q+\lambda-2}(0<p<1,\ 2-p<\lambda<2-q)$, the reverse of (2.2.31) reduces to the result of (Xi JIPAM 2008) [15]; for $r=q, s=p$, inequality (2.2.35) reduces to Hardy-Hilbert's integral inequality (1.1.22); for $r=p, s=q$, (2.2.35) and (2.2.36) reduce to the dual forms of (1.1.22) and (1.1.23) as

$$\int_0^\infty\int_0^\infty \frac{1}{x+y}f(x)g(y)dxdy<\frac{\pi}{\sin(\pi/p)}$$
$$\times\left\{\int_0^\infty x^{p-2}f^p(x)dx\right\}^{\frac{1}{p}}\left\{\int_0^\infty x^{q-2}g^q(x)dx\right\}^{\frac{1}{q}}, \quad (2.2.37)$$

$$\int_0^\infty y^{p-2}\left(\int_0^\infty \frac{f(x)}{x+y}dx\right)^p dy$$
$$<\left[\frac{\pi}{\sin(\pi/p)}\right]^p \int_0^\infty x^{p-2}f^p(x)dx. \quad (2.2.38)$$

Example 2.2.10 If $r,s>1,\alpha<1,\lambda>0,k_\lambda(x,y)=\frac{1}{|x-y|^\alpha(\max\{x,y\})^{\lambda-\alpha}}$, by (1.2.17), we obtain

$$k_\lambda(r)=\int_0^\infty k_\lambda(u,1)u^{\frac{\lambda}{r}-1}du$$
$$=\int_0^\infty \frac{1}{|u-1|^\alpha(\max\{u,1\})^{\lambda-\alpha}}u^{\frac{\lambda}{r}-1}du$$
$$=\int_0^1(1-u)^{(1-\alpha)-1}u^{\frac{\lambda}{r}-1}du+\int_1^\infty \frac{(u-1)^{(1-\alpha)-1}}{u^{\frac{\lambda}{s}+1-\alpha}}du$$
$$=B(1-\alpha,\tfrac{\lambda}{r})+B(1-\alpha,\tfrac{\lambda}{s}),$$

Then by Theorem 2.2.3 and Theorem 2.2.7, (1) for $p>1$, we have the following equivalent inequalities:

$$\int_0^\infty\int_0^\infty \frac{f(x)g(y)}{|x-y|^\alpha(\max\{x,y\})^{\lambda-\alpha}}dxdy$$
$$<[B(1-\alpha,\tfrac{\lambda}{r})+B(1-\alpha,\tfrac{\lambda}{s})]\,||f||_{p,\phi}||g||_{q,\psi}, \quad (2.2.39)$$

$$\int_0^\infty y^{\frac{p\lambda}{s}-1}\left[\int_0^\infty \frac{f(x)}{|x-y|^\alpha(\max\{x,y\})^{\lambda-\alpha}}dx\right]^p dy$$
$$<[B(1-\alpha,\tfrac{\lambda}{r})+B(1-\alpha,\tfrac{\lambda}{s})]^p\,||f||_{p,\phi}^p; \quad (2.2.40)$$

(2) for $0<p<1$, we have the equivalent reverses of (2.2.29) and (2.2.30). In particular, for $\alpha=0$, (i) if $p>1$, we have the following equivalent inequalities (Zhong JJU 2007) [16]:

$$\int_0^\infty\int_0^\infty \frac{f(x)g(y)}{(\max\{x,y\})^\lambda}dxdy<\frac{rs}{\lambda}\,||f||_{p,\phi}||g||_{q,\psi}, \quad (2.2.41)$$

$$\int_0^\infty y^{\frac{p\lambda}{s}-1}\left[\int_0^\infty \frac{f(x)}{(\max\{x,y\})^\lambda}dx\right]^p dy<(\tfrac{rs}{\lambda})^p\,||f||_{p,\phi}^p; \quad (2.2.42)$$

if $0<p<1$, we have the equivalent reverses of (2.2.41) and (2.2.42) (Zhong CM 2008) [17]; (ii) for $0<\alpha=\lambda<1$, if $p>1$, we have the following equivalent inequalities (Zhong JJNU 2007) [18]:

$$\int_0^\infty\int_0^\infty \frac{1}{|x-y|^\lambda}f(x)g(y)dxdy$$
$$<[B(1-\lambda,\tfrac{\lambda}{r})+B(1-\lambda,\tfrac{\lambda}{s})]\,||f||_{p,\phi}||g||_{q,\psi}, \quad (2.2.43)$$

$$\int_0^\infty y^{\frac{p\lambda}{s}-1}\left(\int_0^\infty \frac{f(x)}{|x-y|^\lambda}dx\right)^p dy$$
$$<[B(1-\lambda,\tfrac{\lambda}{r})+B(1-\lambda,\tfrac{\lambda}{s})]^p\,||f||_{p,\phi}^p, \quad (2.2.44)$$

if $0<p<1$, then we have the equivalent reverses of (2.2.43) and (2.2.44).

Remark 2.2.11 (1) Inequalities (2.2.41) and (2.2.42) are the best extensions of (2.1.30) and (2.1.31), and for $\lambda=1$, they reduce to the following equivalent inequalities:

$$\int_0^\infty\int_0^\infty \frac{1}{\max\{x,y\}}f(x)g(y)dxdy<rs$$
$$\times\left\{\int_0^\infty x^{\frac{p}{s}-1}f^p(x)dx\right\}^{\frac{1}{p}}\left\{\int_0^\infty x^{\frac{q}{r}-1}g^q(x)dx\right\}^{\frac{1}{q}}, \quad (2.2.45)$$

$$\int_0^\infty y^{\frac{p}{s}-1}\left(\int_0^\infty \frac{f(x)}{\max\{x,y\}}dx\right)^p dy<(rs)^p$$
$$\times\int_0^\infty x^{\frac{p}{s}-1}f^p(x)dx, \quad (2.2.46)$$

which are the best extensions of (2.1.30) and (2.1.31) with two pairs of conjugate exponents. (2) inequalities (2.2.43) and (2.2.44) are the best extensions of (2.1.50) and (2.1.51) with two pairs of conjugate exponents.

We have the following formula of Gamma function (Wang SP 1979) [19]

$$\Gamma(b):=\int_0^\infty e^{-t}t^{b-1}dt,\ b>0. \quad (2.2.47)$$

By (2.2.47) and simplifications, we still can obtain

$$\int_0^1 x^{a-1}(-\ln x)^{b-1}dx=\int_1^\infty y^{-a-1}(\ln y)^{b-1}dy$$
$$=\frac{1}{a^b}\Gamma(b)\ (a,b>0). \quad (2.2.48)$$

Example 2.2.12 If $\beta>-1,r,s>1,\lambda>0$, $k_\lambda(x,y)=\frac{|\ln(x/y)|^\beta}{(\max\{x,y\})^\lambda}$, then by (2.2.50), we obtain

$$k_\lambda(r)=\int_0^\infty k_\lambda(u,1)u^{\frac{\lambda}{r}-1}du=\int_0^\infty \frac{|\ln u|^\beta u^{\frac{\lambda}{r}-1}}{(\max\{u,1\})^\lambda}du$$

$$= \int_0^1 (-\ln u)^{(\beta+1)-1} u^{\frac{\lambda}{r}-1} du$$

$$+ \int_1^\infty (\ln u)^{(\beta+1)-1} u^{-\frac{\lambda}{s}-1} du$$

$$= \Gamma(\beta+1)[(\tfrac{r}{\lambda})^{\beta+1} + (\tfrac{s}{\lambda})^{\beta+1}].$$

Then by Theorem 2.2.3 and Theorem 2.2.7, (1) for $p>1$, we have the following equivalent inequalities (Yang JIPAM 2008) [20]:

$$\int_0^\infty \int_0^\infty \frac{|\ln(x/y)|^\beta}{(\max\{x,y\})^\lambda} f(x)g(y)dxdy$$
$$< \frac{\Gamma(\beta+1)}{\lambda^{\beta+1}}(r^{\beta+1}+s^{\beta+1}) \| f \|_{p,\phi} \| g \|_{q,\psi}, \quad (2.2.49)$$

$$\int_0^\infty y^{\frac{p\lambda}{s}-1}[\int_0^\infty \frac{|\ln(x/y)|^\beta f(x)}{(\max\{x,y\})^\lambda}dx]^p dy$$
$$< [\frac{\Gamma(\beta+1)}{\lambda^{\beta+1}}(r^{\beta+1}+s^{\beta+1})]^p \| f \|_{p,\phi}^p; \quad (2.2.50)$$

(2) for $0<p<1$, we have the reverses of (2.2.49) and (2.2.50). In particular, for $\beta=1$, if $p>1$, we have the following equivalent inequalities (Yang JSCNU 2007) [7]:

$$\int_0^\infty \int_0^\infty \frac{|\ln(x/y)|}{(\max\{x,y\})^\lambda} f(x)g(y)dxdy$$
$$< \frac{1}{\lambda^2}(r^2+s^2) \| f \|_{p,\phi} \| g \|_{q,\psi}, \quad (2.2.51)$$

$$\int_0^\infty y^{\frac{p\lambda}{s}-1}[\int_0^\infty \frac{|\ln(x/y)| f(x)}{(\max\{x,y\})^\lambda}dx]^p dy$$
$$< [\frac{1}{\lambda^2}(r^2+s^2)]^p \| f \|_{p,\phi}^p; \quad (2.2.52)$$

if $0<p<1$, we have the equivalent reverses of (2.2.51) and (2.2.52).

Remark 2.2.13 Inequalities (2.2.51) and (2.2.52) are the best extensions of (2.1.36) and (2.1.37), and for $\lambda=1$, they reduce to

$$\int_0^\infty \int_0^\infty \frac{|\ln(x/y)|}{\max\{x,y\}} f(x)g(y)dxdy < (r^2+s^2)$$
$$\times\{\int_0^\infty x^{\frac{p}{s}-1} f^p(x)dx\}^{\frac{1}{p}}\{\int_0^\infty x^{\frac{q}{r}-1} g^q(x)dx\}^{\frac{1}{q}}, \quad (2.2.53)$$

$$\int_0^\infty y^{\frac{p}{s}-1}(\int_0^\infty \frac{|\ln(x/y)|}{\max\{x,y\}} f(x)dx)^p dy$$
$$< (r^2+s^2)^p \int_0^\infty x^{\frac{p}{s}-1} f^p(x)dx, \quad (2.2.54)$$

which are the best extensions of (2.1.38) and (2.1.39) with two pairs of conjugate exponents.

Example 2.2.14 If $r,s \neq 0, \lambda \in \mathbf{R}$,

$$\alpha > \max\{0,\tfrac{\lambda}{r},\tfrac{\lambda}{s}\}, |\eta| < \alpha - \max\{\tfrac{\lambda}{r},\tfrac{\lambda}{s}\},$$

$k_\lambda(x,y) = \frac{(\min\{x,y\})^{\alpha-\lambda}}{x^\alpha+y^\alpha}$, then we obtain

$$0 < K_\lambda(\tfrac{\lambda}{r}+\eta) = \int_0^\infty k_\lambda(u,1)u^{\frac{\lambda}{r}+\eta-1}du$$

$$= \int_0^\infty \frac{(\min\{u,1\})^{\alpha-\lambda}}{1+u^\alpha} u^{\frac{\lambda}{r}+\eta-1}du$$

$$= \int_0^1 \frac{1}{1+u^\alpha} u^{\alpha-\lambda+\frac{\lambda}{r}+\eta-1}du + \int_1^\infty \frac{1}{1+u^{-\alpha}} u^{-\alpha+\frac{\lambda}{r}+\eta-1}du$$

$$= \int_0^1 \sum_{k=0}^\infty (-1)^k u^{\alpha(k+1)-\frac{\lambda}{s}+\eta-1}du$$

$$+ \int_1^\infty \sum_{k=0}^\infty (-1)^k u^{-\alpha(k+1)+\frac{\lambda}{r}+\eta-1}du$$

$$= \sum_{k=0}^\infty (-1)^k [\int_0^1 u^{\alpha(k+1)-\frac{\lambda}{s}+\eta-1}du$$

$$+ \int_1^\infty u^{-\alpha(k+1)+\frac{\lambda}{r}+\eta-1}du]$$

$$= \sum_{k=0}^\infty (-1)^k [\frac{1}{\alpha(k+1)-\frac{\lambda}{s}+\eta} + \frac{1}{\alpha(k+1)-\frac{\lambda}{r}-\eta}] < \infty,$$

$$k_{\lambda,\alpha}(r) := \sum_{k=1}^\infty (-1)^{k-1}[\frac{1}{\alpha k-\frac{\lambda}{s}} + \frac{1}{\alpha k-\frac{\lambda}{r}}].$$

Then by Theorem 2.2.3 and Theorem 2.2.7, (1) for $p>1$, we have the following equivalent inequalities:

$$\int_0^\infty \int_0^\infty \frac{(\min\{x,y\})^{\alpha-\lambda}}{x^\alpha+y^\alpha} f(x)g(y)dxdy$$
$$< k_{\lambda,\alpha}(r) \| f \|_{p,\phi} \| g \|_{q,\psi}, \quad (2.2.55)$$

$$\int_0^\infty y^{\frac{p\lambda}{s}-1}[\int_0^\infty \frac{(\min\{x,y\})^{\alpha-\lambda}}{x^\alpha+y^\alpha} f(x)dx]^p dy$$
$$< [k_{\lambda,\alpha}(r)]^p \| f \|_{p,\phi}^p; \quad (2.2.56)$$

(2) for $0<p<1$, we have the equivalent reverses of (2.2.55) and (2.2.56). In particular, for $\lambda=0$, $r=p=2$ in (2.2.55) and (2.2.56), we have (2.1.59) and (2.1.60).

Example 2.2.15 If $r,s \neq 0, \lambda \in \mathbf{R}$,

$$\alpha > \max\{\tfrac{\lambda}{r},\tfrac{\lambda}{s}\}, |\eta| < \alpha - \max\{\tfrac{\lambda}{r},\tfrac{\lambda}{s}\},$$

$$k_\lambda(x,y) = \frac{(\min\{x,y\})^{\alpha-\lambda}}{(\max\{x,y\})^\alpha},$$

then we obtain

$$0 < K_\lambda(\tfrac{\lambda}{r}+\eta) = \int_0^\infty k_\lambda(u,1)u^{\frac{\lambda}{r}+\eta-1}du$$

$$= \int_0^\infty \frac{(\min\{u,1\})^{\alpha-\lambda}}{(\max\{u,1\})^\alpha} u^{\frac{\lambda}{r}+\eta-1}du$$

$$= \int_0^1 u^{\alpha-\frac{\lambda}{s}+\eta-1}du + \int_1^\infty u^{-\alpha+\frac{\lambda}{r}+\eta-1}du$$

$$= \frac{1}{\alpha-\frac{\lambda}{s}+\eta} + \frac{1}{\alpha-\frac{\lambda}{r}-\eta} < \infty,$$

$$k_\lambda(r) = \frac{1}{\alpha-\frac{\lambda}{s}} + \frac{1}{\alpha-\frac{\lambda}{r}}.$$

Then by Theorem 2.2.3 and Theorem 2.2.7, (1) for $p>1$, we have the following equivalent inequalities:

$$\int_0^\infty\int_0^\infty \frac{(\min\{x,y\})^{\alpha-\lambda}}{(\max\{x,y\})^{\alpha}} f(x)g(y)dxdy < \left(\frac{1}{\alpha-\frac{\lambda}{r}}+\frac{1}{\alpha-\frac{\lambda}{s}}\right)\|f\|_{p,\phi}\|g\|_{q,\psi}, \quad (2.2.57)$$

$$\int_0^\infty y^{\frac{p\lambda}{s}-1}\left[\int_0^\infty \frac{(\min\{x,y\})^{\alpha-\lambda}}{(\max\{x,y\})^{\alpha}} f(x)dx\right]^p dy < \left(\frac{1}{\alpha-\frac{\lambda}{r}}+\frac{1}{\alpha-\frac{\lambda}{s}}\right)^p\|f\|_{p,\phi}^p; \quad (2.2.58)$$

(2) for $0<p<1$, we have the equivalent reverses of (2.2.57) and (2.2.58). In particular, for $\alpha=0$, $\lambda<0$, $r>1$ in (2.2.57) and (2.2.58), replacing $-\lambda$ by $\lambda>0$, we have

$$\int_0^\infty\int_0^\infty (\min\{x,y\})^{\lambda} f(x)g(y)dxdy < \frac{rs}{\lambda}\left\{\int_0^\infty x^{p(1+\frac{\lambda}{r})-1}f^p(x)dx\right\}^{\frac{1}{p}} \times\left\{\int_0^\infty x^{q(1+\frac{\lambda}{s})-1}g^q(x)dx\right\}^{\frac{1}{q}}, \quad (2.2.59)$$

$$\int_0^\infty y^{\frac{-p\lambda}{s}-1}\left[\int_0^\infty (\min\{x,y\})^{\lambda} f(x)dx\right]^p dy < \left(\frac{rs}{\lambda}\right)^p\int_0^\infty x^{p(1+\frac{\lambda}{r})-1}f^p(x)dx. \quad (2.2.60)$$

Example 2.2.16 If $r,s\neq 0,\lambda\in\mathbf{R}$,

$$\max\{\tfrac{\lambda}{r},\tfrac{\lambda}{s}\}<\alpha<1,\ |\eta|<\alpha-\max\{\tfrac{\lambda}{r},\tfrac{\lambda}{s}\},$$

$k_\lambda(x,y)=\frac{(\min\{x,y\})^{\alpha-\lambda}}{|x-y|^\alpha}$, then by (1.2.17), we obtain

$$0<K_\lambda(\tfrac{\lambda}{r}+\eta)=\int_0^\infty k_\lambda(u,1)u^{\frac{\lambda}{r}+\eta-1}du$$
$$=\int_0^\infty \frac{(\min\{u,1\})^{\alpha-\lambda}}{|u-1|^\alpha}u^{\frac{\lambda}{r}+\eta-1}du$$
$$=\int_0^1(1-u)^{(1-\alpha)-1}u^{(\alpha-\frac{\lambda}{s}+\eta)-1}du+\int_1^\infty\frac{(u-1)^{(1-\alpha)-1}}{u^{1-\frac{\lambda}{r}-\eta}}du$$
$$=B(1-\alpha,\alpha-\tfrac{\lambda}{s}+\eta)+B(1-\alpha,\alpha-\tfrac{\lambda}{r}-\eta),$$
$$k_\lambda(r)=B(1-\alpha,\alpha-\tfrac{\lambda}{s})+B(1-\alpha,\alpha-\tfrac{\lambda}{r}).$$

Then by Theorem 2.2.3 and Theorem 2.2.7, (1) for $p>1$, we have the following equivalent inequalities:

$$\int_0^\infty\int_0^\infty \frac{(\min\{x,y\})^{\alpha-\lambda}}{|x-y|^{\alpha}} f(x)g(y)dxdy < [B(1-\alpha,\alpha-\tfrac{\lambda}{s})+B(1-\alpha,\alpha-\tfrac{\lambda}{r})] \times\|f\|_{p,\phi}\|g\|_{q,\psi}, \quad (2.2.61)$$

$$\int_0^\infty y^{\frac{p\lambda}{s}-1}\left[\int_0^\infty \frac{(\min\{x,y\})^{\alpha-\lambda}}{|x-y|^{\alpha}} f(x)dx\right]^p dy < [B(1-\alpha,\alpha-\tfrac{\lambda}{s})+B(1-\alpha,\alpha-\tfrac{\lambda}{r})]^p\|f\|_{p,\phi}^p; \quad (2.2.62)$$

(2) for $0<p<1$, we have the equivalent reverses of (2.2.61) and (2.2.62). In particular, for $0<\alpha=\lambda<1$, $r>1$ in (2.2.61) and (2.2.62), we have (2.2.43) and (2.2.44); for $\lambda=0, p=q=2$ in (2.2.61) and (2.2.62), replacing α by $\lambda(\in(0,1))$, we have (2.1.63) and (2.1.64).

2.2.4. SOME HILBERT-TYPE INTEGRAL INEQUALITIES WITH VARIABLES AS PARAMETERS AND CONJUGATE EXPONENTS

Theorem 2.2.17 Suppose that $p>0(p\neq 1)$, $r,s\neq 0$, $\lambda\in\mathbf{R}$, $k_\lambda(x,y)(\geq 0)$ is a homogeneous function of $-\lambda$-degree in $(0,\infty)\times(0,\infty)$, and there exists a neighborhood I_λ of $\frac{\lambda}{r}$, such that

$$0<K_\lambda(x)=\int_0^\infty k_\lambda(u,1)u^{x-1}du<\infty, x\in I_\lambda.$$

If $u(x)$ and $v(x)$ are strict increasing deliverable functions in (a,b) with $u(a^+)=v(a^+)=0$, $u(b^-)=v(b^-)=\infty$, $f(x),g(x)\geq 0$,

$$0<\int_a^b\frac{(u(x))^{p(1-\frac{\lambda}{r})-1}}{(u'(x))^{p-1}}f^p(x)dx<\infty$$

and $0<\int_a^b\frac{(v(x))^{q(1-\frac{\lambda}{s})-1}}{(v'(x))^{q-1}}g^q(x)dx<\infty$, then (1) for $p>1$, we have the following equivalent inequalities:

$$\int_a^b\int_a^b k_\lambda(u(x),v(y))f(x)g(y)dxdy < k_\lambda(r)\left\{\int_a^b\frac{(u(x))^{p(1-\frac{\lambda}{r})-1}}{(u'(x))^{p-1}}f^p(x)dx\right\}^{\frac{1}{p}} \times\left\{\int_a^b\frac{(v(x))^{q(1-\frac{\lambda}{s})-1}}{(v'(x))^{q-1}}g^q(x)dx\right\}^{\frac{1}{q}}, \quad (2.2.63)$$

$$\int_a^b\frac{v'(y)}{(v(y))^{1-(p\lambda)/s}}\left[\int_a^b k_\lambda(u(x),v(y))f(x)dx\right]^p dy < k_\lambda^p(r)\int_a^b\frac{(u(x))^{p(1-\frac{\lambda}{r})-1}}{(u'(x))^{p-1}}f^p(x)dx; \quad (2.2.64)$$

(2) for $0<p<1$, we have the equivalent reverses of (2.2.63) and (2.2.64) with the best constant factors $k_\lambda(r)$ and $k_\lambda^p(r)$.

Proof (1) For $p>1$, replacing x and y respectively to $u(x)$ and $v(y)$ in (2.2.19) and (2.2.20), by simplifications, then replacing $u'(x)f(u(x))$ and $v'(y)g(v(y))$ respectively by $f(x)$ and $g(y)$, we find (2.2.63) and (2.2.64). It is obvious that inequalities (2.2.19) and (2.2.63) are equivalent; so are (2.2.20) and

(2.22.64). Hence inequalities (2.2.63) and (2.2.64) are equivalent. We can prove that the constant factors in (2.2.63) and (2.2.64) are the best possible by using the equivalent relationship of them. (2) for $0<p<1$, by using the above way in (2.2.21) and (2.2.22), we cane obtain the equivalent reverses of (2.2.63) and (2.2.64) with the best constant factors.

Example 2.2.18 If $u(x)=x^{\alpha}, v(x)=x^{\beta}$ $(\alpha,\beta>0; x\in(0,\infty))$, then for $p>1$ in (2.2.63) and (2.2.64), we have the following equivalent inequalities:

$$\int_0^{\infty}\int_0^{\infty} k_{\lambda}(x^{\alpha},y^{\beta})f(x)g(y)dxdy < \frac{k_{\lambda}(r)}{\alpha^{1/q}\beta^{1/p}}\{\int_0^{\infty} x^{p(1-\frac{\alpha\lambda}{r})-1}f^p(x)dx\}^{\frac{1}{p}} \times\{\int_0^{\infty} x^{q(1-\frac{\beta\lambda}{s})-1}g^q(x)dx\}^{\frac{1}{q}}, \quad (2.2.65)$$

$$\int_0^{\infty} y^{\frac{p\beta\lambda}{s}-1}[\int_0^{\infty} k_{\lambda}(x^{\alpha},y^{\beta})f(x)dx]^p dy < (\frac{k_{\lambda}(r)}{\alpha^{1/q}\beta^{1/p}})^p\int_0^{\infty} x^{p(1-\frac{\alpha\lambda}{r})-1}f^p(x)dx; \quad (2.2.66)$$

for $0<p<1$, we have the equivalent reverses of (2.2.65) and (2.2.66).

Example 2.2.19 If $u(x)=v(x)=\ln x (x\in(1,\infty))$, then for $p>1$ in (2.2.63) and (2.2.64), we have the following equivalent inequalities (Yang JM 2007) [6]:

$$\int_1^{\infty}\int_1^{\infty} k_{\lambda}(\ln x,\ln y)f(x)g(y)dxdy < k_{\lambda}(r)\{\int_1^{\infty} x^{p-1}(\ln x)^{p(1-\frac{\lambda}{r})-1}f^p(x)dx\}^{\frac{1}{p}} \times\{\int_1^{\infty} x^{q-1}(\ln x)^{q(1-\frac{\lambda}{s})-1}g^q(x)dx\}^{\frac{1}{q}}, \quad (2.2.67)$$

$$\int_1^{\infty}\frac{(\ln y)^{(p\lambda/s)-1}}{y}[\int_1^{\infty} k_{\lambda}(\ln x,\ln y)f(x)dx]^p dy < k_{\lambda}^p(r)\int_1^{\infty} x^{p-1}(\ln x)^{p(1-\frac{\lambda}{r})-1}f^p(x)dx; \quad (2.2.68)$$

for $0<p<1$, we have the equivalent reverses of (2.2.67) and (2.2.68). In particular, setting

$$k_{\lambda}(x,y)=\frac{1}{(x+y)^{\lambda}} (r,s>1,\lambda>0),$$

for $p>1$, we have the following equivalent inequalities:

$$\int_1^{\infty}\int_1^{\infty}\frac{1}{(\ln xy)^{\lambda}}f(x)g(y)dxdy < B(\frac{\lambda}{r},\frac{\lambda}{s})\{\int_1^{\infty} x^{p-1}(\ln x)^{p(1-\frac{\lambda}{r})-1}f^p(x)dx\}^{\frac{1}{p}} \times\{\int_1^{\infty} x^{q-1}(\ln x)^{q(1-\frac{\lambda}{s})-1}g^q(x)dx\}^{\frac{1}{q}}, \quad (2.2.69)$$

$$\int_1^{\infty}\frac{(\ln y)^{(p\lambda/s)-1}}{y}[\int_1^{\infty}\frac{f(x)}{(\ln xy)^{\lambda}}dx]^p dy < [B(\frac{\lambda}{r},\frac{\lambda}{s})]^p \times\int_1^{\infty} x^{p-1}(\ln x)^{p(1-\frac{\lambda}{r})-1}f^p(x)dx; \quad (2.2.70)$$

for $0<p<1$, we have the equivalent reverses of (2.2.69) and (2.2.70).

Example 2.2.20 If $u(x)=v(x)=e^x, x\in(-\infty,\infty)$, then for $p>1$ in (2.2.63) and (2.2.64), we have the following equivalent inequalities:

$$\int_{-\infty}^{\infty}\int_{-\infty}^{\infty} k_{\lambda}(e^x,e^y)f(x)g(y)dxdy < k_{\lambda}(r)(\int_{-\infty}^{\infty} e^{\frac{-p\lambda}{r}x}f^p(x)dx)^{\frac{1}{p}}(\int_{-\infty}^{\infty} e^{\frac{-q\lambda}{s}x}g^q(x)dx)^{\frac{1}{q}}, \quad (2.2.71)$$

$$\int_{-\infty}^{\infty} e^{\frac{p\lambda}{s}y}[\int_{-\infty}^{\infty} k_{\lambda}(e^x,e^y)f(x)dx]^p dy < k_{\lambda}^p(r)\int_{-\infty}^{\infty} e^{-\frac{p\lambda}{r}x}f^p(x)dx; \quad (2.2.72)$$

for $0<p<1$, we have the equivalent reverses of (2.2.71) and (2.2.72).

Example 2.2.21 If $u(x)=v(x)=\tan x$ $(x\in(0,\frac{\pi}{2}))$, then for $p>1$ in (2.2.63) and (2.2.64), we have the following equivalent inequalities:

$$\int_0^{\frac{\pi}{2}}\int_0^{\frac{\pi}{2}} k_{\lambda}(\tan x,\tan y)f(x)g(y)dxdy < k_{\lambda}(r)\{\int_0^{\frac{\pi}{2}}\frac{(\tan x)^{p(1-\frac{\lambda}{r})-1}}{(\sec x)^{2(p-1)}}f^p(x)dx\}^{\frac{1}{p}} \times\{\int_0^{\frac{\pi}{2}}\frac{(\tan x)^{q(1-\frac{\lambda}{s})-1}}{(\sec x)^{2(q-1)}}g^q(x)dx\}^{\frac{1}{q}}, \quad (2.2.73)$$

$$\int_0^{\frac{\pi}{2}}\frac{\sec^2 y}{(\tan y)^{1-(p\lambda)/s}}[\int_0^{\frac{\pi}{2}} k_{\lambda}(\tan x,\tan y)f(x)dx]^p dy < k_{\lambda}^p(r)\int_0^{\frac{\pi}{2}}\frac{(\tan x)^{p(1-\frac{\lambda}{r})-1}}{(\sec x)^{2(p-1)}}f^p(x)dx; \quad (2.2.74)$$

for $0<p<1$, we have the equivalent reverses of (2.2.73) and (2.2.74).

Example 2.2.22 If $u(x)=v(x)=\sec x-1$ $(x\in(0,\frac{\pi}{2}))$, then for $p>1$ in (2.2.63) and (2.2.64), we have the following equivalent inequalities:

$$\int_0^{\frac{\pi}{2}}\int_0^{\frac{\pi}{2}} k_{\lambda}(\sec x-1,\sec y-1)f(x)g(y)dxdy < k_{\lambda}(r)\{\int_0^{\frac{\pi}{2}}\frac{(\sec x-1)^{p(1-\lambda/r)-1}}{(\sec x\tan x)^{p-1}}f^p(x)dx\}^{\frac{1}{p}}$$

$$\times\{\int_0^{\frac{\pi}{2}}\frac{(\sec x-1)^{q(1-\lambda/s)-1}}{(\sec x\tan x)^{q-1}}g^q(x)dx\}^{\frac{1}{q}}, \quad (2.2.75)$$

$$\int_0^{\frac{\pi}{2}}\frac{\sec y\tan y}{(\sec y-1)^{1-(p\lambda)/s}}$$

$$\times[\int_0^{\frac{\pi}{2}}k_\lambda(\sec x-1,\sec y-1)f(x)dx]^p dy$$

$$<k_\lambda^p(r)\int_0^{\frac{\pi}{2}}\frac{(\sec x-1)^{p(1-\lambda/r)-1}}{(\sec x\tan x)^{p-1}}f^p(x)dx; \quad (2.2.76)$$

for $0<p<1$, we have the equivalent reverses of (2.2.75) and (2.2.76).

2.2.4. SOME HARDY-TYPE INTEGRAL INEQUALITIES

Corollary 2.2.23 As the assumption of Theorem 2.2.3, if $\lambda\in\mathbf{R}$, $\tilde{k}_\lambda(x,y)(\geq 0)$ is a homogeneous function of $-\lambda$-degree,

$$k_\lambda(x,y)=\tilde{k}_\lambda(x,y),0<y\leq x;$$

$k_\lambda(x,y)=0, y>x$, and

$$0<\tilde{k}_\lambda(r):=\int_0^1\tilde{k}_\lambda(1,u)u^{\frac{\lambda}{r}-1}du<\infty,$$

then we have the following equivalent inequalities:

$$\int_0^\infty[\int_0^x\tilde{k}_\lambda(x,y)g(y)dy]f(x)dx$$

$$=\int_0^\infty[\int_y^\infty\tilde{k}_\lambda(x,y)f(x)dx]g(y)dy$$

$$<\tilde{k}_\lambda(r)\|f\|_{p,\phi}\|g\|_{q,\psi}, \quad (2.2.77)$$

$$\int_0^\infty y^{\frac{p\lambda}{s}-1}(\int_y^\infty\tilde{k}_\lambda(x,y)f(x)dx)^p dy$$

$$<\tilde{k}_\lambda^p(r)\|f\|_{p,\phi}^p, \quad (2.2.78)$$

$$\int_0^\infty x^{\frac{q\lambda}{r}-1}(\int_0^x\tilde{k}_\lambda(x,y)g(y)dy)^q dx$$

$$<\tilde{k}_\lambda^q(r)\|g\|_{q,\psi}^q, \quad (2.2.79)$$

where the constant factors $\tilde{k}_\lambda(r)$ and $\tilde{k}_\lambda^\rho(r)$ $(\rho=p,q)$ are the best possible.

Corollary 2.2.24 As the assumption of Corollary 2.2.23, if $0<p<1$, if there exists a neighborhood I_λ of $\frac{\lambda}{r}$, such that for $x\in I_\lambda$,

$$0<H_\lambda(x)=\int_0^1\tilde{k}_\lambda(u,1)u^{x-1}du<\infty,$$

then we have the following equivalent inequalities:

$$\int_0^\infty[\int_0^x\tilde{k}_\lambda(x,y)g(y)dy]f(x)dx$$

$$=\int_0^\infty[\int_y^\infty\tilde{k}_\lambda(x,y)f(x)dx]g(y)dy$$

$$>\tilde{k}_\lambda(r)\|f\|_{p,\phi}\|g\|_{q,\psi}, \quad (2.2.80)$$

$$\int_0^\infty y^{\frac{p\lambda}{s}-1}(\int_y^\infty\tilde{k}_\lambda(x,y)f(x)dx)^p dy$$

$$>\tilde{k}_\lambda^p(r)\|f\|_{p,\phi}^p, \quad (2.2.81)$$

$$\int_0^\infty x^{\frac{q\lambda}{r}-1}(\int_0^x\tilde{k}_\lambda(x,y)g(y)dy)^q dx$$

$$<\tilde{k}_\lambda^q(r)\|g\|_{q,\psi}^q, \quad (2.2.82)$$

where the constant factors $\tilde{k}_\lambda(r)$ and $\tilde{k}_\lambda^\rho(r)$ $(\rho=p,q)$ are the best possible.

In the following examples, the conditions

$$\phi(x)=x^{p(1-\frac{\lambda}{r})-1},\psi(x)=x^{q(1-\frac{\lambda}{s})-1},f(x),g(x)\geq 0,$$

$$0<\|f\|_{p,\phi}=\{\int_0^\infty x^{p(1-\frac{\lambda}{r})-1}f^p(x)dx\}^{\frac{1}{p}}<\infty,$$

$$0<\|g\|_{q,\psi}=\{\int_0^\infty x^{q(1-\frac{\lambda}{s})-1}g^q(x)dx\}^{\frac{1}{q}}<\infty$$

and the conclusions that the constant factors are the best possible are omitted.

Example 2.2.25 If $\beta>-1,r,s>1,\lambda>0$,

$$\tilde{k}_\lambda(x,y)=\frac{|\ln(x/y)|^\beta}{(\max\{x,y\})^\lambda},$$

then by (2.2.48), we obtain

$$\tilde{k}_\lambda(r)=\int_0^1 k_\lambda(u,1)u^{\frac{\lambda}{r}-1}du=\int_0^1\frac{|\ln u|^\beta u^{\frac{\lambda}{r}-1}}{(\max\{u,1\})^\lambda}du$$

$$=\int_0^1(-\ln u)^{(\beta+1)-1}u^{\frac{\lambda}{r}-1}du=\Gamma(\beta+1)(\tfrac{r}{\lambda})^{\beta+1}.$$

Then by Corollary 2.2.23 and Corollary 2.2.24, (1) for $p>1$, we have the following equivalent inequalities:

$$\int_0^\infty[\int_0^x\ln^\beta(x/y)g(y)dy]\frac{1}{x^\lambda}f(x)dx$$

$$=\int_0^\infty[\int_y^\infty\frac{\ln^\beta(x/y)}{x^\lambda}f(x)dx]g(y)dy$$

$$<(\tfrac{r}{\lambda})^{\beta+1}\Gamma(\beta+1)\|f\|_{p,\phi}\|g\|_{q,\psi}, \quad (2.2.83)$$

$$\int_0^\infty y^{\frac{p\lambda}{s}-1}[\int_y^\infty\frac{\ln^\beta(x/y)f(x)}{x^\lambda}dx]^p dy$$

$$<[(\tfrac{r}{\lambda})^{\beta+1}\Gamma(\beta+1)]^p\|f\|_{p,\phi}^p, \quad (2.2.84)$$

$$\int_0^\infty\frac{1}{x^{\frac{q\lambda}{s}+1}}[\int_0^x\ln^\beta(x/y)g(y)dy]^q dx$$

$$<[(\tfrac{r}{\lambda})^{\beta+1}\Gamma(\beta+1)]^q\|g\|_{q,\psi}; \quad (2.2.85)$$

(2) for $0<p<1$, we have the equivalent reverses of (2.2.83) and (2.2.84) (but the reverse of (2.2.85) keeps the same form). In particular, for $\beta=0,\lambda>0$, $r=q>1$ in (2.2.84), and (2.2.85), we have the

following extensions of Hardy's integral inequalities (Hardy CP 1934) [4]:

$$\int_0^\infty y^{\lambda-1}\left(\int_y^\infty \frac{1}{x^\lambda} f(x)dx\right)^p dy < \left(\frac{q}{\lambda}\right)^p \int_0^\infty x^{(p-1)(1-\lambda)} f^p(x)dx, \tag{2.2.86}$$

$$\int_0^\infty \frac{1}{x^{(q-1)\lambda+1}}\left(\int_y^\infty g(y)dy\right)^q dx < \left(\frac{q}{\lambda}\right)^q \int_0^\infty y^{(q-1)(1-\lambda)} g^q(y)dy. \tag{2.2.87}$$

Example 2.2.26 If $r,s \neq 0, \lambda \in \mathbf{R}, \frac{\lambda}{s} < \alpha < 1$, $\tilde{k}_\lambda(x,y) = \frac{(\min\{x,y\})^{\alpha-\lambda}}{|x-y|^\alpha}$, then by (1.2.17), we obtain

$$\tilde{k}_\lambda(r) = \int_0^1 k_\lambda(u,1) u^{\frac{\lambda}{r}-1} du = \int_0^1 \frac{(\min\{u,1\})^{\alpha-\lambda}}{|u-1|^\alpha} u^{\frac{\lambda}{r}-1} du = \int_0^1 (1-u)^{(1-\alpha)-1} u^{(\alpha-\frac{\lambda}{s})-1} du = B(1-\alpha, \alpha-\tfrac{\lambda}{s}),$$

Hence by Corollary 2.2.23 and Corollary 2.2.24, (1) for $p>1$, we have the following equivalent inequalities:

$$\int_0^\infty \left(\int_y^\infty \frac{1}{|x-y|^\alpha} f(x)dx\right) y^{\alpha-\lambda} g(y)dy = \int_0^\infty \left(\int_0^x \frac{y^{\alpha-\lambda}}{|x-y|^\alpha} g(y)dy\right) f(x)dx < B(1-\alpha, \alpha-\tfrac{\lambda}{s}) \| f\|_{p,\phi} \| g\|_{q,\psi}, \tag{2.2.88}$$

$$\int_0^\infty y^{p(\alpha-\frac{\lambda}{r})-1}\left(\int_y^\infty \frac{1}{|x-y|^\alpha} f(x)dx\right)^p dy < [B(1-\alpha, \alpha-\tfrac{\lambda}{s})]^p \| f\|_{p,\phi}^p, \tag{2.2.89}$$

$$\int_0^\infty x^{\frac{q\lambda}{r}-1}\left(\int_0^x \frac{y^{\alpha-\lambda}}{|x-y|^\alpha} g(y)dy\right)^q dx < [B(1-\alpha, \alpha-\tfrac{\lambda}{s})]^q \| g\|_{q,\psi}; \tag{2.2.90}$$

(2) for $0<p<1$, we have the equivalent reverses of (2.2.88) and (2.2.89) (but the reverse of (2.2.90) keeps the same form). In particular, (i) for $0<\alpha=\lambda<1, r,s>1$ if $p>1$, we have the following equivalent inequalities:

$$\int_0^\infty \left(\int_y^\infty \frac{1}{|x-y|^\lambda} f(x)dx\right) g(y)dy = \int_0^\infty \left(\int_0^x \frac{1}{|x-y|^\lambda} g(y)dy\right) f(x)dx < B(1-\lambda, \tfrac{\lambda}{r}) \times \| f\|_{p,\phi} \| g\|_{q,\psi}, \tag{2.2.91}$$

$$\int_0^\infty y^{\frac{p\lambda}{s}-1}\left(\int_y^\infty \frac{1}{|x-y|^\lambda} f(x)dx\right)^p dy < [B(1-\lambda, \tfrac{\lambda}{r})]^p \| f\|_{p,\phi}^p, \tag{2.2.92}$$

$$\int_0^\infty x^{\frac{q\lambda}{r}-1}\left(\int_0^x \frac{1}{|x-y|^\lambda} g(y)dy\right)^q dx < [B(1-\lambda, \tfrac{\lambda}{r})]^q \| g\|_{q,\psi}; \tag{2.2.93}$$

if $0<p<1$, we have the equivalent reverses of (2.2.91) and (2.2.92) (but the reverse of (2.2.93) keeps the same form); (ii) for $\alpha=0$, $\frac{\lambda}{s}<0$, if $p>1$, we have the following equivalent inequalities:

$$\int_0^\infty \left(\int_y^\infty f(x)dx\right) y^{-\lambda} g(y)dy = \int_0^\infty \left[\int_0^x y^{-\lambda} g(y)dy\right] f(x)dx < \frac{-s}{\lambda} \| f\|_{p,\phi} \| g\|_{q,\psi}, \tag{2.2.94}$$

$$\int_0^\infty y^{-\frac{p\lambda}{r}-1}\left(\int_y^\infty f(x)dx\right)^p dy < \left(\frac{-s}{\lambda}\right)^p \| f\|_{p,\phi}^p, \tag{2.2.95}$$

$$\int_0^\infty x^{\frac{q\lambda}{r}-1}\left(\int_0^x y^{-\lambda} g(y)dy\right)^q dx < \left(\frac{-s}{\lambda}\right)^q \| g\|_{q,\psi}. \tag{2.2.96}$$

if $0<p<1$, we have the equivalent reverses of (2.2.94) and (2.2.95) (but the reverse of (2.2.96) keeps the same form).

2.3. SOME SUFFICIENT CONDITIONS OF BOUNDEDNESS IN HILBERT-TYPE INTEGRAL OPERATOR AND APPLICATIONS

If $r,s \neq 0, \lambda \in \mathbf{R}$, $k_\lambda(x,y) (\geq 0)$ is a homogeneous function of $-\lambda$-degree in $(0,\infty)\times(0,\infty)$,

$$k_\lambda(r) = \int_0^\infty k_\lambda(u,1) u^{\frac{\lambda}{r}-1} du$$

is a finite number, then setting $u=\frac{1}{v}$, we find

$$\int_1^\infty k_\lambda(u,1) u^{\frac{\lambda}{r}-1} du = \int_0^1 k_\lambda(1,v) v^{\frac{\lambda}{s}-1} dv,$$

and then

$$\int_0^\infty k_\lambda(u,1) u^{\frac{\lambda}{r}-1} du = \int_0^1 k_\lambda(u,1) u^{\frac{\lambda}{r}-1} du + \int_0^1 k_\lambda(1,u) u^{\frac{\lambda}{s}-1} du. \tag{2.3.1}$$

In this section, by using (2.3.1), we give some sufficient conditions that make $k_\lambda(r) = \int_0^\infty k_\lambda(u,1) u^{\frac{\lambda}{r}-1} du$ as a finite number. As applications, we consider some new equivalent inequalities (cf. (Yang TJM 2008) [21], (Yang JIA 2007) [22]).

In the following examples, the conditions $\phi(x)=x^{p(1-\frac{\lambda}{r})-1}$, $\psi(x)=x^{q(1-\frac{\lambda}{s})-1}$, $f,g\geq 0$,

$$0<\|f\|_{p,\phi}=\{\int_0^\infty x^{p(1-\frac{\lambda}{r})-1}f^p(x)dx\}^{\frac{1}{p}}<\infty,$$

$$0<\|g\|_{q,\psi}=\{\int_0^\infty x^{q(1-\frac{\lambda}{s})-1}g^q(x)dx\}^{\frac{1}{q}}<\infty$$

and the conclusions that the constant factors are the best possible are omitted.

2.3.1. THE CASE THAT THE KERNELS WITH SINGLE VARIABLE ARE BOUNDED IN $(0,1)$

Theorem 2.3.1 If $\lambda\in\mathbf{R}, r,s\neq 0$, both $k_\lambda(u,1)$ and $k_\lambda(1,u)$ are bounded and not zero almost everywhere in the interval $(0,1)$, there exist $L>0$, $\sigma>\max\{\frac{-\lambda}{r},\frac{-\lambda}{s}\}$, such that

$$0\leq k_\lambda(u,1)\leq Lu^\sigma,$$

$$0\leq k_\lambda(1,u)\leq Lu^\sigma \quad (u\in(0,1)).$$

then we have

$$0<k_\lambda(r)=\int_0^\infty k_\lambda(u,1)u^{\frac{\lambda}{r}-1}du<\infty.$$

Proof By (2.3.1), we find

$$0<\int_0^\infty k_\lambda(u,1)u^{\frac{\lambda}{r}-1}du$$

$$=\int_0^1 k_\lambda(u,1)u^{\frac{\lambda}{r}-1}du+\int_0^1 k_\lambda(1,u)u^{\frac{\lambda}{s}-1}du$$

$$\leq L\int_0^1(u^{\sigma+\frac{\lambda}{r}-1}+u^{\sigma+\frac{\lambda}{s}-1})du=L(\frac{1}{\sigma+\frac{\lambda}{r}}+\frac{1}{\sigma+\frac{\lambda}{s}}).$$

Hence $0<k_\lambda(r)=\int_0^\infty k_\lambda(u,1)u^{\frac{\lambda}{r}-1}du<\infty$.□

Lemma 2.3.2 If $f_k(x)(k\in\mathbf{N}_0)$ are non-negative measurable functions, $g(x):=\sum_{k=0}^\infty b_k f_k(x)$ is a measurable function and the integral is determinate in E, then we have

$$\int_E g(x)dx=\sum_{k=0}^\infty b_k\int_E f_k(x)dx. \quad (2.3.2)$$

Proof Setting $\mathbf{N}_1=\{k\in\mathbf{N}_0|\, b_k\geq 0\}$,

$$\mathbf{N}_2=\{k\in\mathbf{N}_0|\, b_k<0\},$$

then both $g^+(x)=\sum_{k\in N_1}b_k f_k(x)$ and

$$g^-(x)=\sum_{k\in N_2}(-b_k)f_k(x)$$

are non-negative measurable. By Lebesgue's term by term integration theorem (Kuang HEP 1996) [2], we have

$$\int_E g(x)dx=\int_E g^+(x)dx-\int_E g^-(x)dx$$

$$=\int_E\sum_{k\in N_1}b_k f_k(x)dx-\int_E\sum_{k\in N_2}(-b_k)f_k(x)dx$$

$$=\sum_{k\in N_1}b_k\int_E f_k(x)dx+\sum_{k\in N_2}b_k\int_E f_k(x)dx$$

$$=\sum_{k=0}^\infty b_k\int_E f_k(x)dx.$$

Hence (2.3.2) is valid.□

Example 2.3.3 If $\lambda>0, r,s>1$,

$$k_\lambda(x,y)=\frac{1}{(x+y)^{\lambda-\alpha}(\max\{x,y\})^\alpha},$$

there exist $\sigma=0$, $L=\max\{1,2^{\alpha-\lambda}\}$, such that

$$k_\lambda(u,1)=k_\lambda(1,u)=\frac{1}{(u+1)^{\lambda-\alpha}}\leq L, u\in(0,1),$$

then by Theorem 2.3.1, we have

$$0<k_\lambda(r)=\int_0^\infty k_\lambda(u,1)u^{\frac{\lambda}{r}-1}du<\infty.$$

By Lemma 2.3.2, there exists a neighborhood I_λ of $\frac{\lambda}{r}$, such that for $\frac{\lambda'}{r}\in I_\lambda$,

$$0<\int_0^1 k_\lambda(u,1)u^{\frac{\lambda'}{r}-1}du+\int_0^1 k_\lambda(1,u)u^{\frac{\lambda'}{s}-1}du$$

$$=\int_0^1(1+u)^{\alpha-\lambda}(u^{\frac{\lambda'}{r}-1}+u^{\frac{\lambda'}{s}-1})du$$

$$=\int_0^1\sum_{k=0}^\infty\binom{\alpha-\lambda}{k}u^k(u^{\frac{\lambda'}{r}-1}+u^{\frac{\lambda'}{s}-1})du$$

$$=\sum_{k=0}^\infty\binom{\alpha-\lambda}{k}\int_0^1(u^{\frac{\lambda'}{r}+k-1}+u^{\frac{\lambda'}{s}+k-1})du$$

$$=\sum_{k=0}^\infty\binom{\alpha-\lambda}{k}\frac{(\lambda'+2k)rs}{(\lambda'+rk)(\lambda'+sk)}<\infty,$$

$$k_\lambda(r)=\sum_{k=0}^\infty\binom{\alpha-\lambda}{k}\frac{(\lambda+2k)rs}{(\lambda+rk)(\lambda+sk)}.$$

By Theorem 2.2.3 and Theorem 2.2.7, for $p>1$, we have the following equivalent inequalities:

$$\int_0^\infty\int_0^\infty\frac{f(x)g(y)}{(x+y)^{\lambda-\alpha}(\max\{x,y\})^\alpha}dxdy$$
$$<k_\lambda(r)\|f\|_{p,\phi}\|g\|_{q,\psi}, \quad (2.3.3)$$

$$\int_0^\infty y^{\frac{p\lambda}{s}-1}[\int_0^\infty\frac{f(x)}{(x+y)^{\lambda-\alpha}(\max\{x,y\})^\alpha}dx]^p dy$$
$$<k_\lambda^p(r)\|f\|_{p,\phi}^p; \quad (2.3.4)$$

for $0<p<1$, we have the equivalent reverses of (2.3.3) and (2.3.4).

Example 2.3.4 If $\lambda > \max\{0,\alpha\}, r, s > 1$,

$$k_\lambda(x,y) = \frac{1}{(x^{\lambda-\alpha}+y^{\lambda-\alpha})(\max\{x,y\})^\alpha},$$

there exist $\sigma = 0,\ L = 1$, such that

$$k_\lambda(u,1) = k_\lambda(1,u) = \frac{1}{u^{\lambda-\alpha}+1} \le 1, u \in (0,1),$$

then by Theorem 2.3.1, we have

$$0 < k_\lambda(r) = \int_0^\infty k_\lambda(u,1)u^{\frac{\lambda}{r}-1}du < \infty.$$

By Lemma 2.3.2, there exists a neighborhood I_λ of $\frac{\lambda}{r}$, such that for $\frac{\lambda'}{r} \in I_\lambda$,

$$0 < \int_0^1 k_\lambda(u,1)u^{\frac{\lambda'}{r}-1}du + \int_0^1 k_\lambda(1,u)u^{\frac{\lambda'}{s}-1}du$$

$$= \int_0^1 (1+u^{\lambda-\alpha})^{-1}(u^{\frac{\lambda'}{r}-1} + u^{\frac{\lambda'}{s}-1})du$$

$$= \int_0^1 \sum_{k=0}^\infty (-1)^k u^{k(\lambda-\alpha)}(u^{\frac{\lambda'}{r}-1} + u^{\frac{\lambda'}{s}-1})du$$

$$= \sum_{k=0}^\infty (-1)^k \int_0^1 (u^{\frac{\lambda'}{r}+k(\lambda-\alpha)-1} + u^{\frac{\lambda'}{s}+k(\lambda-\alpha)-1})du$$

$$= \sum_{k=0}^\infty (-1)^k [\frac{1}{k(\lambda-\alpha)+\frac{\lambda'}{r}} + \frac{1}{k(\lambda-\alpha)+\frac{\lambda'}{s}}] < \infty,$$

$$k_\lambda(r) = \sum_{k=0}^\infty \frac{(-1)^k[(2k+1)\lambda - 2k\alpha]rs}{[(rk+1)\lambda - rk\alpha][(sk+1)\lambda - sk\alpha]}.$$

By Theorem 2.2.3 and Theorem 2.2.7, for $p > 1$, we have the following equivalent inequalities:

$$\int_0^\infty \int_0^\infty \frac{f(x)g(y)}{(x^{\lambda-\alpha}+y^{\lambda-\alpha})(\max\{x,y\})^\alpha}dxdy < k_\lambda(r)\,\|f\|_{p,\phi}\|g\|_{q,\psi}, \quad (2.3.5)$$

$$\int_0^\infty y^{\frac{p\lambda}{s}-1}[\int_0^\infty \frac{f(x)}{(x^{\lambda-\alpha}+y^{\lambda-\alpha})(\max\{x,y\})^\alpha}dx]^p dy < k_\lambda^p(r)\,\|f\|_{p,\phi}^p; \quad (2.3.6)$$

for $0 < p < 1$, we have the equivalent reverses of (2.3.5) and (2.3.6). In particular, for $r = s = 2$, $\lambda = 1$, $\alpha = \frac{1}{2}$, $k_1(2) = 4\ln 2$, if $p > 1$, then we have the following equivalent inequalities:

$$\int_0^\infty \int_0^\infty \frac{f(x)g(y)}{(\sqrt{x}+\sqrt{y})\sqrt{\max\{x,y\}}}dxdy < 4\ln 2\{\int_0^\infty x^{\frac{p}{2}-1}f^p(x)dx\}^{\frac{1}{p}}\{\int_0^\infty x^{\frac{q}{2}-1}g^q(x)dx\}^{\frac{1}{q}}, \quad (2.3.7)$$

$$\int_0^\infty y^{\frac{p}{2}-1}[\int_0^\infty \frac{f(x)}{(\sqrt{x}+\sqrt{y})\sqrt{\max\{x,y\}}}dx]^p dy < (4\ln 2)^p \int_0^\infty x^{\frac{p}{2}-1}f^p(x)dx; \quad (2.3.8)$$

if $0 < p < 1$, then we have the equivalent reverses of (2.3.7) and (2.3.8).

Example 2.3.5 If $0 < \lambda < \alpha, r, s > 1$,

$$k_\lambda(x,y) = \frac{1}{(x^{\lambda-\alpha}+y^{\lambda-\alpha})(\max\{x,y\})^\alpha},$$

there exist $\sigma = 0,\ L = 1$, such that

$$k_\lambda(u,1) = k_\lambda(1,u) = \frac{u^{\alpha-\lambda}}{u^{\alpha-\lambda}+1} \le 1\ (u \in (0,1)),$$

then by Theorem 2.3.1, we have

$$0 < k_\lambda(r) = \int_0^\infty k_\lambda(u,1)u^{\frac{\lambda}{r}-1}du < \infty.$$

By Lemma 2.3.2, there exists a neighborhood I_λ of $\frac{\lambda}{r}$, such that for $\frac{\lambda'}{r} \in I_\lambda$,

$$0 < \int_0^1 k_\lambda(u,1)u^{\frac{\lambda'}{r}-1}du + \int_0^1 k_\lambda(1,u)u^{\frac{\lambda'}{s}-1}du$$

$$= \int_0^1 (1+u^{\alpha-\lambda})^{-1}(u^{-\lambda+\frac{\lambda'}{r}+\alpha-1} + u^{-\lambda+\frac{\lambda'}{s}+\alpha-1})du$$

$$= \int_0^1 \sum_{k=0}^\infty (-1)^k u^{k(\alpha-\lambda)}(u^{-\lambda+\frac{\lambda'}{r}+\alpha-1} + u^{-\lambda+\frac{\lambda'}{s}+\alpha-1})du$$

$$= \sum_{k=0}^\infty (-1)^k \int_0^1 (u^{\frac{\lambda'}{r}+(k+1)(\alpha-\lambda)-1} + u^{\frac{\lambda'}{s}+(k+1)(\alpha-\lambda)-1})du$$

$$= \sum_{k=0}^\infty (-1)^k [\frac{1}{(k+1)(\alpha-\lambda)+\frac{\lambda'}{r}} + \frac{1}{(k+1)(\alpha-\lambda)+\frac{\lambda'}{s}}] < \infty,$$

$$k_\lambda(r) = \sum_{k=1}^\infty (-1)^{k-1}[\frac{1}{k(\alpha-\lambda)+\frac{\lambda}{r}} + \frac{1}{k(\alpha-\lambda)+\frac{\lambda}{s}}].$$

By Theorem 2.2.3 and Theorem 2.2.7, for $p > 1$, we have the following equivalent inequalities:

$$\int_0^\infty \int_0^\infty \frac{f(x)g(y)}{(x^{\lambda-\alpha}+y^{\lambda-\alpha})(\max\{x,y\})^\alpha}dxdy < k_\lambda(r)\,\|f\|_{p,\phi}\|g\|_{q,\psi}, \quad (2.3.9)$$

$$\int_0^\infty y^{\frac{p\lambda}{s}-1}[\int_0^\infty \frac{f(x)}{(x^{\lambda-\alpha}+y^{\lambda-\alpha})(\max\{x,y\})^\alpha}dx]^p dy < k_\lambda^p(r)\,\|f\|_{p,\phi}^p; \quad (2.3.10)$$

for $0 < p < 1$, we have the equivalent reverses of (2.3.9) and (2.3.10). In particular, for $r = s = 2$, $\lambda = \frac{1}{2}, \alpha = 1$, $k_{\frac{1}{2}}(2) = 2(4-\pi)$, if $p > 1$, then we have the following equivalent inequalities:

$$\int_0^\infty \int_0^\infty \frac{f(x)g(y)}{(\sqrt{x^{-1}}+\sqrt{y^{-1}})\max\{x,y\}}dxdy < 2(4-\pi) \times\{\int_0^\infty x^{\frac{p}{4}-1}f^p(x)dx\}^{\frac{1}{p}}\{\int_0^\infty x^{\frac{q}{4}-1}g^q(x)dx\}^{\frac{1}{q}}, \quad (2.3.11)$$

$$\int_0^\infty y^{\frac{p}{4}-1}[\int_0^\infty \frac{f(x)dx}{(\sqrt{x^{-1}}+\sqrt{y^{-1}})\max\{x,y\}}]^p dy < [2(4-\pi)]^p \int_0^\infty x^{\frac{p}{4}-1}f^p(x)dx; \quad (2.3.12)$$

if $0 < p < 1$, then we have the equivalent reverses of (2.3.11) and (2.3.12).

Example 2.3.6 If $\lambda \in \mathbf{R}$, $r, s \neq 0$,

$$\alpha > \max\{\tfrac{-\lambda}{r}, \tfrac{-\lambda}{s}\},\ k_\lambda(x,y) = \tfrac{(\min\{x,y\})^\alpha}{(x+y)^{\lambda+\alpha}},$$

there exist $\sigma = \alpha$, $L = \max\{1, (\frac{1}{2})^{\lambda+\alpha}\}$, such that

$$k_\lambda(u,1) = k_\lambda(1,u) \le Lu^\sigma \ (u \in (0,1)),$$

then by Theorem 2.3.1, we have

$$0 < k_\lambda(r) = \int_0^\infty k_\lambda(u,1) u^{\frac{\lambda}{r}-1} du < \infty.$$

By Lemma 2.3.2, there exists a neighborhood I_λ of $\frac{\lambda}{r}$, such that for $\frac{\lambda'}{r} \in I_\lambda$,

$$0 < \int_0^1 k_\lambda(u,1) u^{\frac{\lambda'}{r}-1} du + \int_0^1 k_\lambda(1,u) u^{\frac{\lambda'}{s}-1} du$$

$$= \int_0^1 \tfrac{1}{(1+u)^{\lambda+\alpha}} u^{\alpha+\frac{\lambda'}{r}-1} du + \int_0^1 \tfrac{1}{(1+u)^{\lambda+\alpha}} u^{\alpha+\frac{\lambda'}{s}-1} du$$

$$= \int_0^1 \sum_{k=0}^\infty \binom{-\lambda-\alpha}{k} [u^{k+\alpha+\frac{\lambda'}{r}-1} + u^{k+\alpha+\frac{\lambda'}{s}-1}] du$$

$$= \sum_{k=0}^\infty \binom{-\lambda-\alpha}{k} \int_0^1 [u^{k+\alpha+\frac{\lambda'}{r}-1} + u^{k+\alpha+\frac{\lambda'}{s}-1}] du$$

$$= \sum_{k=0}^\infty \binom{-\lambda-\alpha}{k} [\tfrac{1}{k+\alpha+\frac{\lambda'}{r}} + \tfrac{1}{k+\alpha+\frac{\lambda'}{s}}] < \infty,$$

$$k_\lambda(r) = \sum_{k=0}^\infty \binom{-\lambda-\alpha}{k} [\tfrac{1}{k+\alpha+\frac{\lambda}{r}} + \tfrac{1}{k+\alpha+\frac{\lambda}{s}}].$$

By Theorem 2.2.3 and Theorem 2.2.7, for $p > 1$, we have the following equivalent inequalities:

$$\int_0^\infty \int_0^\infty \tfrac{(\min\{x,y\})^\alpha}{(x+y)^{\lambda+\alpha}} f(x) g(y) dx dy$$

$$< k_\lambda(r) \| f \|_{p,\phi} \| g \|_{q,\psi}, \quad (2.3.13)$$

$$\int_0^\infty y^{\frac{p\lambda}{s}-1} [\int_0^\infty \tfrac{(\min\{x,y\})^\alpha f(x)}{(x+y)^{\lambda+\alpha}} dx]^p dy$$

$$< k_\lambda^p(r) \| f \|_{p,\phi}^p; \quad (2.3.14)$$

for $0 < p < 1$, we have the equivalent reverses of (2.3.13) and (2.3.14).

2.3.2. THE CASE THAT THE KERNELS WITH SINGLE VARIABLE ARE BOUNDED IN $[\delta,1)(0<\delta<1)$

Theorem 2.3.7 If $\lambda \in \mathbf{R}$, $r, s \neq 0$, for $0 < \delta < 1$, both $k_\lambda(u,1)$ and $k_\lambda(1,u)$ are bounded in $[\delta,1)$ and are not zero almost everywhere in $(0,1)$, there exists constants $\eta < \min\{\frac{\lambda}{r}, \frac{\lambda}{s}\}, C_1$ and C_2, such that

$$\lim_{u\to 0^+} u^\eta k_\lambda(u,1) = C_1,$$

$$\lim_{u\to 0^+} u^\eta k_\lambda(1,u) = C_2, \quad (2.3.15)$$

then $0 < k_\lambda(r) = \int_0^\infty k_\lambda(u,1) u^{\frac{\lambda}{r}-1} du < \infty$.

Proof There exists $L > 0$, such that for $u \in (0,1)$,

$$0 \le u^\eta k_\lambda(u,1) \le L, 0 \le u^\eta k_\lambda(1,u) \le L.$$

By (2.3.1), we have

$$0 < \int_0^\infty k_\lambda(u,1) u^{\frac{\lambda}{r}-1} du = \int_0^1 u^\eta k_\lambda(u,1) u^{\frac{\lambda}{r}-\eta-1} du$$

$$+ \int_0^1 u^\eta k_\lambda(1,u) u^{\frac{\lambda}{s}-\eta-1} du$$

$$\le L[\int_0^1 u^{\frac{\lambda}{r}-\eta-1} du + \int_0^1 u^{\frac{\lambda}{s}-\eta-1} du] = L(\tfrac{1}{\frac{\lambda}{r}-\eta} + \tfrac{1}{\frac{\lambda}{s}-\eta}).$$

Hence $0 < k_\lambda(r) = \int_0^\infty k_\lambda(u,1) u^{\frac{\lambda}{r}-1} du < \infty$.

Example 2.3.8 If $\beta > -1, \lambda \in \mathbf{R}, r, s \neq 0$,

$$\alpha < \min\{\tfrac{\lambda}{r}, \tfrac{\lambda}{s}\},\ k_\lambda(x,y) = \tfrac{|\ln(x/y)|^\beta}{(x+y)^{\lambda-\alpha}(\min\{x,y\})^\alpha},$$

then $k_\lambda(u,1) = k_\lambda(1,u) = \frac{(-\ln u)^\beta}{(u+1)^{\lambda-\alpha} u^\alpha}$ are bounded in $[\delta,1)$. Setting $\eta = \alpha + \sigma$ $(\alpha + \sigma < \min\{\frac{\lambda}{r}, \frac{\lambda}{s}\}$, $\sigma > 0)$, then we obtain

$$\lim_{u\to 0^+} u^\eta k_\lambda(u,1)$$

$$= \lim_{u\to 0^+} u^\eta k_\lambda(1,u) = \lim_{u\to 0^+} \tfrac{(-\ln u)^\beta u^\sigma}{(u+1)^{\lambda-\alpha}} = 0.$$

In view of Theorem 2.3.6, we have

$$0 < k_\lambda(r) = \int_0^\infty k_\lambda(u,1) u^{\frac{\lambda}{r}-1} du < \infty.$$

By Lemma 2.3.2 and (2.2.48), for $\lambda' > 0$, $\alpha < \min\{\frac{\lambda'}{r}, \frac{\lambda'}{s}\}$, we obtain

$$0 < \int_0^1 k_\lambda(u,1) u^{\frac{\lambda'}{r}-1} du + \int_0^1 k_\lambda(1,u) u^{\frac{\lambda'}{s}-1} du$$

$$= \int_0^1 (1+u)^{\alpha-\lambda} (-\ln u)^\beta (u^{\frac{\lambda'}{r}-\alpha-1} + u^{\frac{\lambda'}{s}-\alpha-1}) du$$

$$= \int_0^1 \sum_{k=0}^\infty \binom{\alpha-\lambda}{k} (-\ln u)^\beta (u^{\frac{\lambda'}{r}-\alpha+k-1} + u^{\frac{\lambda'}{s}-\alpha+k-1}) du$$

$$= \sum_{k=0}^\infty \binom{\alpha-\lambda}{k} \int_0^1 (-\ln u)^\beta (u^{\frac{\lambda'}{r}-\alpha+k-1} + u^{\frac{\lambda'}{s}-\alpha+k-1}) du$$

$$= \Gamma(\beta+1) \sum_{k=0}^\infty \binom{\alpha-\lambda}{k} [\tfrac{1}{(\frac{\lambda'}{r}-\alpha+k)^{\beta+1}} + \tfrac{1}{(\frac{\lambda'}{s}-\alpha+k)^{\beta+1}}] < \infty,$$

$$k_\lambda(r) = \Gamma(\beta+1)$$

$$\times \sum_{k=0}^\infty \binom{\alpha-\lambda}{k} [\tfrac{1}{(\frac{\lambda}{r}-\alpha+k)^{\beta+1}} + \tfrac{1}{(\frac{\lambda}{s}-\alpha+k)^{\beta+1}}].$$

By Theorem 2.2.3 and Theorem 2.2.7, for $p > 1$, we have the following equivalent inequalities:

$$\int_0^\infty\int_0^\infty \frac{|\ln(x/y)|^\beta}{(x+y)^{\lambda-\alpha}(\min\{x,y\})^\alpha} f(x)g(y)dxdy$$

$$< k_\lambda(r)\,\|f\|_{p,\phi}\|g\|_{q,\psi}, \qquad (2.3.16)$$

$$\int_0^\infty y^{\frac{p\lambda}{s}-1}\left[\int_0^\infty \frac{|\ln(x/y)|^\beta f(x)}{(x+y)^{\lambda-\alpha}(\min\{x,y\})^\alpha}dx\right]^p dy$$

$$< k_\lambda^p(r)\,\|f\|_{p,\phi}^p; \qquad (2.3.17)$$

for $0<p<1$, we have the equivalent reverses of (2.3.16) and (2.3.17).

Example 2.3.9 If $\lambda\in\mathbf{R}, r,s\neq 0, \alpha<\min\{\frac{\lambda}{r},\frac{\lambda}{s}\}$,

$$k_\lambda(x,y)=\frac{1}{(x^{\lambda-\alpha}+y^{\lambda-\alpha})(\min\{x,y\})^\alpha},$$

then $k_\lambda(u,1)=k_\lambda(1,u)=\frac{1}{(u^{\lambda-\alpha}+1)u^\alpha}$ are bounded in $[\delta,1)$. Setting $\eta=\alpha$ and $C_1=C_2=1$, we have (2.3.13) and then

$$0<k_\lambda(r)=\int_0^\infty k_\lambda(u,1)u^{\frac{\lambda}{r}-1}du<\infty.$$

By Lemma 2.3.2, for $\lambda'>0, \alpha<\min\{\frac{\lambda'}{r},\frac{\lambda'}{s}\}$, we obtain

$$0<\int_0^1 k_\lambda(u,1)u^{\frac{\lambda'}{r}-1}du+\int_0^1 k_\lambda(1,u)u^{\frac{\lambda'}{s}-1}du$$

$$=\int_0^1(1+u^{\lambda-\alpha})^{-1}(u^{\frac{\lambda'}{r}-\alpha-1}+u^{\frac{\lambda'}{s}-\alpha-1})du$$

$$=\int_0^1\sum_{k=0}^\infty(-1)^k u^{k(\lambda-\alpha)}(u^{\frac{\lambda'}{r}-\alpha-1}+u^{\frac{\lambda'}{s}-\alpha-1})du$$

$$=\sum_{k=0}^\infty(-1)^k\int_0^1(u^{\frac{\lambda'}{r}+k(\lambda-\alpha)-\alpha-1}+u^{\frac{\lambda'}{s}+k(\lambda-\alpha)-\alpha-1})du$$

$$=\sum_{k=0}^\infty(-1)^k\left[\frac{1}{\frac{\lambda'}{r}+k(\lambda-\alpha)-\alpha}+\frac{1}{\frac{\lambda'}{s}+k(\lambda-\alpha)-\alpha}\right]<\infty,$$

$$k_\lambda(r)=\sum_{k=0}^\infty\frac{(-1)^k[(2k+1)\lambda-2(k+1)\alpha]}{[(\frac{1}{r}+k)\lambda-(k+1)\alpha][(\frac{1}{s}+k)\lambda-(k+1)\alpha]}.$$

By Theorem 2.2.3 and Theorem 2.2.7, for $p>1$, we have the following equivalent inequalities:

$$\int_0^\infty\int_0^\infty\frac{f(x)g(y)}{(x^{\lambda-\alpha}+y^{\lambda-\alpha})(\min\{x,y\})^\alpha}dxdy$$

$$<k_\lambda(r)\,\|f\|_{p,\phi}\|g\|_{q,\psi}, \qquad (2.3.18)$$

$$\int_0^\infty y^{\frac{p\lambda}{s}-1}\left[\int_0^\infty\frac{f(x)}{(x^{\lambda-\alpha}+y^{\lambda-\alpha})(\min\{x,y\})^\alpha}dx\right]^p dy$$

$$<k_\lambda^p(r)\,\|f\|_{p,\phi}^p; \qquad (2.3.19)$$

for $0<p<1$, we have the equivalent reverses of (2.3.18) and (2.3.19).

Example 2.3.10 If $\lambda>0, r,s>1, \alpha\in\mathbf{R}, \beta>-1$,

$$k_\lambda(x,y)=\frac{|\ln(x/y)|^\beta}{(x+y)^{\lambda-\alpha}(\max\{x,y\})^\alpha},$$

then $k_\lambda(u,1)=k_\lambda(1,u)=\frac{(-\ln u)^\beta}{(u+1)^{\lambda-\alpha}}$ are bounded in $[\delta,1)$. Setting $0<\eta<\min\{\frac{\lambda}{r},\frac{\lambda}{s}\}$, $C_1=C_2=0$, we have (2.3.13) and then

$$0<k_\lambda(r)=\int_0^\infty k_\lambda(u,1)u^{\frac{\lambda}{r}-1}du<\infty.$$

By Lemma 2.3.2, for $\lambda'>0$, we obtain

$$0<\int_0^1 k_\lambda(u,1)u^{\frac{\lambda'}{r}-1}du+\int_0^1 k_\lambda(1,u)u^{\frac{\lambda'}{s}-1}du$$

$$=\int_0^1(1+u)^{\alpha-\lambda}(-\ln u)^\beta(u^{\frac{\lambda'}{r}-1}+u^{\frac{\lambda'}{s}-1})du$$

$$=\int_0^1\sum_{k=0}^\infty\binom{\alpha-\lambda}{k}u^k(-\ln u)^\beta(u^{\frac{\lambda'}{r}-1}+u^{\frac{\lambda'}{s}-1})du$$

$$=\sum_{k=0}^\infty\binom{\alpha-\lambda}{k}\int_0^1(-\ln u)^\beta(u^{\frac{\lambda'}{r}+k-1}+u^{\frac{\lambda'}{s}+k-1})du$$

$$=\Gamma(\beta+1)\sum_{k=0}^\infty\binom{\alpha-\lambda}{k}\left[\frac{1}{(\frac{\lambda'}{r}+k)^{\beta+1}}+\frac{1}{(\frac{\lambda'}{s}+k)^{\beta+1}}\right]<\infty,$$

$$k_\lambda(r)=\Gamma(\beta+1)\sum_{k=0}^\infty\binom{\alpha-\lambda}{k}\left[\frac{1}{(\frac{\lambda}{r}+k)^{\beta+1}}+\frac{1}{(\frac{\lambda}{s}+k)^{\beta+1}}\right].$$

By Theorem 2.2.3 and Theorem 2.2.7, for $p>1$, we have the following equivalent inequalities:

$$\int_0^\infty\int_0^\infty\frac{|\ln(x/y)|^\beta}{(x+y)^{\lambda-\alpha}(\max\{x,y\})^\alpha}f(x)g(y)dxdy$$

$$<k_\lambda(r)\,\|f\|_{p,\phi}\|g\|_{q,\psi}, \qquad (2.3.20)$$

$$\int_0^\infty y^{\frac{p\lambda}{s}-1}\left[\int_0^\infty\frac{|\ln(x/y)|^\beta f(x)}{(x+y)^{\lambda-\alpha}(\max\{x,y\})^\alpha}dx\right]^p dy$$

$$<k_\lambda^p(r)\,\|f\|_{p,\phi}^p; \qquad (2.3.21)$$

for $0<p<1$, we have the equivalent reverses of (2.3.20) and (2.3.21).

Example 2.3.11 If $\lambda>0, r,s>1, \alpha>0$,

$$k_\lambda(x,y)=\frac{\ln(x/y)}{(x^\alpha-y^\alpha)(\max\{x,y\})^{\lambda-\alpha}},$$

then $k_\lambda(u,1)=k_\lambda(1,u)=\frac{\ln u}{u^\alpha-1}$ are bounded in $[\delta,1)$. Setting $0<\eta<\min\{\frac{\lambda}{r},\frac{\lambda}{s}\}$ and $C_1=C_2=0$, we have (2.3.13) and then

$$0<k_\lambda(r)=\int_0^\infty k_\lambda(u,1)u^{\frac{\lambda}{r}-1}du<\infty.$$

By Lemma 2.3.2, for $\lambda'>0$, we obtain

$$0<\int_0^1\frac{-\ln u}{1-u^\alpha}u^{\frac{\lambda'}{r}-1}du+\int_0^1\frac{-\ln u}{1-u^\alpha}u^{\frac{\lambda'}{s}-1}du$$

$$=-\int_0^1\ln u\sum_{k=0}^\infty u^{k\alpha+\frac{\lambda'}{r}-1}du-\int_0^1\ln u\sum_{k=0}^\infty u^{k\alpha+\frac{\lambda'}{s}-1}du$$

$$=\sum_{k=0}^{\infty}\frac{1}{k\alpha+\frac{\lambda'}{r}}\int_0^1(-\ln u)du^{k\alpha+\frac{\lambda'}{r}}$$

$$+\sum_{k=0}^{\infty}\frac{1}{k\alpha+\frac{\lambda'}{s}}\int_0^1(-\ln u)du^{k\alpha+\frac{\lambda'}{s}}$$

$$=\sum_{k=0}^{\infty}\Big[\frac{1}{(k\alpha+\frac{\lambda'}{r})^2}+\frac{1}{(k\alpha+\frac{\lambda'}{s})^2}\Big]<\infty,$$

$$k_\lambda(r)=\sum_{k=0}^{\infty}\Big[\frac{1}{(k\alpha+\frac{\lambda}{r})^2}+\frac{1}{(k\alpha+\frac{\lambda}{s})^2}\Big].$$

By Theorem 2.2.3 and Theorem 2.2.7, for $p>1$, we have the following equivalent inequalities:

$$\int_0^\infty\int_0^\infty\frac{\ln(x/y)}{(x^\alpha-y^\alpha)(\max\{x,y\})^{\lambda-\alpha}}f(x)g(y)dxdy$$
$$<k_{\lambda,\alpha}(r)\,\|f\|_{p,\phi}\|g\|_{q,\psi},\qquad(2.3.22)$$

$$\int_0^\infty y^{\frac{p\lambda}{s}-1}\Big[\int_0^\infty\frac{\ln(x/y)f(x)}{(x^\alpha-y^\alpha)(\max\{x,y\})^{\lambda-\alpha}}dx\Big]^p dy$$
$$<k_{\lambda,\alpha}^p(r)\,\|f\|_{p,\phi}^p;\qquad(2.3.23)$$

for $0<p<1$, we have the equivalent reverses of (2.3.22) and (2.3.23).

Example 2.3.12 If $\lambda\in\mathbf{R}, r,s\neq 0,$

$$\alpha>\max\{0,\tfrac{\lambda}{r},\tfrac{\lambda}{s}\},$$

$$k_\lambda(x,y)=\frac{\ln(x/y)}{(x^\alpha-y^\alpha)(\min\{x,y\})^{\lambda-\alpha}},$$

then $k_\lambda(u,1)=k_\lambda(1,u)=\frac{\ln u}{(u^\alpha-1)u^{\lambda-\alpha}}$ are bounded in $[\delta,1)$. Setting $\lambda-\alpha<\eta<\min\{\frac{\lambda}{r},\frac{\lambda}{s}\}$ and $C_1=C_2=0$, we have (2.3.13) and then

$$0<k_\lambda(r)=\int_0^\infty k_\lambda(u,1)u^{\frac{\lambda}{r}-1}du<\infty.$$

By Lemma 2.3.2, for $\lambda'\in\mathbf{R}$, $\alpha>\max\{0,\frac{\lambda'}{r},\frac{\lambda'}{s}\}$, we obtain

$$0<\int_0^1\frac{-\ln u}{1-u^\alpha}u^{\alpha-\frac{\lambda'}{s}-1}du+\int_0^1\frac{-\ln u}{1-u^\alpha}u^{\alpha-\frac{\lambda'}{r}-1}du$$

$$=\int_0^1(-\ln u)\sum_{k=0}^{\infty}u^{(k+1)\alpha-\frac{\lambda'}{s}-1}du$$

$$+\int_0^1(-\ln u)\sum_{k=0}^{\infty}u^{(k+1)\alpha-\frac{\lambda'}{r}-1}du$$

$$=\sum_{k=0}^{\infty}\frac{1}{(k+1)\alpha-\frac{\lambda'}{s}}\int_0^1(-\ln u)du^{(k+1)\alpha-\frac{\lambda'}{s}}$$

$$+\sum_{k=0}^{\infty}\frac{1}{(k+1)\alpha-\frac{\lambda'}{r}}\int_0^1(\ln u)du^{(k+1)\alpha-\frac{\lambda'}{r}}$$

$$=\sum_{k=0}^{\infty}\Big\{\frac{1}{[(k+1)\alpha-\frac{\lambda}{s}]^2}+\frac{1}{[(k+1)\alpha-\frac{\lambda}{r}]^2}\Big\}<\infty,$$

$$k_\lambda(r)=\sum_{k=1}^{\infty}\Big[\frac{1}{(k\alpha-\frac{\lambda}{s})^2}+\frac{1}{(k\alpha-\frac{\lambda}{r})^2}\Big].$$

By Theorem 2.2.3 and Theorem 2.2.7, for $p>1$, we have the following equivalent inequalities:

$$\int_0^\infty\int_0^\infty\frac{\ln(x/y)f(x)g(y)}{(x^\alpha-y^\alpha)(\min\{x,y\})^{\lambda-\alpha}}dxdy$$
$$<k_\lambda(r)\,\|f\|_{p,\phi}\|g\|_{q,\psi},\qquad(2.3.24)$$

$$\int_0^\infty y^{\frac{p\lambda}{s}-1}\Big[\int_0^\infty\frac{\ln(x/y)f(x)}{(x^\alpha-y^\alpha)(\min\{x,y\})^{\lambda-\alpha}}dx\Big]^p dy$$
$$<k_\lambda^p(r)\,\|f\|_{p,\phi}^p;\qquad(2.3.25)$$

for $0<p<1$, we have the equivalent reverses of (2.3.24) and (2.3.25).

Example 2.3.13 If $\lambda\in\mathbf{R}, r,s\neq 0$,

$$\alpha>\max\{\tfrac{-\lambda}{r},\tfrac{-\lambda}{s}\},$$

$$k_\lambda(x,y)=\frac{(\min\{x,y\})^\alpha}{(x+y)^{\lambda+\alpha}}\,|\ln(\tfrac{x}{y})|,$$

then

$$k_\lambda(u,1)=k_\lambda(1,u)=\frac{u^\alpha(-\ln u)}{(1+u)^{\lambda+\alpha}}$$

are bounded in $[\delta,1)$. Setting

$$\sigma>0,\eta=-\alpha+\sigma<\min\{\tfrac{\lambda}{r},\tfrac{\lambda}{s}\}$$

and $C_1=C_2=0$, we have (2.3.13) and then

$$0<k_\lambda(r)=\int_0^\infty k_\lambda(u,1)u^{\frac{\lambda}{r}-1}du<\infty.$$

By Lemma 2.3.2, there exists a neighborhood I_λ of $\frac{\lambda}{r}$, such that for $\frac{\lambda'}{r}\in I_\lambda$,

$$0<\int_0^1 k_\lambda(u,1)u^{\frac{\lambda'}{r}-1}du+\int_0^1 k_\lambda(1,u)u^{\frac{\lambda'}{s}-1}du$$

$$=\int_0^1\frac{(-\ln u)}{(1+u)^{\lambda+\alpha}}u^{\alpha+\frac{\lambda'}{r}-1}du+\int_0^1\frac{(-\ln u)}{(1+u)^{\lambda+\alpha}}u^{\alpha+\frac{\lambda'}{s}-1}du$$

$$=\int_0^1\sum_{k=0}^{\infty}\binom{-\lambda-\alpha}{k}(-\ln u)[u^{k+\alpha+\frac{\lambda'}{r}-1}+u^{k+\alpha+\frac{\lambda'}{s}-1}]du$$

$$=\sum_{k=0}^{\infty}\binom{-\lambda-\alpha}{k}\int_0^1(-\ln u)[u^{k+\alpha+\frac{\lambda'}{r}-1}+u^{k+\alpha+\frac{\lambda'}{s}-1}]du$$

$$=\sum_{k=0}^{\infty}\binom{-\lambda-\alpha}{k}\Big[\frac{1}{(k+\alpha+\frac{\lambda'}{r})^2}+\frac{1}{(k+\alpha+\frac{\lambda'}{s})^2}\Big]<\infty,$$

$$k_\lambda(r)=\sum_{k=0}^{\infty}\binom{-\lambda-\alpha}{k}\Big[\frac{1}{(k+\alpha+\frac{\lambda}{r})^2}+\frac{1}{(k+\alpha+\frac{\lambda}{s})^2}\Big].$$

By Theorem 2.2.3 and Theorem 2.2.7, for $p>1$, we have the following equivalent inequalities:

$$\int_0^\infty\int_0^\infty\frac{(\min\{x,y\})^\alpha}{(x+y)^{\lambda+\alpha}}|\ln(\tfrac{x}{y})|f(x)g(y)dxdy$$
$$<k_\lambda(r)\,\|f\|_{p,\phi}\|g\|_{q,\psi},\qquad(2.3.26)$$

$$\int_0^\infty y^{\frac{p\lambda}{s}-1}\left[\int_0^\infty \frac{(\min\{x,y\})^\alpha |\ln(\frac{x}{y})|}{(x+y)^{\lambda+\alpha}} f(x)dx\right]^p dy$$
$$< k_\lambda^p(r)\,\|f\|_{p,\phi}^p; \qquad (2.3.27)$$

for $0<p<1$, we have the equivalent reverses of (2.3.26) and (2.3.27).

2.3.3. THE CASE THAT THE KERNELS WITH SINGLE VARIABLE ARE BOUNDED IN $(0,1-\delta](0<\delta<1)$

Theorem 2.3.14 If $\lambda\in\mathbf{R}$, $r,s\neq 0$, for any $0<\delta<1$, both $k_\lambda(u,1)$ and $k_\lambda(1,u)$ are bounded in $(0,1-\delta]$ and are not zero almost everywhere in $(0,1)$, there exist constants $\sigma>\max\{\frac{-\lambda}{r},\frac{-\lambda}{s}\}$, $\theta<1, C_1$ and C_2, such that

$$\lim_{u\to1^-}(1-u)^\theta u^{-\sigma}k_\lambda(u,1)=C_1,$$
$$\lim_{u\to1^-}(1-u)^\theta u^{-\sigma}k_\lambda(1,u)=C_2, \qquad (2.3.28)$$

then $0<k_\lambda(r)=\int_0^\infty k_\lambda(u,1)u^{\frac{\lambda}{r}-1}du<\infty$.

Proof There exists $L>0$, such that

$$0\le(1-u)^\theta u^{-\sigma}k(u,1)\le L,$$
$$0\le(1-u)^\theta u^{-\sigma}k(1,u)\le L(u\in(0,1)).$$

Then by (2.3.1) and (1.2.17), we have

$$0<\int_0^\infty k_\lambda(u,1)u^{\frac{\lambda}{r}-1}du$$
$$=\int_0^1[(1-u)^\theta u^{-\sigma}k_\lambda(u,1)](1-u)^{-\theta}u^{\sigma+\frac{\lambda}{r}-1}du$$
$$+\int_0^1[(1-u)^\theta u^{-\sigma}k_\lambda(1,u)](1-u)^{-\theta}u^{\sigma+\frac{\lambda}{s}-1}du$$
$$\le L\left[\int_0^1(1-u)^{(1-\theta)-1}u^{\sigma+\frac{\lambda}{r}-1}du\right.$$
$$\left.+\int_0^1(1-u)^{(1-\theta)-1}u^{\sigma+\frac{\lambda}{s}-1}du\right]$$
$$=L[B(1-\theta,\sigma+\tfrac{\lambda}{r})+B(1-\theta,\sigma+\tfrac{\lambda}{s})].$$

Hence $0<k_\lambda(r)=\int_0^\infty k_\lambda(u,1)u^{\frac{\lambda}{r}-1}du<\infty$.□

Example 2.3.15 If $0<\lambda<1+\alpha, r,s>1$,

$$k_\lambda(x,y)=\frac{1}{|x-y|^{\lambda-\alpha}(\max\{x,y\})^\alpha}$$

then

$$k_\lambda(u,1)=k_\lambda(1,u)=\frac{1}{(1-u)^{\lambda-\alpha}}$$

are bounded in $(0,1-\delta]$. Setting $\sigma=0$, $\theta=\lambda-\alpha<1$, $C_1=C_2=1$, we have (2.3.28) and

$$0<k_\lambda(r)=\int_0^\infty k_\lambda(u,1)u^{\frac{\lambda}{r}-1}du<\infty.$$

For $0<\lambda'<1+\alpha$, we obtain

$$0<\int_0^1 k_\lambda(u,1)u^{\frac{\lambda'}{r}-1}du+\int_0^1 k_\lambda(1,u)u^{\frac{\lambda'}{s}-1}du$$
$$=\int_0^1(1-u)^{\alpha-\lambda}u^{\frac{\lambda'}{r}-1}du+\int_0^1(1-u)^{\alpha-\lambda}u^{\frac{\lambda'}{s}-1}du$$
$$=B(1+\alpha-\lambda,\tfrac{\lambda'}{r})+B(1+\alpha-\lambda,\tfrac{\lambda'}{s})<\infty,$$
$$k_\lambda(r)=B(1+\alpha-\lambda,\tfrac{\lambda}{r})+B(1+\alpha-\lambda,\tfrac{\lambda}{s}).$$

By Theorem 2.2.3 and Theorem 2.2.7, for $p>1$, we have the following equivalent inequalities:

$$\int_0^\infty\int_0^\infty \frac{f(x)g(y)}{|x-y|^{\lambda-\alpha}(\max\{x,y\})^\alpha}dxdy$$
$$<k_\lambda(r)\,\|f\|_{p,\phi}\|g\|_{q,\psi}, \qquad (2.3.29)$$
$$\int_0^\infty y^{\frac{p\lambda}{s}-1}\left[\int_0^\infty \frac{f(x)}{|x-y|^{\lambda-\alpha}(\max\{x,y\})^\alpha}dx\right]^p dy$$
$$<k_\lambda{}^p(r)\,\|f\|_{p,\phi}^p; \qquad (2.3.30)$$

for $0<p<1$, we have the equivalent reverses of (2.3.29) and (2.3.30).

Example 2.3.16 If $0<\lambda<1+\alpha, r,s>1$, $k_\lambda(x,y)=\frac{1}{|x-y|^{\lambda-\alpha}(x+y)^\alpha}$, then

$$k_\lambda(u,1)=k_\lambda(1,u)=\frac{1}{(1-u)^{\lambda-\alpha}(1+u)^\alpha}$$

are bounded in $(0,1-\delta]$. Setting $\sigma=0$, $\theta=\lambda-\alpha$, $C_1=C_2=1$, we have (2.3.28) and then

$$0<k_\lambda(r)=\int_0^\infty k_\lambda(u,1)u^{\frac{\lambda}{r}-1}du<\infty.$$

By Lemma 2.3.2, for $\lambda'>0$, we obtain

$$0<\int_0^1 k_\lambda(u,1)u^{\frac{\lambda'}{r}-1}du+\int_0^1 k_\lambda(1,u)u^{\frac{\lambda'}{s}-1}du$$
$$=\int_0^1(1+u)^{-\alpha}(1-u)^{\alpha-\lambda}u^{\frac{\lambda'}{r}-1}du$$
$$+\int_0^1(1+u)^{-\alpha}(1-u)^{\alpha-\lambda}u^{\frac{\lambda'}{s}-1}du$$
$$=\int_0^1\sum_{k=0}^\infty\binom{-\alpha}{k}u^k(1-u)^{\alpha-\lambda}u^{\frac{\lambda'}{r}-1}du$$
$$+\int_0^1\sum_{k=0}^\infty\binom{-\alpha}{k}u^k(1-u)^{\alpha-\lambda}u^{\frac{\lambda'}{s}-1}du$$
$$=\sum_{k=0}^\infty\binom{-\alpha}{k}\left[\int_0^1(1-u)^{\alpha-\lambda}u^{\frac{\lambda'}{r}+k-1}du\right.$$
$$\left.+\int_0^1(1-u)^{\alpha-\lambda}u^{\frac{\lambda'}{s}+k-1}du\right]$$
$$=\sum_{k=0}^\infty\binom{-\alpha}{k}[B(1+\alpha-\lambda,\tfrac{\lambda'}{r}+k)$$

$$+B(1+\alpha-\lambda,\tfrac{\lambda'}{s}+k)]<\infty,$$

$$k_\lambda(r)=\sum_{k=0}^{\infty}\binom{-\alpha}{k}[B(1+\alpha-\lambda,\tfrac{\lambda}{r}+k)$$
$$+B(1+\alpha-\lambda,\tfrac{\lambda}{s}+k)].$$

By Theorem 2.2.3 and Theorem 2.2.7, for $p>1$, we have the following equivalent inequalities:

$$\int_0^\infty\int_0^\infty \frac{f(x)g(y)}{|x-y|^{\lambda-\alpha}(x+y)^\alpha}dxdy$$
$$<k_\lambda(r)\,\|f\|_{p,\phi}\|g\|_{q,\psi}, \tag{2.3.31}$$

$$\int_0^\infty y^{\frac{p\lambda}{s}-1}\Big[\int_0^\infty \frac{f(x)}{|x-y|^{\lambda-\alpha}(x+y)^\alpha}dx\Big]^p dy$$
$$<k_\lambda^p(r)\,\|f\|_{p,\phi}^p, \tag{2.3.32}$$

for $0<p<1$, we have the equivalent reverses of (2.3.31) and (2.3.32).

Example 2.3.17 If $\alpha>0, 0<\lambda<1+\alpha, r,s>1$,

$$k_\lambda(x,y)=\frac{1}{|x-y|^{\lambda-\alpha}(x^\alpha+y^\alpha)},$$

then

$$k_\lambda(u,1)=k_\lambda(1,u)=\frac{1}{(1-u)^{\lambda-\alpha}(1+u^\alpha)}$$

are bounded in $(0,1-\delta]$. Setting $\sigma=0$, $\theta=\lambda-\alpha$, $C_1=C_2=\frac{1}{2}$, we have (2.3.28) and then

$$0<k_\lambda(r)=\int_0^\infty k_\lambda(u,1)u^{\frac{\lambda}{r}-1}du<\infty.$$

By Lemma 2.3.2, for $\lambda'>0$, we obtain

$$0<\int_0^1 k_\lambda(u,1)u^{\frac{\lambda'}{r}-1}du+\int_0^1 k_\lambda(1,u)u^{\frac{\lambda'}{s}-1}du$$
$$=\int_0^1(1+u^\alpha)^{-1}(1-u)^{\alpha-\lambda}u^{\frac{\lambda'}{r}-1}du$$
$$+\int_0^1(1+u^\alpha)^{-1}(1-u)^{\alpha-\lambda}u^{\frac{\lambda'}{s}-1}du$$
$$=\int_0^1\sum_{k=0}^{\infty}(-1)^k u^{k\alpha}(1-u)^{\alpha-\lambda}u^{\frac{\lambda'}{r}-1}du$$
$$+\int_0^1\sum_{k=0}^{\infty}(-1)^k u^{k\alpha}(1-u)^{\alpha-\lambda}u^{\frac{\lambda'}{s}-1}du$$
$$=\sum_{k=0}^{\infty}(-1)^k\Big[\int_0^1(1-u)^{\alpha-\lambda}u^{\frac{\lambda'}{r}+k\alpha-1}du$$
$$+\int_0^1(1-u)^{\alpha-\lambda}u^{\frac{\lambda'}{s}+k\alpha-1}du\Big]$$
$$=\sum_{k=0}^{\infty}(-1)^k[B(1+\alpha-\lambda,\tfrac{\lambda'}{r}+k\alpha)$$
$$+B(1+\alpha-\lambda,\tfrac{\lambda'}{s}+k\alpha)]<\infty$$

,

$$k_\lambda(r)=\sum_{k=0}^{\infty}(-1)^k[B(1+\alpha-\lambda,\tfrac{\lambda}{r}+k\alpha)$$
$$+B(1+\alpha-\lambda,\tfrac{\lambda}{s}+k\alpha)].$$

By Theorem 2.2.3 and Theorem 2.2.7, for $p>1$, we have the following equivalent inequalities:

$$\int_0^\infty\int_0^\infty \frac{f(x)g(y)}{|x-y|^{\lambda-\alpha}(x^\alpha+y^\alpha)}dxdy$$
$$<k_\lambda(r)\,\|f\|_{p,\phi}\|g\|_{q,\psi}, \tag{2.3.33}$$

$$\int_0^\infty y^{\frac{p\lambda}{s}-1}\Big[\int_0^\infty \frac{f(x)}{|x-y|^{\lambda-\alpha}(x^\alpha+y^\alpha)}dx\Big]^p dy$$
$$<k_\lambda^p(r)\,\|f\|_{p,\phi}^p; \tag{2.3.34}$$

for $0<p<1$, we have the equivalent reverses of (2.3.33) and (2.3.34).

Example 2.3.18 If $\lambda\in\mathbf{R}, r,s\neq 0$,

$$\max\{0,\tfrac{-\lambda}{r},\tfrac{-\lambda}{s}\}<\alpha<1,$$

$$k_\lambda(x,y)=\frac{(\min\{x,y\})^\alpha}{(x+y)^\lambda|x-y|^\alpha},$$

then

$$k_\lambda(u,1)=k_\lambda(1,u)=\frac{u^\alpha}{(1+u)^\lambda(1-u)^\alpha}$$

are bounded in $(0,1-\delta]$. Setting $0<\sigma=\theta<1$ and $C_1=C_2=\frac{1}{2^\lambda}$, we have (2.3.28) and then

$$0<k_\lambda(r)=\int_0^\infty k_\lambda(u,1)u^{\frac{\lambda}{r}-1}du<\infty.$$

By Lemma 2.3.2, there exists a neighborhood I_λ of $\frac{\lambda}{r}$, such that for $\frac{\lambda'}{r}\in I_\lambda$,

$$0<\int_0^1 k_\lambda(u,1)u^{\frac{\lambda'}{r}-1}du+\int_0^1 k_\lambda(1,u)u^{\frac{\lambda'}{s}-1}du$$
$$=\int_0^1\frac{(1-u)^{-\alpha}}{(1+u)^\lambda}u^{\alpha+\frac{\lambda'}{r}-1}du+\int_0^1\frac{(1-u)^{-\alpha}}{(1+u)^\lambda}u^{\alpha+\frac{\lambda'}{s}-1}du$$
$$=\int_0^1\sum_{k=0}^{\infty}\binom{-\lambda}{k}(1-u)^{-\alpha}[u^{k+\alpha+\frac{\lambda'}{r}-1}+u^{k+\alpha+\frac{\lambda'}{s}-1}]du$$
$$=\sum_{k=0}^{\infty}\binom{-\lambda}{k}\int_0^1(1-u)^{-\alpha}[u^{k+\alpha+\frac{\lambda'}{r}-1}+u^{k+\alpha+\frac{\lambda'}{s}-1}]du$$
$$=\sum_{k=0}^{\infty}\binom{-\lambda}{k}[B(1-\alpha,k+\alpha+\tfrac{\lambda'}{r})$$
$$+B(1-\alpha,k+\alpha+\tfrac{\lambda'}{s})]<\infty,$$

$$k_\lambda(r)=\sum_{k=0}^{\infty}\binom{-\lambda}{k}[B(1-\alpha,k+\alpha+\tfrac{\lambda}{r})$$
$$+B(1-\alpha,k+\alpha+\tfrac{\lambda}{s})].$$

By Theorem 2.2.3 and Theorem 2.2.7, for $p>1$, we have the following equivalent inequalities:

$$\int_0^\infty\int_0^\infty \frac{(\min\{x,y\})^\alpha}{(x+y)^\lambda|x-y|^\alpha} f(x)g(y)dxdy$$
$$< k_\lambda(r)\,||\, f\,||_{p,\phi}||\, g\,||_{q,\psi}, \qquad (2.3.35)$$
$$\int_0^\infty y^{\frac{p\lambda}{s}-1}[\int_0^\infty \frac{(\min\{x,y\})^\alpha}{(x+y)^\lambda|x-y|^\alpha} f(x)dx]^p dy$$
$$< k_\lambda^p(r)\,||\, f\,||_{p,\phi}^p; \qquad (2.3.36)$$

for $0<p<1$, we have the equivalent reverses of (2.3.35) and (2.3.36).

2.3.4. THE CASE THAT THE KERNELS WITH SINGLE VARIABLE ARE BOUNDED IN $[\delta,1-\delta](0<\delta<\frac{1}{2})$

Theorem 2.3.19 If $\lambda\in\mathbf{R}$, $r,s\neq 0$, for any $0<\delta<\frac{1}{2}$, both $k_\lambda(u,1)$ and $k_\lambda(1,u)$ are bounded in $[\delta,1-\delta]$ and are not zero almost everywhere in $(0,1)$, there exists constants $\eta<\min\{\frac{\lambda}{r},\frac{\lambda}{s}\}$, $\theta<1$ and $C_i(i=1,2,3,4)$, such that

$$\lim_{u\to 0^+} u^\eta k_\lambda(u,1)=C_1,\ \lim_{u\to 0^+} u^\eta k_\lambda(1,u)=C_2,$$
$$\lim_{u\to 1^-}(1-u)^\theta k_\lambda(u,1)=C_3,$$
$$\lim_{u\to 1^-}(1-u)^\theta k_\lambda(1,u)=C_4, \qquad (2.3.37)$$

then $0<k_\lambda(r)=\int_0^\infty k_\lambda(u,1)u^{\frac{\lambda}{r}-1}du<\infty$.

Proof There exists $L>0$, such that

$$0\le(1-u)^\theta u^\eta k(u,1)\le L,$$
$$0\le(1-u)^\theta u^\eta k(1,u)\le L\ (u\in(0,1)).$$

Then by (2.3.1) and (1.2.17), we have

$$0<\int_0^\infty k_\lambda(u,1)u^{\frac{\lambda}{r}-1}du$$
$$=\int_0^1[(1-u)^\theta u^\eta k_\lambda(u,1)](1-u)^{-\theta}u^{\frac{\lambda}{r}-\eta-1}du$$
$$+\int_0^1[(1-u)^\theta u^\eta k_\lambda(1,u)](1-u)^{-\theta}u^{\frac{\lambda}{s}-\eta-1}du$$
$$\le L[\int_0^1(1-u)^{(1-\theta)-1}u^{\frac{\lambda}{r}-\eta-1}du$$
$$+\int_0^1(1-u)^{(1-\theta)-1}u^{\frac{\lambda}{s}-\eta-1}du]$$
$$=L[B(1-\theta,\tfrac{\lambda}{r}-\eta)+B(1-\theta,\tfrac{\lambda}{s}-\eta)].$$

Hence $0<k_\lambda(r)=\int_0^\infty k_\lambda(u,1)u^{\frac{\lambda}{r}-1}du<\infty$.□

Example 2.3.20 If $r,s\neq 0,\beta>-1,\lambda\in\mathbf{R}$,

$$\max\{\tfrac{\lambda}{r},\tfrac{\lambda}{s}\}<1,\ \max\{\tfrac{-\lambda}{r},\tfrac{-\lambda}{s}\}<\alpha<1-\lambda,$$
$$k_\lambda(x,y)=\frac{|\ln(x/y)|^\beta}{|x-y|^{\lambda+\alpha}}(\min\{x,y\})^\alpha,$$

then

$$k_\lambda(u,1)=k_\lambda(1,u)=\frac{(-\ln u)^\beta u^\alpha}{(1-u)^{\lambda+\alpha}}$$

are bounded in $[\delta,1-\delta]$. Setting $\theta=\lambda+\alpha<1$, $\eta=-\alpha+\sigma$ $(0<\sigma<\min\{\frac{\lambda}{r},\frac{\lambda}{s}\}+\alpha)$, $C_i=0$ $(i=1,\cdots,4)$), then we have (2.3.37) and

$$0<k_\lambda(r)=\int_0^\infty k_\lambda(u,1)u^{\frac{\lambda}{r}-1}du<\infty.$$

By Lemma 2.3.2, for $\lambda'\in\mathbf{R}, \max\{\frac{\lambda'}{r},\frac{\lambda'}{s}\}<1$, $\max\{\frac{-\lambda'}{r},\frac{-\lambda'}{s}\}<\alpha<1-\lambda'$, we obtain

$$0<\int_0^1 k_\lambda(u,1)u^{\frac{\lambda'}{r}-1}du+\int_0^1 k_\lambda(1,u)u^{\frac{\lambda'}{s}-1}du$$
$$=\int_0^1(1-u)^{-\alpha-\lambda}(-\ln u)^\beta u^{\frac{\lambda'}{r}+\alpha-1}du$$
$$+\int_0^1(1-u)^{-\alpha-\lambda}(-\ln u)^\beta u^{\frac{\lambda'}{s}+\alpha-1}du$$
$$=\int_0^1\sum_{k=0}^\infty\binom{-\alpha-\lambda}{k}(-1)^k(-\ln u)^\beta u^{\frac{\lambda'}{r}+\alpha+k-1}du$$
$$+\int_0^1\binom{-\alpha-\lambda}{k}(-1)^k(-\ln u)^\beta u^{\frac{\lambda'}{s}+\alpha+k-1}du$$
$$=\sum_{k=0}^\infty\binom{-\alpha-\lambda}{k}(-1)^k[\int_0^1(-\ln u)^\beta u^{\frac{\lambda'}{r}+\alpha+k-1}du$$
$$+\int_0^1(-\ln u)^\beta u^{\frac{\lambda'}{s}+\alpha+k-1}du]$$
$$=\Gamma(\beta+1)$$
$$\times\sum_{k=0}^\infty\binom{-\alpha-\lambda}{k}(-1)^k[\frac{1}{(\frac{\lambda'}{r}+\alpha+k)^{\beta+1}}+\frac{1}{(\frac{\lambda'}{s}+\alpha+k)^{\beta+1}}]<\infty,$$
$$k_\lambda(r)=\Gamma(\beta+1)$$
$$\times\sum_{k=0}^\infty\binom{-\alpha-\lambda}{k}(-1)^k[\frac{1}{(\frac{\lambda}{r}+\alpha+k)^{\beta+1}}+\frac{1}{(\frac{\lambda}{s}+\alpha+k)^{\beta+1}}].$$

By Theorem 2.2.3 and Theorem 2.2.7, for $p>1$, we have the following equivalent inequalities:

$$\int_0^\infty\int_0^\infty \frac{|\ln(x/y)|^\beta}{|x-y|^{\lambda+\alpha}}(\min\{x,y\})^\alpha f(x)g(y)dxdy$$
$$< k_\lambda(r)\,||\, f\,||_{p,\phi}||\, g\,||_{q,\psi}, \qquad (2.3.38)$$
$$\int_0^\infty y^{\frac{p\lambda}{s}-1}[\int_0^\infty \frac{|\ln(x/y)|^\beta}{|x-y|^{\lambda+\alpha}}(\min\{x,y\})^\alpha f(x)dx]^p dy$$
$$< k_\lambda^p(r)\,||\, f\,||_{p,\phi}^p; \qquad (2.3.39)$$

for $0<p<1$, we have the equivalent reverses of (2.3.38) and (2.3.39).

Example 2.3.21 If $r,s>1,\beta>-1,0<\lambda<1+\alpha$,

$$k_\lambda(x,y)=\frac{|\ln(x/y)|^\beta}{|x-y|^{\lambda-\alpha}(\max\{x,y\})^\alpha},$$

then $k_\lambda(u,1) = k_\lambda(1,u) = \frac{(-\ln u)^\beta}{(1-u)^{\lambda-\alpha}}$ are bounded in $[\delta, 1-\delta]$.

Setting $\theta = \lambda - \alpha$, $0 < \eta < \min\{\frac{\lambda}{r}, \frac{\lambda}{s}\}$, $C_i = 0$ or 1, then we have (2.3.37), and then

$$0 < k_\lambda(r) = \int_0^\infty k_\lambda(u,1) u^{\frac{\lambda}{r}-1} du < \infty.$$

By Lemma 2.3.2, for $0 < \lambda' < 1 + \alpha$, we obtain

$$\begin{aligned} 0 &< \int_0^1 k_\lambda(u,1) u^{\frac{\lambda'}{r}-1} du + \int_0^1 k_\lambda(1,u) u^{\frac{\lambda'}{s}-1} du \\ &= \int_0^1 (1-u)^{\alpha-\lambda} (-\ln u)^\beta u^{\frac{\lambda'}{r}-1} du \\ &\qquad + \int_0^1 (1-u)^{\alpha-\lambda} (-\ln u)^\beta u^{\frac{\lambda'}{s}-1} du \\ &= \int_0^1 \sum_{k=0}^\infty \binom{\alpha-\lambda}{k} (-1)^k (-\ln u)^\beta u^{\frac{\lambda'}{r}+k-1} du \\ &\qquad + \int_0^1 \binom{\alpha-\lambda}{k} (-1)^k (-\ln u)^\beta u^{\frac{\lambda'}{s}+k-1} du \\ &= \sum_{k=0}^\infty \binom{\alpha-\lambda}{k} (-1)^k \Big[\int_0^1 (-\ln u)^\beta u^{\frac{\lambda'}{r}+k-1} du \\ &\qquad + \int_0^1 (-\ln u)^\beta u^{\frac{\lambda'}{s}+k-1} du \Big] \\ &= \Gamma(\beta+1) \sum_{k=0}^\infty \binom{\alpha-\lambda}{k} (-1)^k \Big[\frac{1}{(\frac{\lambda'}{r}+k)^{\beta+1}} + \frac{1}{(\frac{\lambda'}{s}+k)^{\beta+1}} \Big], \end{aligned}$$

$$k_\lambda(r) = \Gamma(\beta+1) \sum_{k=0}^\infty \binom{\alpha-\lambda}{k} (-1)^k \Big[\frac{1}{(\frac{\lambda}{r}+k)^{\beta+1}} + \frac{1}{(\frac{\lambda}{s}+k)^{\beta+1}} \Big].$$

By Theorem 2.2.3 and Theorem 2.2.7, for $p > 1$, we have the following equivalent inequalities:

$$\int_0^\infty \int_0^\infty \frac{|\ln(x/y)|^\beta}{|x-y|^{\lambda-\alpha} (\max\{x,y\})^\alpha} f(x) g(y) dx dy < k_\lambda(r) \| f \|_{p,\phi} \| g \|_{q,\psi}, \quad (2.3.40)$$

$$\int_0^\infty y^{\frac{p\lambda}{s}-1} \Big[\int_0^\infty \frac{|\ln(x/y)|^\beta f(x)}{|x-y|^{\lambda-\alpha} (\max\{x,y\})^\alpha} dx \Big]^p dy < k_\lambda^p(r) \| f \|_{p,\phi}^p; \quad (2.3.41)$$

for $0 < p < 1$, we have the equivalent reverses of (2.3.40) and (2.3.41).

2.4. REFERENCES

1. Kuang JC. Applied inequalities. Jinan: Shandong Science Technic Press, 2004.
2. Kuang JC. Introduction to real analysis. Changsha: Hunan Education Press, 1996.
3. Yang BC. A note on Hilbert's integral inequalities. Chinese Quarterly Journal of Mathematics, 1998; 13(4): 83-86.
4. Hardy GH, Littlewood JE, Polya G. Inequalities. Cambridge : Cambridge University Press, 1934.
5. Yang BC. On a generalization of the Hilbert's type inequality and its applications. Chinese Journal of Engineering Mathematics, 2004; 21(5): 821-824.
6. Yang BC. An extension of the Hilbert's type integral inequality and its applications. J.Math., 2007; 27(3): 285-290.
7. Yang BC. On a Hilbert-type integral inequality with multiple-parameters. Journal of Southwest China Normal University (Natural Science), 2007; 32(5): 33-38.
8. Yang BC. A basic Hilbert-type integral inequality. College Mathematics, 2008; 24(1): 87-92.
9. Yang BC. A Basic Hilbert-type integral inequality with the homogeneous kernel of -1-degree and extensions. Journal of Guangdong Education Institute, 2008; 28(3): 5-10.
10. Yang BC. A new Hilbert-type integral inequality and its generalization. J. Jilin Univ. , 2005; 43(5): 580-584.
11. Taylor AE, Lay DC. Introduction to functional analysis. New York: Hohn Wiley & Sons, 1980.
12. Yang BC. A relation to Hardy-Hilbert's integral inequality and Mulholland's inequality. Journal of Ineq. in Pure & Applied Math., 2005; 6(4), Art.112: 1-12.
13. Yang BC, Brnetc I, Krnic M, Pecaric J. Generalization of Hilbert and Hardy-Hilbert integral inequalities. Math. Ineq. & Appl., 2005; 8(2): 259-272.
14. Yang BC. On an extension of Hilbert's integral inequality with some parameters. The Australian Journal of Mathematical Analysis and Appl., 2004; 1(1), Art.11: 1-8.
15. Xi GW. A reverse Hardy-Hilbert-type integral inequality. J. Ineq. in Pure and Appl. Math., 2008; 9(2), Article 49: 1-9.
16. Zhong WY, Yang BC. A best extension of Hilbert inequality involving several parameters. Journal of Jinan University (Natural Science), 2007; 28(1): 20-23.
17. Zhong WY, Yang BC. A reverse Hilbert's type integral inequality and its equivalent forms. College Mathematics, 2008; 24(5): 89-93.
18. Zhong WY, Yang BC. A best extension of new Hilbert's type integral inequality with some parameters. Journal of Jianxi Normal Univ. (Natural Science), 2007; 31(4): 410-414.
19. Wang DC, Guo DR. Introduction to special functions. Beijing: Science Press, 1979.
20. Yang BC. A relation to Hilbert's integral inequality and some basic Hilbert-type inequalities. Journal of Inequalities in Pure and Applied Mathematics, 2008; 9(2), Article 59: 1-8.
21. Yang BC. On the norm of a certain self-adjoint integral operator and applications to bilinear integral inequalities. Taiwan Journal of Mathematics, 2008; 12(2): 315-324.
22. Yang BC. On a Hilbert-type operator with a symmetric homogeneous kernel of -1-order and applications. Journal of Inequalities and Applications, Volume 2007; Article ID 47812, 9 pages, doi: 10.1155/2007/47812.

3. Hilbert-Type Integral Inequalities Restricted in the Subintervals

Bicheng Yang

Department of Mathematics, Guangdong Education Institute, Guangzhou, Guangdong 510303, P. R. China; E-mail: bcyang@pub.guangzhou.gd.cn

Abstract: In this chapter, by using the way of weight functions and the technique of real analysis, we introduce multi-parameters and give some extended Hilbert -type integral inequalities restricted respectively in the subintervals (a,∞), $(0,b)$ and (a,b) $(0<a<b)$. The strengthened versions, the equivalent forms and some reverses are also considered.

3.1. SOME GENERAL RESULTS RESTRICTED IN THE SUBINTERVALS AND LEMMAS

3.1.1. TWO EQUIVALENT INTEGRAL INEQUALITIES

Lemma 3.1.1 Suppose that $\lambda\in\mathbf{R}$, $r,s\neq 0$, $k_\lambda(x,y)(\geq 0)$ is a homogeneous function of $-\lambda$-degree in $(0,\infty)\times(0,\infty)$, such that both $k_\lambda(u,1)$ and $k_\lambda(1,u)$ are positive a.e. in $(0,1]$ and $k_\lambda(r)=\int_0^\infty k_\lambda(u,1)u^{\frac{\lambda}{r}-1}du$ is a positive constant. If $(a,b)\subset(0,\infty)$, define the weight functions $\varpi_\lambda(s,x,a,b)$ and $\omega_\lambda(r,y,a,b)$ as

$$\varpi_\lambda(s,x,a,b):=x^{\frac{\lambda}{r}}\int_a^b k_\lambda(x,y)y^{\frac{\lambda}{s}-1}dy,x\in(a,b),\qquad(3.1.1)$$

$$\omega_\lambda(r,y,a,b):=y^{\frac{\lambda}{s}}\int_a^b k_\lambda(x,y)x^{\frac{\lambda}{r}-1}dx,y\in(a,b).\qquad(3.1.2)$$

Then for $x,y\in(a,b)$, both $\varpi_\lambda(s,x,a,b)$ and $\omega_\lambda(r,y,a,b)$ possess the positive lower bounds

Proof For fixed $x\in(a,b)$ ($0<a<b<\infty$), setting $u=\frac{y}{x}$ and $u=\frac{x}{y}$ respectively in the following two integrals, we find

$$\varpi_\lambda(s,x,a,b)$$
$$=x^{\frac{\lambda}{r}}[\int_a^x k_\lambda(x,y)y^{\frac{\lambda}{s}-1}dy+\int_x^b k_\lambda(x,y)y^{\frac{\lambda}{s}-1}dy]$$
$$=\int_{\frac{a}{x}}^1 k_\lambda(1,u)u^{\frac{\lambda}{s}-1}du+\int_{\frac{x}{b}}^1 k_\lambda(u,1)u^{\frac{\lambda}{r}-1}du.$$

Since $\varpi_\lambda(s,x,a,b)$ is continuous and positive in $x\in[a,b]$, then $\varpi_\lambda(s,x,a,b)$ possesses the positive lower bound in $x\in(a,b)$. We find

$$\varpi_\lambda(s,x,0,b)$$
$$=\int_0^1 k_\lambda(1,u)u^{\frac{\lambda}{s}-1}du+\int_{\frac{x}{b}}^1 k_\lambda(u,1)u^{\frac{\lambda}{r}-1}du$$
$$\geq\int_0^1 k_\lambda(1,u)u^{\frac{\lambda}{s}-1}du>0,x\in(0,b),$$
$$\varpi_\lambda(s,x,a,\infty)$$
$$=\int_{\frac{a}{x}}^1 k_\lambda(1,u)u^{\frac{\lambda}{s}-1}du+\int_0^1 k_\lambda(u,1)u^{\frac{\lambda}{r}-1}du$$
$$\geq\int_0^1 k_\lambda(u,1)u^{\frac{\lambda}{r}-1}du>0,x\in(a,\infty),$$
$$\varpi_\lambda(s,x,0,\infty)=k_\lambda(r)>0,\ x\in(0,\infty).$$

By the same way, we can prove that $\omega_\lambda(r,y,a,b)$ possesses the positive lower bound in $y\in(a,b)$. □

Note 3.1.2 By the assumption of Lemma 3.1.1, we can conform that for fixed $y\in(a,b)$, $k_\lambda(x,y)>0$ *a.e.* in $x\in(a,b)$. In fact,

$$k_\lambda(x,y)=y^{-\lambda}k_\lambda(\tfrac{x}{y},1)>0 \text{ a.e. in } x\leq y;$$
$$k_\lambda(x,y)=x^{-\lambda}k_\lambda(1,\tfrac{y}{x})>0 \text{ a.e. in } x>y.$$

By the same way, we can show that for fixed $x\in(a,b)$, $k_\lambda(x,y)>0$ *a.e.* in $y\in(a,b)$.

Theorem 3.1.3 Suppose that $\lambda\in\mathbf{R}$, $r,s\neq 0$, $p>1$, $k_\lambda(x,y)(\geq 0)$ is a homogeneous function of $-\lambda$-degree in $(0,\infty)\times(0,\infty)$, such that $k_\lambda(u,1)$ and $k_\lambda(1,u)>0$ a.e. in $(0,1]$, and

$$k_\lambda(r)=\int_0^\infty k_\lambda(u,1)u^{\frac{\lambda}{r}-1}du$$

is a positive constant. If $a=0$ or $b=\infty$, $\tilde{\kappa}(x)$ and $\kappa(y)$ are measurable functions in $(a,b)(\subset(0,\infty))$, satisfying $0<\kappa(y),\tilde{\kappa}(x)\le 1$ and

$$\varpi_\lambda(s,x,a,b)\le\tilde{\kappa}(x)k_\lambda(r), \quad (3.1.3)$$

$$\omega_\lambda(r,y,a,b)\le\kappa(y)k_\lambda(r), x,y\in(a,b), \quad (3.1.4)$$

$f(x)$ and $g(x)$ are non-negative measurable functions in (a,b), $0<\int_a^b x^{p(1-\frac{\lambda}{r})-1}f^p(x)dx<\infty$ and $0<\int_a^b y^{q(1-\frac{\lambda}{s})-1}g^q(y)dy<\infty$, then we have the following equivalent inequalities:

$$I_\lambda(a,b):=\int_a^b\int_a^b k_\lambda(x,y)f(x)g(y)dxdy$$
$$<k_\lambda(r)\{\int_a^b\tilde{\kappa}(x)x^{p(1-\frac{\lambda}{r})-1}f^p(x)dx\}^{\frac{1}{p}}$$
$$\times\{\int_a^b\kappa(y)y^{q(1-\frac{\lambda}{s})-1}g^q(y)dy\}^{\frac{1}{q}}, \quad (3.1.5)$$

$$J_\lambda(a,b):=\int_a^b\frac{y^{p\lambda/s-1}}{\kappa^{p-1}(y)}(\int_a^b k_\lambda(x,y)f(x)dx)^p dy$$
$$<k_\lambda^p(r)\int_a^b\tilde{\kappa}(x)x^{p(1-\frac{\lambda}{r})-1}f^p(x)dx. \quad (3.1.6)$$

Proof By Lemma 3.1.1 and (3.1.3), we conform that both $\varpi_\lambda(s,x,a,b)$ and $\omega_\lambda(r,y,a,b)$ possess the same upper bound $k_\lambda(r)>0$ and the same lower bound $m(r)>0$. Hence, it is obvious that the following two inequalities

$$0<\int_a^b\tilde{\kappa}(x)x^{p(1-\frac{\lambda}{r})-1}f^p(x)dx<\infty$$

and $0<\int_a^b\kappa(y)y^{q(1-\frac{\lambda}{s})-1}g^q(y)dy<\infty$ are equivalent to $0<\int_a^b x^{p(1-\frac{\lambda}{r})-1}f^p(x)dx<\infty$ and

$$0<\int_a^b y^{q(1-\frac{\lambda}{s})-1}g^q(y)dy<\infty.$$

By Hölder's inequality with weight (Kuang SSTP 2004) [1], (3.1.1) and (3.1.2), it follows

$$I_\lambda(a,b)=\int_a^b\int_a^b k_\lambda(x,y)[\frac{x^{(1-\lambda/r)/q}}{y^{(1-\lambda/s)/p}}f(x)]$$
$$\times[\frac{y^{(1-\lambda/s)/p}}{x^{(1-\lambda/r)/q}}g(y)]dxdy$$
$$\le\{\int_a^b\int_a^b k_\lambda(x,y)\frac{x^{(1-\lambda/r)(p-1)}}{y^{1-\lambda/s}}f^p(x)dxdy\}^{\frac{1}{p}}$$
$$\times\{\int_a^b\int_a^b k_\lambda(x,y)\frac{y^{(1-\lambda/s)(q-1)}}{x^{1-\lambda/r}}g^q(y)dxdy\}^{\frac{1}{q}}$$
$$=\{\int_a^b\varpi_\lambda(s,x,a,b)x^{p(1-\frac{\lambda}{r})-1}f^p(x)dx\}^{\frac{1}{p}}$$
$$\times\{\int_a^b\omega_\lambda(r,y,a,b)y^{q(1-\frac{\lambda}{s})-1}g^q(y)dy\}^{\frac{1}{q}}. \quad (3.1.7)$$

If (3.1.7) takes the form of equality, then by Note 3.1.2, there exist constants A and B, which are not all zero and (Kuang SSTP 2004) [1]

$$A\frac{x^{(1-\lambda/r)(p-1)}}{y^{1-\lambda/s}}f^p(x)=B\frac{y^{(1-\lambda/s)(q-1)}}{x^{1-\lambda/r}}g^q(y)$$
a.e. in $(a,b)\times(a,b)$.

e.t. $Ax^{p(1-\frac{\lambda}{r})}f^p(x)=By^{q(1-\frac{\lambda}{s})}g^q(y)$ a.e. in $(a,b)\times(a,b)$ .Without lose of generality, suppose $A\ne 0$ .Then for a fixed $y\in(a,b)$, we have

$$x^{p(1-\frac{\lambda}{r})-1}f^p(x)=[By^{q(1-\frac{\lambda}{s})}g^q(y)]/(Ax)$$
a.e. in $x\in(a,b)$,

which contradicts the fact that

$$0<\int_a^b x^{p(1-\frac{\lambda}{r})-1}f^p(x)dx<\infty$$

due to $a=0$ or $b=\infty$. Hence (3.1.7) takes the strict sign-inequality and by (3.1.3), we have (3.1.5).

Setting $[f(x)]_n=f(x), f(x)<n; [f(x)]_n=n$, $f(x)\ge n$ $(n\in\mathbf{N}; x\in(a,b))$, there exist $[a_n,b_n]\subset(a,b), \lim_{n\to\infty}[a_n,b_n]=(a,b)$ and $n_0\in\mathbf{N}$, such that

$$\int_{a_n}^{b_n}\tilde{\kappa}(x)x^{p(1-\frac{\lambda}{r})-1}[f(x)]_n^p dx>0\,(n\ge n_0).$$

Setting

$$g_n(y):=\frac{y^{p\lambda/s-1}}{\kappa^{p-1}(y)}(\int_{a_n}^{b_n}k_\lambda(x,y)[f(x)]_n dx)^{p-1},$$

$y\in(a_n,b_n), n\ge n_0$, there exists $M>0$, such that $[f(x)]_n\le n\le Mx^{\frac{\lambda}{r}-1}$, $x\in[a_n,b_n]$, and

$$0<\int_{a_n}^{b_n}\kappa(y)y^{q(1-\frac{\lambda}{s})-1}g_n^q(y)dy$$
$$=\int_{a_n}^{b_n}\frac{y^{p\lambda/s-1}}{\kappa^{p-1}(y)}(\int_{a_n}^{b_n}k_\lambda(x,y)[f(x)]_n dx)^p dy$$
$$\le M^p\frac{1}{m^{p-1}}\int_{a_n}^{b_n}y^{\frac{p\lambda}{s}-1}(\int_0^\infty k_\lambda(x,y)x^{\frac{\lambda}{r}-1}dx)^p dy$$
$$=[k_\lambda(r)M]^p\frac{1}{m^{p-1}}\ln(\frac{b_n}{a_n})<\infty.$$

Then for $n\ge n_0$, using (3.1.5), we find

$$0<\int_{a_n}^{b_n}\kappa(y)y^{q(1-\frac{\lambda}{s})-1}g_n^q(y)dy$$
$$=\int_{a_n}^{b_n}\frac{y^{p\lambda/s-1}}{\kappa^{p-1}(y)}(\int_{a_n}^{b_n}k_\lambda(x,y)[f(x)]_n dx)^p dy$$
$$=\int_{a_n}^{b_n}\int_{a_n}^{b_n}k_\lambda(x,y)[f(x)]_n g_n(y)dxdy$$

$$< k_\lambda(r)\{\int_{a_n}^{b_n} \tilde{\kappa}(x)x^{p(1-\frac{\lambda}{r})-1}[f(x)]_n^p dx\}^{\frac{1}{p}}$$
$$\times\{\int_{a_n}^{b_n} \kappa(y)y^{q(1-\frac{\lambda}{s})-1}g_n^q(y)dy\}^{\frac{1}{q}} < \infty ,$$
$$0 < \int_{a_n}^{b_n} \kappa(y)y^{q(1-\frac{\lambda}{s})-1}g_n^q(y)dy$$
$$< k_\lambda^p(r)\int_a^b \tilde{\kappa}(x)x^{p(1-\frac{\lambda}{r})-1}f^p(x)dx < \infty .$$

It follows

$$0 < \int_a^b \kappa(y)y^{q(1-\frac{\lambda}{s})-1}g_\infty^q(y)dy < \infty$$

and then $0 < \int_a^b y^{q(1-\frac{\lambda}{s})-1}g_\infty^q(y)dy < \infty$. Hence for $n \to \infty$, using (3.1.5), both of the above inequalities still take the strict sign-inequalities, and we have (3.1.6).

On the other hand, suppose that (3.1.6) is valid. By Hölder's inequality (Kuang SSTP 2004) [1], we find

$$I_\lambda(a,b) = \int_a^b [\frac{y^{-1/p+\lambda/s}}{\kappa^{1/q}(y)}\int_a^b k_\lambda(x,y)f(x)dx]$$
$$\times[\kappa^{\frac{1}{q}}(y)y^{\frac{1}{p}-\frac{\lambda}{s}}g(y)]dy$$
$$\le [J_\lambda(a,b)]^{\frac{1}{p}}\{\int_a^b \kappa(y)y^{q(1-\frac{\lambda}{s})-1}g^q(y)dy\}^{\frac{1}{q}}. \tag{3.1.8}$$

By (3.1.6), since

$$0 < \int_a^b \kappa(y)y^{q(1-\frac{\lambda}{s})-1}g^q(y)dy < \infty ,$$

we have (3.1.5). Hence inequalities (3.1.5) and (3.1.6) are equivalent. □

Note 3.1.4 For $0 < a < b < \infty$, under the assumption of Theorem 3.1.2, inequalities (3.1.5) and (3.1.6) may possibly keep the forms of equality. It is obvious that if at least one of (3.1.3) and (3.1.4) keeps the strict sign-inequality *a.e.* in (a,b), then we still have (3.1.5) and (3.1.6).

3.1.2. TWO LEMMAS

Lemma 3.1.5 Suppose $\lambda \in \mathbf{R}$, $k_\lambda(x,y) \ge 0$ is a homogeneous function of $-\lambda$-degree. If $k_\lambda(u,1) \in C^1(0,1]$, $k_\lambda'(u,1) \le 0$, and $\sigma_\lambda > 0$,

$$h_\lambda(y) := y^{-\sigma_\lambda}\int_0^y k_\lambda(u,1)u^{\sigma_\lambda-1}du \in \mathbf{R}$$

($y \in (0,1]$), then

$$h_\lambda(y) \ge h_\lambda(1) = \int_0^1 k_\lambda(u,1)u^{\sigma_\lambda-1}du, y \in (0,1].$$

Proof Since $k_\lambda'(u,1) \le 0$, $y \in (0,1]$, then

$$h_\lambda'(y) = -\sigma_\lambda y^{-\sigma_\lambda-1}\int_0^y k_\lambda(u,1)u^{\sigma_\lambda-1}du$$
$$+y^{-\sigma_\lambda}k(y,1)y^{\sigma_\lambda-1}$$
$$= -y^{-\sigma_\lambda-1}\int_0^y k_\lambda(u,1)du^{\sigma_\lambda} + y^{-\sigma_\lambda}k(y,1)y^{\sigma_\lambda-1}$$
$$= -y^{-\sigma_\lambda-1}k_\lambda(y,1)y^{\sigma_\lambda}$$
$$+y^{-\sigma_\lambda-1}\int_0^y u^{\sigma_\lambda}dk_\lambda(u,1) + k_\lambda(y,1)y^{-1}$$
$$= y^{-\sigma_\lambda-1}\int_0^y u^{\sigma_\lambda}k_\lambda'(u,1)du \le 0, \ y \in (0,1].$$

Hence $h_\lambda(y) \ge h_\lambda(1)$ $(y \in (0,1])$.

By the same way, we still have

Lemma 3.1.6 Suppose $\lambda \in \mathbf{R}$, $k_\lambda(x,y) \ge 0$ is a homogeneous function of $-\lambda$-degree. If $k_\lambda(1,u) \in C^1(0,1]$, $k_\lambda'(1,u) \le 0$, and $\sigma_\lambda > 0$,

$$\tilde{h}_\lambda(y) := y^{-\sigma_\lambda}\int_0^y k_\lambda(1,u)u^{\sigma_\lambda-1}du \in \mathbf{R}$$

($y \in (0,1]$), then

$$\tilde{h}_\lambda(y) \ge \tilde{h}_\lambda(1) = \int_0^1 k_\lambda(1,u)u^{\sigma_\lambda-1}du, y \in (0,1].$$

3.2. SOME HILBERT-TYPE INTEGRAL INEQUALITIES RESTICTED IN THE SUBINTERVAL (a,∞) $(a>0)$

3.2.1. SOME RESULTS

Theorem 3.2.1 As the assumption of Theorem 3.1.3, for $\lambda \in \mathbf{R}$, $r,s \ne 0, a > 0, b = \infty$, we have the following equivalent inequalities:

$$\int_a^\infty\int_a^\infty k_\lambda(x,y)f(x)g(y)dxdy$$
$$< k_\lambda(r)\{\int_a^\infty \tilde{\kappa}(x)x^{p(1-\frac{\lambda}{r})-1}f^p(x)dx\}^{\frac{1}{p}}$$
$$\times\{\int_a^\infty \kappa(y)y^{q(1-\frac{\lambda}{s})-1}g^q(y)dy\}^{\frac{1}{q}}, \tag{3.2.1}$$
$$\int_a^\infty \frac{y^{p\lambda/s-1}}{\kappa^{p-1}(y)}(\int_a^\infty k_\lambda(x,y)f(x)dx)^p dy$$
$$< k_\lambda^p(r)\int_a^\infty \tilde{\kappa}(x)x^{p(1-\frac{\lambda}{r})-1}f^p(x)dx, \tag{3.2.2}$$

where the constant factors

$$k_\lambda(r) = \int_0^\infty k_\lambda(u,1)u^{\frac{\lambda}{r}-1}du$$

and $k_\lambda^p(r)$ are the best possible. In particular, for $\tilde{\kappa}(x) = \kappa(y) = 1$, we have the following equivalent inequalities:

$$\int_a^\infty \int_a^\infty k_\lambda(x,y)f(x)g(y)dxdy$$
$$< k_\lambda(r)\{\int_a^\infty x^{p(1-\frac{\lambda}{r})-1} f^p(x)dx\}^{\frac{1}{p}}$$
$$\times\{\int_a^\infty y^{q(1-\frac{\lambda}{s})-1} g^q(y)dy\}^{\frac{1}{q}}, \tag{3.2.3}$$
$$\int_a^\infty y^{\frac{p\lambda}{s}-1}(\int_a^\infty k_\lambda(x,y)f(x)dx)^p dy$$
$$< k_\lambda^p(r)\int_a^\infty x^{p(1-\frac{\lambda}{r})-1} f^p(x)dx. \tag{3.2.4}$$

Proof By Theorem 3.1.3, for $a>0, b=\infty$, we have equivalent inequalities (3.2.1) and (3.2.2). For $n\in \mathbf{N}$, we set

$$f_n(x) := x^{\frac{\lambda}{r}-1-\frac{1}{np}},\ g_n(x) := x^{\frac{\lambda}{s}-1-\frac{1}{nq}},\ x\in(a,\infty).$$

If there exists a positive number $k \le k_\lambda(r)$, such that (3.2.1) is still valid as we replace $k_\lambda(r)$ by k, then in particular, for $0<\kappa(y), \tilde{\kappa}(x)\le 1$, it follows

$$I_n := \int_a^\infty \int_a^\infty k_\lambda(x,y)f_n(x)g_n(y)dxdy$$
$$< k\{\int_a^\infty x^{p(1-\frac{\lambda}{r})-1} f_n^{\ p}(x)dx\}^{\frac{1}{p}}$$
$$\times\{\int_a^\infty y^{q(1-\frac{\lambda}{s})-1} g_n^{\ q}(y)dy\}^{\frac{1}{q}} = \frac{n}{a^{1/n}}k. \tag{3.2.5}$$

For fixed $y>a$, setting $u=\frac{y}{x}$, by Fubini theorem, we find

$$I_n = \int_a^\infty y^{\frac{\lambda}{s}-1-\frac{1}{nq}}[\int_a^\infty k_\lambda(x,y)x^{\frac{\lambda}{r}-1-\frac{1}{np}}dx]dy$$
$$= \int_a^\infty y^{-1-\frac{1}{n}}[\int_0^{\frac{y}{a}} k_\lambda(1,u)u^{\frac{\lambda}{s}-1+\frac{1}{np}}du]dy$$
$$= \int_a^\infty y^{-1-\frac{1}{n}}[\int_0^1 k_\lambda(1,u)u^{\frac{\lambda}{s}-1+\frac{1}{np}}du]dy$$
$$+\int_a^\infty y^{-1-\frac{1}{n}}[\int_1^{\frac{y}{a}} k_\lambda(1,u)u^{\frac{\lambda}{s}-1+\frac{1}{np}}du]dy$$
$$= \frac{n}{a^{1/n}}\int_0^1 k_\lambda(1,u)u^{\frac{\lambda}{s}-1+\frac{1}{np}}du$$
$$+\int_1^\infty (\int_{au}^\infty y^{-1-\frac{1}{n}}dy)k_\lambda(1,u)u^{\frac{\lambda}{s}-1+\frac{1}{np}}du$$
$$= \frac{n}{a^{1/n}}[\int_0^1 k_\lambda(1,u)u^{\frac{\lambda}{s}-1+\frac{1}{np}}du$$
$$+\int_1^\infty k_\lambda(1,u)u^{\frac{\lambda}{s}-1-\frac{1}{nq}}du]. \tag{3.2.6}$$

Hence in view of (3.2.5) and (3.2.6), it follows

$$\int_0^1 k_\lambda(1,u)u^{\frac{\lambda}{s}-1+\frac{1}{np}}du$$
$$+\int_1^\infty k_\lambda(1,u)u^{\frac{\lambda}{s}-1-\frac{1}{nq}}du < k.$$

For $n\to\infty$, by Fatou lemma, we obtain

$$k_\lambda(r) = \tilde{k}_\lambda(s) = \int_0^1 \lim_{n\to\infty} k_\lambda(1,u)u^{\frac{\lambda}{s}-1+\frac{1}{np}}du$$
$$+\int_1^\infty \lim_{n\to\infty} k_\lambda(1,u)u^{\frac{\lambda}{s}-1-\frac{1}{nq}}du$$
$$\le \underline{\lim}_{n\to\infty}[\int_0^1 k_\lambda(1,u)u^{\frac{\lambda}{s}-1+\frac{1}{np}}du$$
$$+\int_1^\infty k_\lambda(1,u)u^{\frac{\lambda}{s}-1-\frac{1}{nq}}du] \le k.$$

Therefore, the constant factor $k = k_\lambda(r)$ in (3.2.1) is the best possible. We conform that the constant factor $k_\lambda^p(r)$ in (3.2.2) is the best possible, otherwise, by (3.1.8) (for $a>0, b=\infty$), we can get a contradiction that the constant factor in (3.2.1) is not the best possible. □

Setting $\phi(x) = x^{p(1-\frac{\lambda}{r})-1}$, $\psi(x) = x^{q(1-\frac{\lambda}{s})-1}$, $x\in(a,\infty)$ and $\psi^{1-p}(x) = x^{\frac{p\lambda}{s}-1}$, define the following real function spaces:

$$L_\phi^p(a,\infty) = \{f; \|f\|_{p,\phi} := \{\int_a^\infty \phi(x)|f(x)|^p dx\}^{\frac{1}{p}} < \infty\},$$
$$L_\psi^q(a,\infty) = \{g; \|g\|_{q,\psi} := \{\int_a^\infty \psi(x)|g(x)|^q dx\}^{\frac{1}{q}} < \infty\},$$

and Hilbert-type operator

$$T_a : L_\phi^p(a,\infty) \to L_{\psi^{1-p}}^p(a,\infty)$$

as: for $f\in L_\phi^p(a,\infty)$,

$$(T_a f)(y) = \int_a^\infty k_\lambda(x,y)f(x)dx,\ y\in(a,\infty).$$

It is obvious that $T_a f \in L_{\psi^{1-p}}^p(a,\infty)$ due to (3.2.4). For $g\in L_\psi^q(a,\infty)$, define the formal inner product of $T_a f$ and g as

$$(T_a f, g) := \int_a^\infty \int_a^\infty k_\lambda(x,y)f(x)g(y)dxdy.$$

We may rewrite inequalities (3.2.3) and (3.2.4) as the following equivalent forms:

$$(T_a f, g) < k_\lambda(r)\|f\|_{p,\phi}\|g\|_{q,\psi}, \tag{3.2.7}$$
$$\|T_a f\|_{p,\psi^{1-p}} < k_\lambda(r)\|f\|_{p,\phi}. \tag{3.2.8}$$

Since the constant factors in (3.2.7) and (3.2.8) are the best possible, we can conclude that T_a is bounded and $\|T_a\| = k_\lambda(r)$.

Corollary 3.2.2 As the assumption of Theorem 3.2.1, if $\lambda>0, r,s>1$, both $k_\lambda(u,1)$ and $k_\lambda(1,u)$ possess the same lower bound $l_\lambda>0$ in $(0,1]$, then we have the following equivalent inequalities:

$$(T_a f,g)$$
$$<k_\lambda(r)\{\int_a^\infty[1-\tfrac{sl_\lambda}{\lambda k_\lambda(r)}(\tfrac{a}{x})^{\frac{\lambda}{s}}]x^{p(1-\frac{\lambda}{r})-1}f^p(x)dx\}^{\frac{1}{p}}$$
$$\times\{\int_a^\infty[1-\tfrac{rl_\lambda}{\lambda k_\lambda(r)}(\tfrac{a}{y})^{\frac{\lambda}{r}}]y^{q(1-\frac{\lambda}{s})-1}g^q(y)dy\}^{\frac{1}{q}}, \tag{3.2.9}$$
$$\int_a^\infty\frac{y^{p\lambda/s-1}}{[1-\frac{rl_\lambda}{\lambda k_\lambda(r)}(\frac{a}{y})^{\lambda/r}]^{p-1}}(\int_a^\infty k_\lambda(x,y)f(x)dx)^p dy$$
$$<k_\lambda^p(r)\int_a^\infty[1-\tfrac{sl_\lambda}{\lambda k_\lambda(r)}(\tfrac{a}{x})^{\frac{\lambda}{s}}]x^{p(1-\frac{\lambda}{r})-1}f^p(x)dx, \tag{3.2.10}$$

where the constant factors $k_\lambda(r)$ and $k_\lambda^p(r)$ are the best possible. In particular, we have the following equivalent inequalities:

$$(T_a f,g)<k_\lambda(r)$$
$$\times\{\int_a^\infty[1-\tfrac{sl_\lambda}{\lambda k_\lambda(r)}(\tfrac{a}{x})^{\frac{\lambda}{s}}]x^{p(1-\frac{\lambda}{r})-1}f^p(x)dx\}^{\frac{1}{p}}\|g\|_{q,\psi}, \tag{3.2.11}$$
$$\|T_a f\|^p_{p,\psi^{1-p}}$$
$$<k_\lambda^p(r)\int_a^\infty[1-\tfrac{sl_\lambda}{\lambda k_\lambda(r)}(\tfrac{a}{x})^{\frac{\lambda}{s}}]x^{p(1-\frac{\lambda}{r})-1}f^p(x)dx. \tag{3.2.12}$$

Proof Since $k_\lambda(u,1)\ge l_\lambda, k_\lambda(1,u)\ge l_\lambda, u\in(0,1]$, we find

$$\varpi_\lambda(s,x,a,\infty)=x^{\frac{\lambda}{r}}\int_a^\infty k_\lambda(x,y)y^{\frac{\lambda}{s}-1}dy$$
$$=\int_{\frac{a}{x}}^\infty k_\lambda(1,u)u^{\frac{\lambda}{s}-1}du$$
$$=k_\lambda(r)-\int_0^{\frac{a}{x}}k_\lambda(1,u)u^{\frac{\lambda}{s}-1}du$$
$$\le k_\lambda(r)-l_\lambda\int_0^{\frac{a}{x}}u^{\frac{\lambda}{s}-1}du$$
$$=k_\lambda(r)[1-\tfrac{sl_\lambda}{\lambda k_\lambda(r)}(\tfrac{a}{x})^{\frac{\lambda}{s}}],\ x\in(a,\infty).$$

By the same way, it follows

$$\omega_\lambda(r,y,a,\infty)$$
$$\le k_\lambda(r)[1-\tfrac{rl_\lambda}{\lambda k_\lambda(r)}(\tfrac{a}{y})^{\frac{\lambda}{r}}], y\in(a,\infty).$$

Setting $\tilde{\kappa}(x)=1-\frac{sl_\lambda}{\lambda k_\lambda(r)}(\frac{a}{x})^{\frac{\lambda}{s}}$ and

$$\kappa(y)=1-\tfrac{rl_\lambda}{\lambda k_\lambda(r)}(\tfrac{a}{y})^{\frac{\lambda}{r}}$$

in (3.2.1) and (3.2.2), we have equivalent inequalities (3.2.9) and (3.2.10). Setting

$$\tilde{\kappa}(x)=1-\tfrac{sl_\lambda}{\lambda k_\lambda(r)}(\tfrac{a}{x})^{\frac{\lambda}{s}}$$

and $\kappa(y)=1$ in (3.2.1) and (3.2.2), we obtain (3.2.11) and (3.2.12). □

For $a>0, b=\infty$, setting $u=\frac{y}{x}$ in (3.1.1), we find

$$\varpi_\lambda(s,x,a,\infty)=x^{\frac{\lambda}{r}}\int_a^\infty k_\lambda(x,y)y^{\frac{\lambda}{s}-1}dy$$
$$=k_\lambda(r)-\int_0^{\frac{a}{x}}k_\lambda(1,u)u^{\frac{\lambda}{s}-1}du$$
$$=k_\lambda(r)-(\tfrac{a}{x})^{\frac{\lambda}{s}}[(\tfrac{a}{x})^{\frac{-\lambda}{s}}\int_0^{\frac{a}{x}}k_\lambda(1,u)u^{\frac{\lambda}{s}-1}du].$$

If $\lambda\in\mathbf{R}, k_\lambda(1,u), k_\lambda(u,1)\in C^1(0,1]$,

$$k_\lambda'(1,u)\le 0, k_\lambda'(u,1)\le 0, u\in(0,1],$$

by Lemma 3.1.5, since $0<\frac{a}{x}\le 1$, we have

$$(\tfrac{a}{x})^{\frac{-\lambda}{s}}\int_0^{\frac{a}{x}}k_\lambda(1,u)u^{\frac{\lambda}{s}-1}du$$
$$\ge\int_0^1 k_\lambda(1,u)u^{\frac{\lambda}{s}-1}du=:\tilde{\theta}_\lambda(s),\quad x>a,$$

and then

$$\varpi_\lambda(s,x,a,\infty)$$
$$\le k_\lambda(r)[1-\tfrac{\tilde{\theta}_\lambda(s)}{k_\lambda(r)}(\tfrac{a}{x})^{\frac{\lambda}{s}}], x>a. \tag{3.2.13}$$

By the same way, setting $\theta_\lambda(r):=\int_0^1 k_\lambda(u,1)u^{\frac{\lambda}{r}-1}du$, we find

$$\omega_\lambda(r,y,a,\infty)$$
$$\le k_\lambda(r)[1-\tfrac{\theta_\lambda(r)}{k_\lambda(r)}(\tfrac{a}{y})^{\frac{\lambda}{r}}], y>a. \tag{3.2.14}$$

By Theorem 3.2.1, it follows

Corollary 3.2.3 As the assumption of Theorem 3.2.1, If $\lambda\in\mathbf{R}, r,s\ne 0, k_\lambda(1,u), k_\lambda(u,1)\in C^1(0,1]$,

$$k_\lambda'(1,u)\le 0, k_\lambda'(u,1)\le 0, u\in(0,1]$$

and $l_\lambda=k_\lambda(1,1)>0$, then we have the following equivalent inequalities:

$$(T_a f,g)$$
$$<k_\lambda(r)\{\int_a^\infty[1-\tfrac{\tilde{\theta}_\lambda(s)}{k_\lambda(r)}(\tfrac{a}{x})^{\frac{\lambda}{s}}]x^{p(1-\frac{\lambda}{r})-1}f^p(x)dx\}^{\frac{1}{p}}$$
$$\times\{\int_a^\infty[1-\tfrac{\theta_\lambda(r)}{k_\lambda(r)}(\tfrac{a}{y})^{\frac{\lambda}{r}}]y^{q(1-\frac{\lambda}{s})-1}g^q(y)dy\}^{\frac{1}{q}}, \tag{3.2.15}$$
$$\int_a^\infty\frac{y^{p\lambda/s-1}}{[1-\frac{\theta_\lambda(r)}{k_\lambda(r)}(\frac{a}{y})^{\lambda/r}]^{p-1}}(\int_a^\infty k_\lambda(x,y)f(x)dx)^p dy$$
$$<k_\lambda^p(r)\int_a^\infty[1-\tfrac{\tilde{\theta}_\lambda(s)}{k_\lambda(r)}(\tfrac{a}{x})^{\frac{\lambda}{s}}]x^{p(1-\frac{\lambda}{r})-1}f^p(x)dx, \tag{3.2.16}$$

where $\tilde{\theta}_\lambda(s)=\int_0^1 k_\lambda(1,u)u^{\frac{\lambda}{s}-1}du$,

$$\theta_\lambda(r)=\int_0^1 k_\lambda(u,1)u^{\frac{\lambda}{r}-1}du,$$

and the constant factors $k_\lambda(r)$ and $k_\lambda^p(r)$ are the best possible. In fact, we still have the following equivalent inequalities:

$$(T_a f, g) < k_\lambda(r) \times \{\int_a^\infty [1 - \tfrac{\tilde{\theta}_\lambda(s)}{k_\lambda(r)} (\tfrac{a}{x})^{\frac{\lambda}{s}}] x^{p(1-\frac{\lambda}{r})-1} f^p(x)dx\}^{\frac{1}{p}} \| g \|_{q,\psi}, \quad (3.2.17)$$

$$\| T_a f \|^p_{p,\psi^{1-p}} < k_\lambda^p(r) \int_a^\infty [1 - \tfrac{\tilde{\theta}_\lambda(s)}{k_\lambda(r)} (\tfrac{a}{x})^{\frac{\lambda}{s}}] x^{p(1-\frac{\lambda}{r})-1} f^p(x)dx. \quad (3.2.18)$$

In particular, for $r = s = 2$, if $k_\lambda(1,u) = k_\lambda(u,1)$, setting $k_\lambda := k_\lambda(2)$, we have $\tilde{\theta}_\lambda(2) = \theta_\lambda(2) = \frac{k_\lambda}{2}$ and the following equivalent inequalities:

$$(T_a f, g) < k_\lambda \{\int_a^\infty [1 - \tfrac{1}{2}(\tfrac{a}{x})^{\frac{\lambda}{2}}] x^{p(1-\frac{\lambda}{2})-1} f^p(x)dx\}^{\frac{1}{p}} \times \{\int_a^\infty [1 - \tfrac{1}{2}(\tfrac{a}{y})^{\frac{\lambda}{2}}] y^{q(1-\frac{\lambda}{2})-1} g^q(y)dy\}^{\frac{1}{q}}, \quad (3.2.19)$$

$$\int_a^\infty \frac{y^{p\lambda/2-1}}{[1-\frac{1}{2}(\frac{a}{y})^{\lambda/2}]^{p-1}} (\int_a^\infty k_\lambda(x,y) f(x)dx)^p dy < k_\lambda^p \int_a^\infty [1 - \tfrac{1}{2}(\tfrac{a}{x})^{\frac{\lambda}{2}}] x^{p(1-\frac{\lambda}{2})-1} f^p(x)dx; \quad (3.2.20)$$

$$(T_a f, g) < k_\lambda \{\int_a^\infty [1 - \tfrac{1}{2}(\tfrac{a}{x})^{\frac{\lambda}{2}}] x^{p(1-\frac{\lambda}{2})-1} f^p(x)dx\}^{\frac{1}{p}} \times \{\int_a^\infty y^{q(1-\frac{\lambda}{2})-1} g^q(y)dy\}^{\frac{1}{q}}, \quad (3.2.21)$$

$$\int_a^\infty y^{\frac{p\lambda}{2}-1} (\int_a^\infty k_\lambda(x,y) f(x)dx)^p dy < k_\lambda^p \int_a^\infty [1 - \tfrac{1}{2}(\tfrac{a}{x})^{\frac{\lambda}{2}}] x^{p(1-\frac{\lambda}{2})-1} f^p(x)dx. \quad (3.2.22)$$

Note 3.2.4 (1) For $a \to 0^+$, inequalities (3.2.3) and (3.2.4) reduce to (2.2.1) and (2.2.2) . It follows that (3.2.3) and (3.2.4) are extensions of (2.2.1) and (2.2.2). (2) Inequalities (3.2.9) and (3.2.11) are strengthened versions of (3.2.3).

3.2.2. SOME PARTICULAR EXAMPLES

In the following examples, we omit to write the assumption of f and g satisfying Theorem 3.2.1 and the conclusions that the constant factors are the best possible.

Example 3.2.5 If $k_\lambda(x,y) = \frac{1}{(x^\alpha + y^\alpha)^{\lambda/\alpha}}$ $(\lambda, \alpha > 0)$, $r, s > 1$, then we find

$$k_\lambda(u,1) = k_\lambda(1,u) = \frac{1}{(1+u^\alpha)^{\lambda/\alpha}},$$

$l_\lambda = \frac{1}{2^{\lambda/\alpha}} > 0$ and $k_\lambda(r) = \frac{1}{\alpha} B(\frac{\lambda}{\alpha r}, \frac{\lambda}{\alpha s})$. By (3.2.9) and (3.2.10), we obtain the following equivalent inequalities:

$$\int_a^\infty \int_a^\infty \frac{1}{(x^\alpha + y^\alpha)^{\lambda/\alpha}} f(x) g(y) dxdy < \frac{1}{\alpha} B(\frac{\lambda}{\alpha r}, \frac{\lambda}{\alpha s}) \times \{\int_a^\infty [1 - \frac{s\alpha}{2^{\lambda/\alpha} \lambda B(\frac{\lambda}{\alpha r}, \frac{\lambda}{\alpha s})} (\tfrac{a}{x})^{\frac{\lambda}{s}}] x^{p(1-\frac{\lambda}{r})-1} f^p(x)dx\}^{\frac{1}{p}} \times \{\int_a^\infty [1 - \frac{r\alpha}{2^{\lambda/\alpha} \lambda B(\frac{\lambda}{\alpha r}, \frac{\lambda}{\alpha s})} (\tfrac{a}{y})^{\frac{\lambda}{r}}] y^{q(1-\frac{\lambda}{s})-1} g^q(y)dy\}^{\frac{1}{q}}, \quad (3.2.23)$$

$$\int_a^\infty \frac{y^{p\lambda/s-1}}{[1 - \frac{r\alpha}{2^{\lambda/\alpha}\lambda B(\frac{\lambda}{\alpha r}, \frac{\lambda}{\alpha s})} (\frac{a}{y})^{\lambda/r}]^{p-1}} [\int_a^\infty \frac{f(x)}{(x^\alpha + y^\alpha)^{\lambda/\alpha}} dx]^p dy < [\frac{1}{\alpha} B(\frac{\lambda}{\alpha r}, \frac{\lambda}{\alpha s})]^p \times \int_a^\infty [1 - \frac{s\alpha}{2^{\lambda/\alpha} \lambda B(\frac{\lambda}{\alpha r}, \frac{\lambda}{\alpha s})} (\tfrac{a}{x})^{\frac{\lambda}{s}}] x^{p(1-\frac{\lambda}{r})-1} f^p(x)dx. \quad (3.2.24)$$

In particular, due to (3.2.19) and (3.2.20), we deduce the following equivalent inequalities:

$$\int_a^\infty \int_a^\infty \frac{1}{(x^\alpha + y^\alpha)^{\lambda/\alpha}} f(x) g(y) dxdy < \frac{1}{\alpha} B(\frac{\lambda}{2\alpha}, \frac{\lambda}{2\alpha}) \{\int_a^\infty [1 - \tfrac{1}{2}(\tfrac{a}{x})^{\frac{\lambda}{2}}] x^{p(1-\frac{\lambda}{2})-1} f^p(x)dx\}^{\frac{1}{p}} \times \{\int_a^\infty [1 - \tfrac{1}{2}(\tfrac{a}{y})^{\frac{\lambda}{2}}] y^{q(1-\frac{\lambda}{2})-1} g^q(y)dy\}^{\frac{1}{q}}, \quad (3.2.25)$$

$$\int_a^\infty \frac{y^{p\lambda/2-1}}{[1-\frac{1}{2}(\frac{a}{y})^{\lambda/2}]^{p-1}} [\int_a^\infty \frac{f(x)}{(x^\alpha + y^\alpha)^{\lambda/\alpha}} dx]^p dy < [\frac{1}{\alpha} B(\frac{\lambda}{2\alpha}, \frac{\lambda}{2\alpha})]^p \int_a^\infty [1 - \tfrac{1}{2}(\tfrac{a}{x})^{\frac{\lambda}{2}}] x^{p(1-\frac{\lambda}{2})-1} f^p(x)dx. \quad (3.2.26)$$

Note 3.2.6 For $\alpha = 1, (r,s)$ takes some particular value, (3.2.23) and (3.2.24) are the results of (Yang AMS 1998) [2], (Yang CMA 2000)[3]; Some other particular results are given in (Yang JMAA 1998)[4], (Yang JNUMB 2000)[5].

Example 3.2.7 If $k_\lambda(x,y) = \frac{\ln(x/y)}{x^\lambda - y^\lambda} (\lambda > 0), r, s > 1$, then we find $k_\lambda(u,1) = k_\lambda(1,u) = \frac{\ln u}{u^\lambda - 1}$, $u \in (0,1]$, $l_\lambda = \frac{1}{\lambda} > 0$ and $k_\lambda(r) = [\frac{1}{\lambda} \frac{\pi}{\sin(\pi/r)}]^2$. By (3.2.9) and (3.2.10), we have the following equivalent inequalities:

$$\int_a^\infty \int_a^\infty \frac{\ln(x/y)}{x^\lambda - y^\lambda} f(x) g(y) dxdy < [\frac{\pi}{\lambda \sin(\pi/r)}]^2 \times \{\int_a^\infty [1 - s(\tfrac{\sin(\pi/r)}{\pi})^2 (\tfrac{a}{x})^{\frac{\lambda}{s}}] x^{p(1-\frac{\lambda}{r})-1} f^p(x)dx\}^{\frac{1}{p}} \times \{\int_a^\infty [1 - r(\tfrac{\sin(\pi/r)}{\pi})^2 (\tfrac{a}{y})^{\frac{\lambda}{r}}] y^{q(1-\frac{\lambda}{s})-1} g^q(y)dy\}^{\frac{1}{q}}, \quad (3.2.27)$$

$$\int_a^\infty \frac{y^{p\lambda/s-1}}{[1 - r(\frac{\sin(\pi/r)}{\pi})^2 (\frac{a}{y})^{\lambda/r}]^{p-1}} [\int_a^\infty \frac{\ln(x/y)}{x^\lambda - y^\lambda} f(x)dx]^p dy$$

$$<[\frac{\pi}{\lambda\sin(\pi/r)}]^{2p}$$
$$\times\int_a^\infty[1-s(\frac{\sin(\pi/r)}{\pi})^2(\frac{a}{x})^{\frac{\lambda}{s}}]x^{p(1-\frac{\lambda}{r})-1}f^p(x)dx. \quad (3.2.28)$$

In particular, due to (3.2.19) and (3.2.20), we deduce the following equivalent inequalities:

$$\int_a^\infty\int_a^\infty\frac{\ln(x/y)}{x^\lambda-y^\lambda}f(x)g(y)dxdy$$
$$<(\frac{\pi}{\lambda})^2\{\int_a^\infty[1-\frac{1}{2}(\frac{a}{x})^{\frac{\lambda}{2}}]x^{p(1-\frac{\lambda}{2})-1}f^p(x)dx\}^{\frac{1}{p}}$$
$$\times\{\int_a^\infty[1-\frac{1}{2}(\frac{a}{y})^{\frac{\lambda}{2}}]y^{q(1-\frac{\lambda}{2})-1}g^q(y)dy\}^{\frac{1}{q}}, \quad (3.2.29)$$
$$\int_a^\infty\frac{y^{p\lambda/2-1}}{[1-\frac{1}{2}(\frac{a}{y})^{\lambda/2}]^{p-1}}[\int_a^\infty\frac{\ln(x/y)}{x^\lambda-y^\lambda}f(x)dx]^p dy$$
$$<(\frac{\pi}{\lambda})^{2p}\int_a^\infty[1-\frac{1}{2}(\frac{a}{x})^{\frac{\lambda}{2}}]x^{p(1-\frac{\lambda}{2})-1}f^p(x)dx. \quad (3.2.30)$$

Example 3.2.8 If $k_\lambda(x,y)=\frac{1}{(\max\{x,y\})^\lambda}(\lambda>0)$, $r,s>1$, then we find $k_\lambda(u,1)=k_\lambda(1,u)=1$, $u\in(0,1]$, $k_\lambda(r)=\frac{rs}{\lambda}$ and

$$\varpi_\lambda(s,x,a,\infty)=\frac{rs}{\lambda}[1-\frac{1}{r}(\frac{a}{x})^{\frac{\lambda}{s}}],$$
$$\omega_\lambda(r,y,a,\infty)=\frac{rs}{\lambda}[1-\frac{1}{s}(\frac{a}{y})^{\frac{\lambda}{r}}].$$

By (3.2.1) and (3.2.2) (for $a>0,b=\infty$), we have the following equivalent inequalities:

$$\int_a^\infty\int_a^\infty\frac{1}{(\max\{x,y\})^\lambda}f(x)g(y)dxdy$$
$$<\frac{rs}{\lambda}\{\int_a^\infty[1-\frac{1}{r}(\frac{a}{x})^{\frac{\lambda}{s}}]x^{p(1-\frac{\lambda}{r})-1}f^p(x)dx\}^{\frac{1}{p}}$$
$$\times\{\int_a^\infty[1-\frac{1}{s}(\frac{a}{y})^{\frac{\lambda}{r}}]y^{q(1-\frac{\lambda}{s})-1}g^q(y)dy\}^{\frac{1}{q}}, \quad (3.2.31)$$
$$\int_a^\infty\frac{y^{p\lambda/s-1}}{[1-\frac{1}{s}(\frac{a}{y})^{\lambda/r}]^{p-1}}[\int_a^\infty\frac{f(x)}{(\max\{x,y\})^\lambda}dx]^p dy$$
$$<(\frac{rs}{\lambda})^p\int_a^\infty[1-\frac{1}{r}(\frac{a}{x})^{\frac{\lambda}{s}}]x^{p(1-\frac{\lambda}{r})-1}f^p(x)dx. \quad (3.2.32)$$

Example 3.2.9 If $k_\lambda(x,y)=\frac{|\ln(x/y)|}{(\max\{x,y\})^\lambda}(\lambda>0)$, $r>1$, then we find $k_\lambda(u,1),k_\lambda(1,u)>0$ a.e. in $(0,1]$, $k_\lambda(r)=\frac{r^2+s^2}{\lambda^2}$ and

$$\varpi_\lambda(s,x,a,\infty)=\frac{r^2+s^2}{\lambda^2}-\int_0^{\frac{a}{x}}(-\ln u)u^{\frac{\lambda}{s}-1}du$$
$$=\frac{r^2+s^2}{\lambda^2}-\frac{s}{\lambda}\int_0^{\frac{a}{x}}(-\ln u)du^{\frac{\lambda}{s}}$$
$$\le\frac{r^2+s^2}{\lambda^2}[1-\frac{s^2}{r^2+s^2}(\frac{a}{x})^{\frac{\lambda}{s}}],$$
$$\omega_\lambda(r,y,a,\infty)\le\frac{r^2+s^2}{\lambda^2}[1-\frac{r^2}{r^2+s^2}(\frac{a}{y})^{\frac{\lambda}{r}}].$$

By (3.2.1) and (3.2.2) (for $a>0,b=\infty$), we have the following equivalent inequalities:

$$\int_a^\infty\int_a^\infty\frac{|\ln(\frac{x}{y})|}{(\max\{x,y\})^\lambda}f(x)g(y)dxdy$$
$$<\frac{r^2+s^2}{\lambda^2}\{\int_a^\infty[1-\frac{s^2}{r^2+s^2}(\frac{a}{x})^{\frac{\lambda}{s}}]x^{p(1-\frac{\lambda}{r})-1}f^p(x)dx\}^{\frac{1}{p}}$$
$$\times\{\int_a^\infty[1-\frac{r^2}{r^2+s^2}(\frac{a}{y})^{\frac{\lambda}{r}}]y^{q(1-\frac{\lambda}{s})-1}g^q(y)dy\}^{\frac{1}{q}}, \quad (3.2.33)$$
$$\int_a^\infty\frac{y^{p\lambda/s-1}}{[1-\frac{r^2}{r^2+s^2}(\frac{a}{y})^{\lambda/r}]^{p-1}}[\int_a^\infty\frac{|\ln(x/y)|}{(\max\{x,y\})^\lambda}f(x)dx]^p dy$$
$$<(\frac{r^2+s^2}{\lambda^2})^p\int_a^\infty[1-\frac{s^2}{r^2+s^2}(\frac{a}{x})^{\frac{\lambda}{s}}]x^{p(1-\frac{\lambda}{r})-1}f^p(x)dx. \quad (3.2.34)$$

Example 3.2.10 If $r,s>1$,

$$k_\lambda(x,y)=\frac{1}{|x^\alpha-y^\alpha|^{\lambda/\alpha}}(\alpha>0,0<\lambda<\alpha),$$

then we find $k_\lambda(u,1)=k_\lambda(1,u)\ge l_\lambda=1, u\in(0,1]$ and $\tilde{k}_\lambda(r)=\frac{1}{\alpha}[B(1-\frac{\lambda}{\alpha},\frac{\lambda}{\alpha r})+B(1-\frac{\lambda}{\alpha},\frac{\lambda}{\alpha s})]$. By Corollary 3.2.2, we have the following equivalent inequalities

$$\int_a^\infty\int_a^\infty\frac{1}{|x^\alpha-y^\alpha|^{\lambda/\alpha}}f(x)g(y)dxdy$$
$$<\tilde{k}_\lambda(r)\{\int_a^\infty[1-\frac{s}{\lambda\tilde{k}_\lambda(r)}(\frac{a}{x})^{\frac{\lambda}{s}}]x^{p(1-\frac{\lambda}{r})-1}f^p(x)dx\}^{\frac{1}{p}}$$
$$\times\{\int_a^\infty[1-\frac{r}{\lambda\tilde{k}_\lambda(r)}(\frac{a}{y})^{\frac{\lambda}{r}}]y^{q(1-\frac{\lambda}{s})-1}g^q(y)dy\}^{\frac{1}{q}}, \quad (3.2.35)$$
$$\int_a^\infty\frac{y^{p\lambda/s-1}}{[1-\frac{r}{\lambda\tilde{k}_\lambda(r)}(\frac{a}{y})^{\lambda/r}]^{p-1}}(\int_a^\infty\frac{f(x)}{|x^\alpha-y^\alpha|^{\lambda/\alpha}}dx)^p dy$$
$$<\tilde{k}_\lambda^p(r)\int_a^\infty[1-\frac{s}{\lambda\tilde{k}_\lambda(r)}(\frac{a}{x})^{\frac{\lambda}{s}}]x^{p(1-\frac{\lambda}{r})-1}f^p(x)dx. \quad (3.2.36)$$

3.3. SOME HILBERT-TYPE INTEGRAL INEQUALITIES RESTRICTED IN THE SUBINTERVAL $(0,b)$ $(b>0)$

3.3.1. SOME RESULTS

Theorem 3.3.1 As the assumption of Theorem 3.1.3, for $\lambda\in\mathbf{R}, r,s\ne0, a=0, 0<b<\infty$, we have the following equivalent inequalities:

$$\int_0^b\int_0^b k_\lambda(x,y)f(x)g(y)dxdy$$
$$<k_\lambda(r)\{\int_0^b\tilde{\kappa}(x)x^{p(1-\frac{\lambda}{r})-1}f^p(x)dx\}^{\frac{1}{p}}$$
$$\times\{\int_0^b\kappa(y)y^{q(1-\frac{\lambda}{s})-1}g^q(y)dy\}^{\frac{1}{q}}, \quad (3.3.1)$$
$$\int_0^b\frac{y^{p\lambda/s-1}}{\kappa^{p-1}(y)}(\int_0^b k_\lambda(x,y)f(x)dx)^p dy$$

$$< k_\lambda^p(r)\int_0^b \tilde{\kappa}(x)x^{p(1-\frac{\lambda}{r})-1}f^p(x)dx, \qquad (3.3.2)$$

where the constant factors

$$k_\lambda(r)=\int_0^\infty k_\lambda(u,1)u^{\frac{\lambda}{r}-1}du$$

and $k_\lambda^p(r)$ are the best possible. In particular, for $\tilde{\kappa}(x) = \kappa(y) = 1$, we have the following equivalent inequalities:

$$\int_0^b\int_0^b k_\lambda(x,y)f(x)g(y)dxdy$$
$$< k_\lambda(r)\{\int_0^b x^{p(1-\frac{\lambda}{r})-1}f^p(x)dx\}^{\frac{1}{p}}$$
$$\times\{\int_0^b y^{q(1-\frac{\lambda}{s})-1}g^q(y)dy\}^{\frac{1}{q}}, \qquad (3.3.3)$$

$$\int_0^b y^{\frac{p\lambda}{s}-1}(\int_0^b k_\lambda(x,y)f(x)dx)^p dy$$
$$< k_\lambda^p(r)\int_0^b x^{p(1-\frac{\lambda}{r})-1}f^p(x)dx, \qquad (3.3.4)$$

where the constant factors $k_\lambda(r)$ and $k_\lambda^p(r)$ are the best possible.

Proof By (3.3.1) and (3.3.2), for $a=0, 0<b<\infty$, we still have the equivalent inequalities (3.3.1) and (3.3.2). For $n\in \mathbf{N}$, we set

$$f_n(x):=x^{\frac{\lambda}{r}-1+\frac{1}{np}},\ g_n(x):=x^{\frac{\lambda}{s}-1+\frac{1}{nq}}, x\in(0,b).$$

If there exists a positive number $k\le k_\lambda(r)$, such that (3.3.1) is still valid as we replace $k_\lambda(r)$ by k, then in particular, we find

$$\tilde{I}_n := \int_0^b\int_0^b k_\lambda(x,y)f_n(x)g_n(y)dxdy$$
$$< k\{\int_0^b x^{p(1-\frac{\lambda}{r})-1}f_n{}^p(x)dx\}^{\frac{1}{p}}$$
$$\times\{\int_0^b y^{q(1-\frac{\lambda}{s})-1}g_n{}^q(y)dy\}^{\frac{1}{q}} = nb^{\frac{1}{n}}k; \qquad (3.3.5)$$

For fixed $y\in(0,b]$, setting $u=\frac{y}{x}$, in view of Fubini theorem, we obtain

$$\tilde{I}_n = \int_0^b y^{\frac{\lambda}{s}-1+\frac{1}{nq}}[\int_0^b k_\lambda(x,y)x^{\frac{\lambda}{r}-1+\frac{1}{np}}dx]dy$$
$$=\int_0^b y^{-1+\frac{1}{n}}[\int_{\frac{y}{b}}^\infty k_\lambda(1,u)u^{\frac{\lambda}{s}-1-\frac{1}{np}}du]dy$$
$$=\int_0^b y^{-1+\frac{1}{n}}[\int_{\frac{y}{b}}^1 k_\lambda(1,u)u^{\frac{\lambda}{s}-1-\frac{1}{np}}du]dy$$
$$+\int_0^b y^{-1+\frac{1}{n}}[\int_1^\infty k_\lambda(1,u)u^{\frac{\lambda}{s}-1-\frac{1}{np}}du]dy$$
$$=\int_0^1(\int_0^{bu} y^{-1+\frac{1}{n}}dy)k_\lambda(1,u)u^{\frac{\lambda}{s}-1-\frac{1}{np}}du$$
$$+nb^{\frac{1}{n}}\int_1^\infty k_\lambda(1,u)u^{\frac{\lambda}{s}-1-\frac{1}{np}}du$$
$$=nb^{\frac{1}{n}}[\int_0^1 k_\lambda(1,u)u^{\frac{\lambda}{s}-1+\frac{1}{nq}}du$$
$$+\int_1^\infty k_\lambda(1,u)u^{\frac{\lambda}{s}-1-\frac{1}{np}}du]. \qquad (3.3.6)$$

By (3.3.5), we have

$$\int_0^1 k_\lambda(1,u)u^{\frac{\lambda}{s}-1+\frac{1}{nq}}du$$
$$+\int_1^\infty k_\lambda(1,u)u^{\frac{\lambda}{s}-1-\frac{1}{np}}du \le k. \qquad (3.3.7)$$

By Fatou lemma, we find

$$k_\lambda(r)=\tilde{k}_\lambda(s)=\int_0^1 \lim_{n\to\infty} k_\lambda(1,u)u^{\frac{\lambda}{s}-1+\frac{1}{nq}}du$$
$$+\int_1^\infty \lim_{n\to\infty} k_\lambda(1,u)u^{\frac{\lambda}{s}-1-\frac{1}{np}}du$$
$$\le \underline{\lim}_{n\to\infty}[\int_0^1 k_\lambda(1,u)u^{\frac{\lambda}{s}-1+\frac{1}{nq}}du$$
$$+\int_1^\infty k_\lambda(1,u)u^{\frac{\lambda}{s}-1-\frac{1}{np}}du]\le k.$$

Therefore, the constant factor $k=k_\lambda(r)$ in (3.3.1) is the best possible. We conform that the constant factor $k_\lambda^p(r)$ in (3.3.2) is the best possible, otherwise, due to (3.1.8) (for $a=0, b<\infty$), we can get a contradiction that the constant factor in (3.3.1) is not the best possible. □

As the assumption of Theorem 3.3.1, setting

$$\phi(x)=x^{p(1-\frac{\lambda}{r})-1}, \psi(x)=x^{q(1-\frac{\lambda}{s})-1}, x\in(0,b),$$

and $\psi^{1-p}(x) = x^{\frac{p\lambda}{s}-1}$, define the following real function spaces:

$$L_\phi^p(0,b)=\{f;\|f\|_{p,\phi}:=\{\int_0^b\phi(x)|f(x)|^p dx\}^{\frac{1}{p}}<\infty\},$$
$$L_\psi^q(0,b)=\{g;\|g\|_{q,\psi}:=\{\int_0^b\psi(x)|g(x)|^q dx\}^{\frac{1}{q}}<\infty\},$$

and Hilbert-type integral operator

$$\tilde{T}_b : L_\phi^p(0,b)\to L_{\psi^{1-p}}^p(0,b)$$

as: for $f\in L_\phi^p(0,b)$,

$$(\tilde{T}_b f)(y)=\int_0^b k_\lambda(x,y)f(x)dx,\ y\in(0,b).$$

By (3.3.2), it follows $\tilde{T}_b f\in L_{\psi^{1-p}}^p(0,b)$. For $g\in L_\psi^q(0,b)$, setting the formal inner product of $\tilde{T}_b f$ and g as

$$(\tilde{T}_b f, g) := \int_0^b\int_0^b k_\lambda(x,y)f(x)g(y)dxdy,$$

we may rewrite inequalities (3.3.3) and (3.3.4) as the following equivalent forms:

$$(\tilde{T}_b f, g) < k_\lambda(r)\|f\|_{p,\phi}\|g\|_{q,\psi}, \qquad (3.3.8)$$
$$\|\tilde{T}_b f\|_{p,\psi^{1-p}} < k_\lambda(r)\|f\|_{p,\phi}. \qquad (3.3.9)$$

Since the constant factors in (3.3.8) and (3.3.9) are the best possible, we can conclude that $\tilde{T}_b$ is bounded and $\|\tilde{T}_b\| = k_\lambda(r)$.

Corollary 3.3.2 As the assumption of Theorem 3.3.1, if $\lambda > 0, r, s > 1$, both $k_\lambda(u,1)$ and $k_\lambda(1,u)$ possess the same lower bound $l_\lambda > 0$ in $(0,1]$, then we find the following equivalent inequalities:

$$(\tilde{T}_b f, g) < k_\lambda(r)\{\int_0^b [1-\tfrac{rl_\lambda}{\lambda k_\lambda(r)}(\tfrac{x}{b})^{\frac{\lambda}{r}}]x^{p(1-\frac{\lambda}{r})-1} f^p(x)dx\}^{\frac{1}{p}} \times\{\int_0^b [1-\tfrac{sl_\lambda}{\lambda k_\lambda(r)}(\tfrac{y}{b})^{\frac{\lambda}{s}}]y^{q(1-\frac{\lambda}{s})-1} g^q(y)dy\}^{\frac{1}{q}}, \quad (3.3.10)$$

$$\int_0^b \frac{y^{p\lambda/s-1}}{[1-\frac{sl_\lambda}{\lambda k_\lambda(r)}(\frac{y}{b})^{\lambda/s}]^{p-1}}(\int_0^b k_\lambda(x,y)f(x)dx)^p dy < k_\lambda^p(r)\int_0^b [1-\tfrac{rl_\lambda}{\lambda k_\lambda(r)}(\tfrac{x}{b})^{\frac{\lambda}{r}}]x^{p(1-\frac{\lambda}{r})-1} f^p(x)dx, \quad (3.3.11)$$

where the constant factors $k_\lambda(r)$ and $k_\lambda^p(r)$ are the best possible. In particular, we have the following equivalent inequalities:

$$(\tilde{T}_b f, g) < k_\lambda(r) \times\{\int_0^b [1-\tfrac{rl_\lambda}{\lambda k_\lambda(r)}(\tfrac{x}{b})^{\frac{\lambda}{r}}]x^{p(1-\frac{\lambda}{r})-1} f^p(x)dx\}^{\frac{1}{p}} \|g\|_{q,\psi}, \quad (3.3.12)$$

$$\|\tilde{T}_b f\|_{p,\psi^{1-p}}^p < k_\lambda^p(r) \times\int_0^b [1-\tfrac{rl_\lambda}{\lambda k_\lambda(r)}(\tfrac{x}{b})^{\frac{\lambda}{r}}]x^{p(1-\frac{\lambda}{r})-1} f^p(x)dx. \quad (3.3.13)$$

Proof we find

$$\varpi_\lambda(s,x,0,b) = x^{\frac{\lambda}{r}}\int_0^b k_\lambda(x,y)y^{\frac{\lambda}{s}-1}dy = \int_{\frac{x}{b}}^\infty k_\lambda(u,1)u^{\frac{\lambda}{r}-1}du$$
$$= k_\lambda(r) - \int_0^{\frac{x}{b}} k_\lambda(u,1)u^{\frac{\lambda}{r}-1}du \le k_\lambda(r) - l_\lambda\int_0^{\frac{x}{b}} u^{\frac{\lambda}{r}-1}du$$
$$= k_\lambda(r)[1-\tfrac{rl_\lambda}{\lambda k_\lambda(r)}(\tfrac{x}{b})^{\frac{\lambda}{r}}],\ x\in(0,b).$$

By the same way, it follows

$$\omega_\lambda(r,y,0,b) \le k_\lambda(r)[1-\tfrac{sl_\lambda}{\lambda k_\lambda(r)}(\tfrac{y}{b})^{\frac{\lambda}{s}}], y\in(0,b).$$

Setting $\tilde{\kappa}(x) = 1-\frac{rl_\lambda}{\lambda k_\lambda(r)}(\frac{x}{b})^{\frac{\lambda}{r}}$ and

$$\kappa(y) = 1-\tfrac{sl_\lambda}{\lambda k_\lambda(r)}(\tfrac{y}{b})^{\frac{\lambda}{s}}$$

in (3.3.1) and (3.3.2), we obtain (3.3.10) and (3.3.11); setting $\tilde{\kappa}(x) = 1-\frac{rl_\lambda}{\lambda k_\lambda(r)}(\frac{x}{b})^{\frac{\lambda}{r}}$ and $\kappa(y)=1$, we obtain (3.3.12) and (3.3.13).

For $0<b<\infty$, setting $u=\frac{x}{y}$, we find

$$\varpi_\lambda(s,x,0,b) := x^{\frac{\lambda}{r}}\int_0^b k_\lambda(x,y)y^{\frac{\lambda}{s}-1}dy = \int_{\frac{x}{b}}^\infty k_\lambda(u,1)u^{\frac{\lambda}{r}-1}du$$
$$= k_\lambda(r) - \int_0^{\frac{x}{b}} k_\lambda(u,1)u^{\frac{\lambda}{r}-1}du = k_\lambda(r) - (\tfrac{x}{b})^{\frac{\lambda}{r}}[(\tfrac{x}{b})^{\frac{-\lambda}{r}}\int_0^{\frac{x}{b}} k_\lambda(u,1)u^{\frac{\lambda}{r}-1}du].$$

If $\lambda>0$, $k_\lambda(1,u), k_\lambda(u,1) \in C^1(0,1]$,

$$k_\lambda'(1,u)\le 0, k_\lambda'(u,1)\le 0, u\in(0,1],$$

by Lemma 3.1.5, we have

$$(\tfrac{x}{b})^{\frac{-\lambda}{r}}\int_0^{\frac{x}{b}} k_\lambda(u,1)u^{\frac{\lambda}{r}-1}du \ge \int_0^1 k_\lambda(u,1)u^{\frac{\lambda}{r}-1}du =: \theta_\lambda(r), 0<x<b,$$

$$\varpi_\lambda(s,x,0,b) \le \tilde{\kappa}(x) := k_\lambda(r) - (\tfrac{x}{b})^{\frac{\lambda}{r}}\theta_\lambda(r),\ 0<x<b. \quad (3.3.14)$$

By the same way, it follows

$$\tilde{\theta}_\lambda(s) = \int_0^1 k_\lambda(1,u)u^{\frac{\lambda}{s}-1}du$$

and

$$\omega_\lambda(r,y,0,b) \le \kappa(y) := k_\lambda(r) - (\tfrac{y}{b})^{\frac{\lambda}{s}}\tilde{\theta}_\lambda(s), 0<y<b. \quad (3.3.15)$$

In view of Theorem 3.3.1, we have

Corollary 3.3.3 As the assumption of Theorem 3.3.1, If $\lambda\in\mathbf{R}, r,s\ne 0$, $k_\lambda(1,u), k_\lambda(u,1)\in C^1(0,1]$,

$$k_\lambda'(1,u)\le 0, k_\lambda'(u,1)\le 0, u\in(0,1]$$

and $l_\lambda = k_\lambda(1,1) > 0$, then we have the following equivalent inequalities:

$$(\tilde{T}_b f, g) < k_\lambda(r)\{\int_0^b [1-\tfrac{\theta_\lambda(r)}{k_\lambda(r)}(\tfrac{x}{b})^{\frac{\lambda}{r}}]x^{p(1-\frac{\lambda}{r})-1} f^p(x)dx\}^{\frac{1}{p}} \times\{\int_0^b [1-\tfrac{\tilde{\theta}_\lambda(s)}{k_\lambda(r)}(\tfrac{y}{b})^{\frac{\lambda}{s}}]y^{q(1-\frac{\lambda}{s})-1} g^q(y)dy\}^{\frac{1}{q}}, \quad (3.3.16)$$

$$\int_0^b \frac{y^{p\lambda/s-1}}{[1-\frac{\tilde{\theta}_\lambda(s)}{k_\lambda(r)}(\frac{y}{b})^{\lambda/s}]^{p-1}}(\int_0^b k_\lambda(x,y)f(x)dx)^p dy < k_\lambda^p(r)\int_0^b [1-\tfrac{\theta_\lambda(r)}{k_\lambda(r)}(\tfrac{x}{b})^{\frac{\lambda}{r}}]x^{p(1-\frac{\lambda}{r})-1} f^p(x)dx, \quad (3.3.17)$$

where the constant factors $k_\lambda(r)$ and $k_\lambda^p(r)$ are the best possible. We still have the following equivalent inequalities:

$$(\tilde{T}_b f, g) < k_\lambda(r) \times \{\int_0^b [1-\tfrac{\theta_\lambda(r)}{k_\lambda(r)}(\tfrac{x}{b})^{\frac{\lambda}{r}}] x^{p(1-\frac{\lambda}{r})-1} f^p(x)dx\}^{\frac{1}{p}} \| g \|_{q,\psi}, \tag{3.3.18}$$

$$\| \tilde{T}_b f \|^p_{p,\psi^{1-p}} < k_\lambda^p(r) \times \int_0^b [1-\tfrac{\theta_\lambda(r)}{k_\lambda(r)}(\tfrac{x}{b})^{\frac{\lambda}{r}}] x^{p(1-\frac{\lambda}{r})-1} f^p(x)dx. \tag{3.3.19}$$

In particular, for $r = s = 2$, if $k_\lambda(1,u) = k_\lambda(u,1)$, $k_\lambda := k_\lambda(2)$, we have $\tilde{\theta}_\lambda(2) = \theta_\lambda(2) = \frac{1}{2}k_\lambda$ and the following two pairs of equivalent inequalities:

$$(\tilde{T}_b f, g) < k_\lambda \{\int_0^b [1-\tfrac{1}{2}(\tfrac{x}{b})^{\frac{\lambda}{2}}] x^{p(1-\frac{\lambda}{2})-1} f^p(x)dx\}^{\frac{1}{p}} \times \{\int_0^b [1-\tfrac{1}{2}(\tfrac{y}{b})^{\frac{\lambda}{2}}] y^{q(1-\frac{\lambda}{2})-1} g^q(y)dy\}^{\frac{1}{q}}, \tag{3.3.20}$$

$$\int_0^b \frac{y^{p\lambda/2-1}}{[1-\frac{1}{2}(\frac{y}{b})^{\lambda/2}]^{p-1}} (\int_0^b k_\lambda(x,y) f(x)dx)^p dy < k_\lambda^p \int_0^b [1-\tfrac{1}{2}(\tfrac{x}{b})^{\frac{\lambda}{2}}] x^{p(1-\frac{\lambda}{2})-1} f^p(x)dx, \tag{3.3.21}$$

$$(\tilde{T}_b f, g) < k_\lambda \{\int_0^b [1-\tfrac{1}{2}(\tfrac{x}{b})^{\frac{\lambda}{2}}] x^{p(1-\frac{\lambda}{2})-1} f^p(x)dx\}^{\frac{1}{p}} \times \{\int_0^b y^{q(1-\frac{\lambda}{2})-1} g^q(y)dy\}^{\frac{1}{q}}, \tag{3.3.22}$$

$$\int_0^b y^{\frac{p\lambda}{2}-1} (\int_0^b k_\lambda(x,y) f(x)dx)^p dy < k_\lambda^p \int_0^b [1-\tfrac{1}{2}(\tfrac{x}{b})^{\frac{\lambda}{2}}] x^{p(1-\frac{\lambda}{2})-1} f^p(x)dx. \tag{3.3.23}$$

Note 3.3.4 (1) For $b \to \infty$, inequalities (3.3.3) and (3.3.4) reduce to (2.2.1) and (2.2.2). It follows that (3.3.3) and (3.3.4) are extensions of (2.2.1) and (2.2.2). (2) For $\lambda = 1$, setting $(r,s) = (q,p)$ in (3.3.16), it deduces to the result of (Yang JMAA 1998) [4]. (3) Inequalities (3.3.10) and (3.3.16) are strengthened versions of (3.3.3).

3.3.2. SOME PARTICULAR EXAMPLES

In the following examples, we omit to write the assumption of f and g satisfying Theorem 3.3.1 and the conclusions that the constant factors are the best possible.

Example 3.3.5 If $k_\lambda(x,y) = \frac{1}{(x^\alpha + y^\alpha)^{\lambda/\alpha}}$ $(\lambda, \alpha > 0)$, $r, s > 1$, then we find

$$k_\lambda(u,1) = k_\lambda(1,u) = \frac{1}{(1+u^\alpha)^{\lambda/\alpha}},\ u \in (0,1],$$

$l_\lambda = \frac{1}{2^{\lambda/\alpha}} > 0$ and $k_\lambda(r) = \frac{1}{\alpha}B(\frac{\lambda}{\alpha r}, \frac{\lambda}{\alpha s})$. Then by (3.3.10) and (3.3.11), we have the following equivalent inequalities:

$$\int_a^\infty \int_a^\infty \frac{1}{(x^\alpha + y^\alpha)^{\lambda/\alpha}} f(x)g(y)dxdy < \frac{1}{\alpha}B(\tfrac{\lambda}{\alpha r}, \tfrac{\lambda}{\alpha s}) \times \{\int_0^b [1-\tfrac{r\alpha}{2^{\lambda/\alpha}\lambda B(\frac{\lambda}{\alpha r},\frac{\lambda}{\alpha s})}(\tfrac{x}{b})^{\frac{\lambda}{r}}] x^{p(1-\frac{\lambda}{r})-1} f^p(x)dx\}^{\frac{1}{p}} \times \{\int_0^b [1-\tfrac{s\alpha}{2^{\lambda/\alpha}\lambda B(\frac{\lambda}{\alpha r},\frac{\lambda}{\alpha s})}(\tfrac{y}{b})^{\frac{\lambda}{s}}] y^{q(1-\frac{\lambda}{s})-1} g^q(y)dy\}^{\frac{1}{q}}, \tag{3.3.24}$$

$$\int_0^b \frac{y^{p\lambda/s-1}}{[1-\frac{s\alpha}{2^{\lambda/\alpha}\lambda B(\frac{\lambda}{\alpha r},\frac{\lambda}{\alpha s})}(\frac{y}{b})^{\lambda/s}]^{p-1}} [\int_0^b \frac{f(x)}{(x^\alpha + y^\alpha)^{\lambda/\alpha}} dx]^p dy < [\tfrac{1}{\alpha}B(\tfrac{\lambda}{\alpha r}, \tfrac{\lambda}{\alpha s})]^p \times \int_0^b [1-\tfrac{r\alpha}{2^{\lambda/\alpha}\lambda B(\frac{\lambda}{\alpha r},\frac{\lambda}{\alpha s})}(\tfrac{x}{b})^{\frac{\lambda}{r}}] x^{p(1-\frac{\lambda}{r})-1} f^p(x)dx. \tag{3.3.25}$$

In particular, due to (3.3.20) and (3.3.21), we have the following equivalent inequalities:

$$\int_0^b \int_0^b \frac{1}{(x^\alpha + y^\alpha)^{\lambda/\alpha}} f(x)g(y)dxdy < \frac{1}{\alpha}B(\tfrac{\lambda}{2\alpha}, \tfrac{\lambda}{2\alpha}) \{\int_0^b [1-\tfrac{1}{2}(\tfrac{x}{b})^{\frac{\lambda}{2}}] x^{p(1-\frac{\lambda}{2})-1} f^p(x)dx\}^{\frac{1}{p}} \times \{\int_0^b [1-\tfrac{1}{2}(\tfrac{y}{b})^{\frac{\lambda}{2}}] y^{q(1-\frac{\lambda}{2})-1} g^q(y)dy\}^{\frac{1}{q}}, \tag{3.3.26}$$

$$\int_0^b \frac{y^{p\lambda/2-1}}{[1-\frac{1}{2}(\frac{y}{b})^{\lambda/2}]^{p-1}} [\int_0^b \frac{f(x)}{(x^\alpha + y^\alpha)^{\lambda/\alpha}} dx]^p dy < [\tfrac{1}{\alpha}B(\tfrac{\lambda}{2\alpha}, \tfrac{\lambda}{2\alpha})]^p \int_0^b [1-\tfrac{1}{2}(\tfrac{x}{b})^{\frac{\lambda}{2}}] x^{p(1-\frac{\lambda}{2})-1} f^p(x)dx. \tag{3.3.27}$$

Note 3.3.6 For $\alpha = 1$, taking some particular value of (r,s) in (3.3.23) and (3.3.24), we obtain some results of (Yang JMAA 2002) [7], (Yang JMAA 2001) [8]. We see inequality (3.3.25) in (Zhong JSHU 2007) [9].

Example 3.3.7 If $k_\lambda(x,y) = \frac{\ln(x/y)}{x^\lambda - y^\lambda}$, $\lambda > 0; r, s > 1$, then we find $k_\lambda(u,1) = k_\lambda(1,u) = \frac{\ln u}{u^\lambda - 1}$, $u \in (0,1]$, $l_\lambda = \frac{1}{\lambda} > 0$ and $k_\lambda(r) = [\frac{1}{\lambda}\frac{\pi}{\sin(\pi/r)}]^2$. By (3.3.10) and (3.3.11), we have the following equivalent inequalities:

$$\int_0^b \int_0^b \frac{\ln(x/y)}{x^\lambda - y^\lambda} f(x)g(y)dxdy < [\tfrac{\pi}{\lambda \sin(\pi/r)}]^2 \times \{\int_0^b [1-r(\tfrac{\sin(\pi/r)}{\pi})^2(\tfrac{x}{b})^{\frac{\lambda}{r}}] x^{p(1-\frac{\lambda}{r})-1} f^p(x)dx\}^{\frac{1}{p}} \times \{\int_0^b [1-s(\tfrac{\sin(\pi/r)}{\pi})^2(\tfrac{y}{b})^{\frac{\lambda}{s}}] y^{q(1-\frac{\lambda}{s})-1} g^q(y)dy\}^{\frac{1}{q}}, \tag{3.3.28}$$

$$\int_0^b \frac{y^{p\lambda/s-1}}{[1-s(\frac{\sin(\pi/r)}{\pi})^2(\frac{y}{b})^{\lambda/s}]^{p-1}} [\int_0^b \frac{\ln(x/y)}{x^\lambda - y^\lambda} f(x)dx]^p dy$$

$$<[\frac{\pi}{\lambda\sin(\pi/r)}]^{2p}$$
$$\times\int_0^b[1-r(\frac{\sin(\pi/r)}{\pi})^2(\frac{x}{b})^{\frac{\lambda}{r}}]x^{p(1-\frac{\lambda}{r})-1}f^p(x)dx. \tag{3.3.29}$$

In particular, due to (3.3.20) and (3.3.21), we have the following equivalent inequalities:

$$\int_0^b\int_0^b\frac{\ln(x/y)}{x^\lambda-y^\lambda}f(x)g(y)dxdy<(\frac{\pi}{\lambda})^2$$
$$\times\{\int_0^b[1-\frac{1}{2}(\frac{x}{b})^{\frac{\lambda}{2}}]x^{p(1-\frac{\lambda}{2})-1}f^p(x)dx\}^{\frac{1}{p}}$$
$$\times\{\int_0^b[1-\frac{1}{2}(\frac{y}{b})^{\frac{\lambda}{2}}]y^{q(1-\frac{\lambda}{2})-1}g^q(y)dy\}^{\frac{1}{q}}, \tag{3.3.30}$$
$$\int_0^b\frac{y^{p/2-1}}{[1-\frac{1}{2}(\frac{y}{b})^{\lambda/2}]^{p-1}}[\int_0^b\frac{\ln(x/y)}{x^\lambda-y^\lambda}f(x)dx]^p dy$$
$$<(\frac{\pi}{\lambda})^{2p}\int_0^b[1-\frac{1}{2}(\frac{x}{b})^{\frac{\lambda}{2}}]x^{p(1-\frac{\lambda}{2})-1}f^p(x)dx. \tag{3.3.31}$$

Example 3.3.8 If $k_\lambda(x,y)=\frac{1}{(\max\{x,y\})^\lambda}(\lambda>0)$, $r,s>1$, then we find $k_\lambda(u,1)=k_\lambda(1,u)=1$, $u\in(0,1]$, $k_\lambda(r)=\frac{rs}{\lambda}$ and

$$\varpi_\lambda(s,x,0,b)=\frac{rs}{\lambda}[1-\frac{1}{s}(\frac{x}{b})^{\frac{\lambda}{r}}],$$
$$\varpi_\lambda(r,y,0,b)=\frac{rs}{\lambda}[1-\frac{1}{r}(\frac{y}{b})^{\frac{\lambda}{s}}].$$

Hence by (3.3.1), (3.3.2), we have the following equivalent inequalities

$$\int_0^b\int_0^b\frac{1}{(\max\{x,y\})^\lambda}f(x)g(y)dxdy$$
$$<\frac{rs}{\lambda}\{\int_0^b[1-\frac{1}{s}(\frac{x}{b})^{\frac{\lambda}{r}}]x^{p(1-\frac{\lambda}{r})-1}f^p(x)dx\}^{\frac{1}{p}}$$
$$\times\{\int_0^b[1-\frac{1}{r}(\frac{y}{b})^{\frac{\lambda}{s}}]y^{q(1-\frac{\lambda}{s})-1}g^q(y)dy\}^{\frac{1}{q}}, \tag{3.3.32}$$
$$\int_0^b\frac{y^{p\lambda/s-1}}{[1-\frac{1}{r}(\frac{y}{b})^{\lambda/s}]^{p-1}}[\int_0^b\frac{f(x)}{(\max\{x,y\})^\lambda}dx]^p dy$$
$$<(\frac{rs}{\lambda})^p\int_0^b[1-\frac{1}{s}(\frac{x}{b})^{\frac{\lambda}{r}}]x^{p(1-\frac{\lambda}{r})-1}f^p(x)dx. \tag{3.3.33}$$

Example 3.3.9 If $k_\lambda(x,y)=\frac{|\ln(x/y)|}{(\max\{x,y\})^\lambda}(\lambda>0)$, $r,s>1$, then we find $k_\lambda(r)=\frac{r^2+s^2}{\lambda^2}$ and

$$\varpi_\lambda(s,x,0,b)=\frac{r^2+s^2}{\lambda^2}-\int_0^{\frac{x}{b}}(-\ln u)u^{\frac{\lambda}{r}-1}du$$
$$=\frac{r^2+s^2}{\lambda^2}-\frac{r}{\lambda}\int_0^{\frac{x}{b}}(-\ln u)du^{\frac{\lambda}{r}}$$
$$\le\frac{r^2+s^2}{\lambda^2}[1-\frac{r^2}{r^2+s^2}(\frac{x}{b})^{\frac{\lambda}{r}}].$$
$$\omega_\lambda(r,y,0,b)\le\frac{r^2+s^2}{\lambda^2}[1-\frac{s^2}{r^2+s^2}(\frac{y}{b})^{\frac{\lambda}{r}}].$$

Hence by (3.3.1) and (3.3.2), we have the following equivalent inequalities:

$$\int_0^b\int_0^b\frac{|\ln(x/y)|}{(\max\{x,y\})^\lambda}f(x)g(y)dxdy$$
$$<\frac{r^2+s^2}{\lambda^2}\{\int_0^b[1-\frac{r^2}{r^2+s^2}(\frac{x}{b})^{\frac{\lambda}{r}}]x^{p(1-\frac{\lambda}{r})-1}f^p(x)dx\}^{\frac{1}{p}}$$
$$\times\{\int_0^b[1-\frac{s^2}{r^2+s^2}(\frac{y}{b})^{\frac{\lambda}{s}}]y^{q(1-\frac{\lambda}{s})-1}g^q(y)dy\}^{\frac{1}{q}}; \tag{3.3.34}$$
$$\int_0^b\frac{y^{p\lambda/s-1}}{[1-\frac{s^2}{r^2+s^2}(\frac{y}{b})^{\lambda/s}]^{p-1}}[\int_0^b\frac{|\ln(x/y)|}{(\max\{x,y\})^\lambda}f(x)dx]^p dy$$
$$<(\frac{r^2+s^2}{\lambda^2})^p\int_0^b[1-\frac{r^2}{r^2+s^2}(\frac{x}{b})^{\frac{\lambda}{r}}]x^{p(1-\frac{\lambda}{r})-1}f^p(x)dx. \tag{3.3.35}$$

Example 3.3.10 If $r,s>1$,

$$k_\lambda(x,y)=\frac{1}{|x^\alpha-y^\alpha|^{\lambda/\alpha}}(\alpha>0,0<\lambda<\alpha),$$

then we find $k_\lambda(u,1)=k_\lambda(1,u)\ge l_\lambda=1$, $u\in(0,1]$ and

$$\tilde{k}_\lambda(r)=\frac{1}{\alpha}[B(1-\frac{\lambda}{\alpha},\frac{\lambda}{\alpha r})+B(1-\frac{\lambda}{\alpha},\frac{\lambda}{\alpha s})].$$

By Corollary 3.3.2, we have the following equivalent inequalities:

$$\int_0^b\int_0^b\frac{1}{|x^\alpha-y^\alpha|^{\lambda/\alpha}}f(x)g(y)dxdy$$
$$<\tilde{k}_\lambda(r)\{\int_0^b[1-\frac{r}{\lambda\tilde{k}_\lambda(r)}(\frac{x}{b})^{\frac{\lambda}{r}}]x^{p(1-\frac{\lambda}{r})-1}f^p(x)dx\}^{\frac{1}{p}}$$
$$\times\{\int_0^b[1-\frac{s}{\lambda\tilde{k}_\lambda(r)}(\frac{y}{b})^{\frac{\lambda}{s}}]y^{q(1-\frac{\lambda}{s})-1}g^q(y)dy\}^{\frac{1}{q}}, \tag{3.3.36}$$
$$\int_0^b\frac{y^{p\lambda/s-1}}{[1-\frac{s}{\lambda\tilde{k}_\lambda(r)}(\frac{y}{b})^{\lambda/s}]^{p-1}}(\int_0^b\frac{f(x)}{|x^\alpha-y^\alpha|^{\lambda/\alpha}}dx)^p dy$$
$$<\tilde{k}_\lambda^p(r)\int_0^b[1-\frac{r}{\lambda\tilde{k}_\lambda(r)}(\frac{x}{b})^{\frac{\lambda}{r}}]x^{p(1-\frac{\lambda}{r})-1}f^p(x)dx. \tag{3.3.37}$$

3.4. SOME HILBERT-TYPE INTEG-RAL INEQUALITIES RESTRICTED IN THE SUBINTERVAL (a,b) $(0<a<b<\infty)$

3.4.1. SOME THEOREMS AND COROLLARIES

In view of Note 3.1.4, for $\kappa(y)=\tilde{\kappa}(x)=1$, we have

Theorem 3.4.1 For $\lambda\in\mathbf{R}, r,s\neq0, 0<a<b<\infty$, as the assumption of Theorem 3.1.3, if at least one of (3.1.3) and (3.1.4) keeps the strict sign-inequality *a.e.* in (a,b), then we have the following equivalent inequalities:

$$\int_a^b\int_a^b k_\lambda(x,y)f(x)g(y)dxdy$$

$$< k_\lambda(r)\{\int_a^b x^{p(1-\frac{\lambda}{r})-1} f^p(x)dx\}^{\frac{1}{p}}$$

$$\times\{\int_a^b y^{q(1-\frac{\lambda}{s})-1} g^q(y)dy\}^{\frac{1}{q}}, \qquad (3.4.1)$$

$$\int_a^b y^{\frac{p\lambda}{s}-1}(\int_a^b k_\lambda(x,y)f(x)dx)^p dy$$

$$< k_\lambda^p(r)\int_a^b x^{p(1-\frac{\lambda}{r})-1} f^p(x)dx. \qquad (3.4.2)$$

If $\lambda>0, r,s>1$, both $k_\lambda(u,1)$ and $k_\lambda(1,u)$ possess the lower bound $l_\lambda>0$ in $(0,1]$, setting $u=\frac{y}{x}$ in the front two integrals and $u=\frac{x}{y}$ in the last integral, we find

$$\varpi_\lambda(s,x,a,b)=x^{\frac{\lambda}{r}}\int_a^b k_\lambda(x,y)y^{\frac{\lambda}{s}-1}dy$$

$$=x^{\frac{\lambda}{r}}[\int_0^\infty k_\lambda(x,y)y^{\frac{\lambda}{s}-1}dy$$

$$-\int_0^a k_\lambda(x,y)y^{\frac{\lambda}{s}-1}dy-\int_b^\infty k_\lambda(x,y)y^{\frac{\lambda}{s}-1}dy]$$

$$=\int_0^\infty k_\lambda(1,u)u^{\frac{\lambda}{s}-1}du$$

$$-[\int_0^{\frac{a}{x}} k_\lambda(1,u)u^{\frac{\lambda}{s}-1}du+\int_0^{\frac{x}{b}} k_\lambda(u,1)u^{\frac{\lambda}{r}-1}du]$$

$$\le k_\lambda(r)-l_\lambda[\int_0^{\frac{a}{x}} u^{\frac{\lambda}{s}-1}du+\int_0^{\frac{x}{b}} u^{\frac{\lambda}{r}-1}du]$$

$$=k_\lambda(r)\{1-\frac{rsl_\lambda}{\lambda k_\lambda(r)}[\frac{1}{r}(\frac{a}{x})^{\frac{\lambda}{s}}+\frac{1}{s}(\frac{x}{b})^{\frac{\lambda}{r}}]\}.$$

In view of the general mean value inequality (Kuang STP 2004) [1], we have

$$\frac{1}{r}(\frac{a}{x})^{\frac{\lambda}{s}}+\frac{1}{s}(\frac{x}{b})^{\frac{\lambda}{r}}\ge[(\frac{a}{x})^{\frac{\lambda}{s}}]^{\frac{1}{r}}[(\frac{x}{b})^{\frac{\lambda}{r}}]^{\frac{1}{s}}=(\frac{a}{b})^{\frac{\lambda}{rs}}$$

and then

$$\varpi_\lambda(s,x,a,b)\le k_\lambda(r)[1-\frac{rsl_\lambda}{\lambda k_\lambda(r)}(\frac{a}{b})^{\frac{\lambda}{rs}}].$$

By the same way, we obtain

$$\omega_\lambda(r,y,a,b)$$

$$\le k_\lambda(r)\{1-\frac{rsl_\lambda}{\lambda k_\lambda(r)}[\frac{1}{s}(\frac{a}{y})^{\frac{\lambda}{r}}+\frac{1}{r}(\frac{y}{b})^{\frac{\lambda}{s}}]\}$$

$$\le k_\lambda(r)[1-\frac{rsl_\lambda}{\lambda k_\lambda(r)}(\frac{a}{b})^{\frac{\lambda}{rs}}].$$

Then by Note 3.1.4, we have

Corollary 3.4.2 As the assumption of Theorem 3.4.1, if $\lambda>0, r,s>1$, both $k_\lambda(u,1)$ and $k_\lambda(1,u)$ possess the lower bound $l_\lambda>0$ in $(0,1]$ and among them, there is a function that dose not equal l_λ a.e. in $(0,1]$, then we have the following equivalent inequalities:

$$\int_a^b\int_a^b k_\lambda(x,y)f(x)g(y)dxdy<k_\lambda(r)$$

$$\times\{\int_a^b[1-\frac{rsl_\lambda}{\lambda k_\lambda(r)}(\frac{1}{r}(\frac{a}{x})^{\frac{\lambda}{s}}+\frac{1}{s}(\frac{x}{b})^{\frac{\lambda}{r}})]x^{p(1-\frac{\lambda}{r})-1}f^p(x)dx\}^{\frac{1}{p}}$$

$$\times\{\int_a^b[1-\frac{srl_\lambda}{\lambda k_\lambda(r)}(\frac{1}{s}(\frac{a}{y})^{\frac{\lambda}{r}}+\frac{1}{r}(\frac{y}{b})^{\frac{\lambda}{s}})]y^{q(1-\frac{\lambda}{s})-1}g^q(y)dy\}^{\frac{1}{q}}, \qquad (3.4.3)$$

$$\int_a^b\frac{y^{p\lambda/s-1}}{\{1-\frac{rsl_\lambda}{\lambda k_\lambda(r)}[\frac{1}{s}(\frac{a}{y})^{\frac{\lambda}{r}}+\frac{1}{r}(\frac{y}{b})^{\frac{\lambda}{s}}]\}^{p-1}}(\int_a^b k_\lambda(x,y)f(x)dx)^p dy$$

$$<k_\lambda^p(r)\int_a^b[1-\frac{srl_\lambda}{\lambda k_\lambda(r)}(\frac{1}{r}(\frac{a}{x})^{\frac{\lambda}{s}}+\frac{1}{s}(\frac{x}{b})^{\frac{\lambda}{r}})]$$

$$\times x^{p(1-\frac{\lambda}{r})-1}f^p(x)dx. \qquad (3.4.4)$$

In particular, we have the following equivalent inequalities:

$$\int_a^b\int_a^b k_\lambda(x,y)f(x)g(y)dxdy$$

$$<[k_\lambda(r)-\frac{rsl_\lambda}{\lambda}(\frac{a}{b})^{\frac{\lambda}{rs}}]\{\int_a^b x^{p(1-\frac{\lambda}{r})-1}f^p(x)dx\}^{\frac{1}{p}}$$

$$\times\{\int_a^b y^{q(1-\frac{\lambda}{s})-1}g^q(y)dy\}^{\frac{1}{q}}, \qquad (3.4.5)$$

$$\int_a^b y^{\frac{p\lambda}{s}-1}(\int_a^b k_\lambda(x,y)f(x)dx)^p dy$$

$$<[k_\lambda(r)-\frac{rsl_\lambda}{\lambda}(\frac{a}{b})^{\frac{\lambda}{rs}}]^p\int_a^b x^{p(1-\frac{\lambda}{r})-1}f^p(x)dx. \qquad (3.4.6)$$

If $\lambda\in\mathbf{R}, r,s\ne0$, $k_\lambda(1,u)$, $k_\lambda(u,1)\in C^1(0,1]$ and $k'_\lambda(1,u)\le0, k'_\lambda(u,1)\le0$, then by Lemma 3.1.5 and Lemma 3.1.6, we have

$$(\frac{a}{x})^{-\frac{\lambda}{s}}\int_0^{\frac{a}{x}}k_\lambda(1,u)u^{\frac{\lambda}{s}-1}du$$

$$\ge\int_0^1 k_\lambda(1,u)u^{\frac{\lambda}{s}-1}du=\tilde{\theta}_\lambda(s),$$

$$(\frac{x}{b})^{\frac{-\lambda}{r}}\int_0^{\frac{x}{b}}k_\lambda(u,1)u^{\frac{\lambda}{r}-1}du$$

$$\ge\int_0^1 k_\lambda(u,1)u^{\frac{\lambda}{r}-1}du=\theta_\lambda(r),\ a<x<b.$$

Hence we find

$$\varpi_\lambda(s,x,a,b)$$

$$\le k_\lambda(r)-[\tilde{\theta}_\lambda(s)(\frac{a}{x})^{\frac{\lambda}{s}}+\theta_\lambda(r)(\frac{x}{b})^{\frac{\lambda}{r}}],\ a<x<b.$$

By the same way, it follows

$$\omega_\lambda(r,y,a,b)$$

$$\le k_\lambda(r)-[\theta_\lambda(r)(\frac{a}{y})^{\frac{\lambda}{r}}+\tilde{\theta}_\lambda(s)(\frac{y}{b})^{\frac{\lambda}{s}}],a<y<b.$$

Corollary 3.4.3 For $0<a<b<\infty$, as the assumption of Theorem 3.1.3, if $\lambda\in\mathbf{R}, r,s\ne0$, $k_\lambda(1,u)$, $k_\lambda(u,1)\in C^1(0,1], k'_\lambda(1,u)\le0$, $k'_\lambda(u,1)\le0$, and among then there is a function that keeps strictly decreasing in a subinterval of $(0,1]$, then we have the following equivalent inequalities:

$$\int_a^b\int_a^b k_\lambda(x,y)f(x)g(y)dxdy < k_\lambda(r)$$
$$\times\{\int_a^b [1-\frac{\tilde{\theta}_\lambda(s)}{k_\lambda(r)}(\frac{a}{x})^{\frac{\lambda}{s}} - \frac{\theta_\lambda(r)}{k_\lambda(r)}(\frac{x}{b})^{\frac{\lambda}{r}}]x^{p(1-\frac{\lambda}{r})-1}f^p(x)dx\}^{\frac{1}{p}}$$
$$\times\{\int_a^b [1-\frac{\theta_\lambda(r)}{k_\lambda(r)}(\frac{a}{y})^{\frac{\lambda}{r}} - \frac{\tilde{\theta}_\lambda(s)}{k_\lambda(r)}(\frac{y}{b})^{\frac{\lambda}{s}}]y^{q(1-\frac{\lambda}{s})-1}g^q(y)dy\}^{\frac{1}{q}}, \quad (3.4.7)$$

$$\int_a^b \frac{y^{p\lambda/s-1}}{[1-\frac{\theta_\lambda(r)}{k_\lambda(r)}(\frac{a}{y})^{\frac{\lambda}{r}}-\frac{\tilde{\theta}_\lambda(s)}{k_\lambda(r)}(\frac{y}{b})^{\frac{\lambda}{s}}]^{p-1}}(\int_a^b k_\lambda(x,y)f(x)dx)^p dy$$
$$< k_\lambda^p(r)\int_a^b [1-\frac{\tilde{\theta}_\lambda(s)}{k_\lambda(r)}(\frac{a}{x})^{\frac{\lambda}{s}} - \frac{\theta_\lambda(r)}{k_\lambda(r)}(\frac{x}{b})^{\frac{\lambda}{r}}]$$
$$\times x^{p(1-\frac{\lambda}{r})-1}f^p(x)dx, \quad (3.4.8)$$

where, $\tilde{\theta}_\lambda(s) = \int_0^1 k_\lambda(1,u)u^{\frac{\lambda}{s}-1}du,$

$$\theta_\lambda(r) = \int_0^1 k_\lambda(u,1)u^{\frac{\lambda}{r}-1}du.$$

In particular, for $r = s = 2$, if

$$k_\lambda(u,1) = k_\lambda(1,u), u\in(0,1],$$

setting $k_\lambda = k_\lambda(2)$, we deduce the following equivalent inequalities:

$$\int_a^b\int_a^b k_\lambda(x,y)f(x)g(y)dxdy$$
$$< k_\lambda\{\int_a^b [1-\frac{1}{2}(\frac{a}{x})^{\frac{\lambda}{2}} - \frac{1}{2}(\frac{x}{b})^{\frac{\lambda}{2}}]x^{p(1-\frac{\lambda}{2})-1}f^p(x)dx\}^{\frac{1}{p}}$$
$$\times\{\int_a^b [1-\frac{1}{2}(\frac{a}{x})^{\frac{\lambda}{2}} - \frac{1}{2}(\frac{x}{b})^{\frac{\lambda}{2}}]x^{q(1-\frac{\lambda}{2})-1}g^q(x)dx\}^{\frac{1}{q}}, \quad (3.4.9)$$

$$\int_a^b \frac{y^{p\lambda/s-1}}{[1-\frac{1}{2}(\frac{a}{y})^{\frac{\lambda}{2}}-\frac{1}{2}(\frac{y}{b})^{\frac{\lambda}{2}}]^{p-1}}(\int_a^b k_\lambda(x,y)f(x)dx)^p dy$$
$$< k_\lambda^p\int_a^b [1-\frac{1}{2}(\frac{a}{x})^{\frac{\lambda}{2}} - \frac{1}{2}(\frac{x}{b})^{\frac{\lambda}{2}}]x^{p(1-\frac{\lambda}{2})-1}f^p(x)dx. \quad (3.4.10)$$

In view of the mean value inequality (Kuang STP 2004) [1], we find

$$\frac{1}{2}[(\frac{a}{x})^{\frac{\lambda}{2}} + (\frac{x}{b})^{\frac{\lambda}{2}}] \ge \sqrt{(\frac{a}{x})^{\frac{\lambda}{2}}(\frac{x}{b})^{\frac{\lambda}{2}}} = (\frac{a}{b})^{\frac{\lambda}{4}}.$$

Due to (3.4.9) and (3.4.10), we have

Corollary 3.4.4 For $0 < a < b < \infty$, as the assumption of Theorem 3.1.3, if $\lambda\in\mathbf{R}$,

$$k_\lambda(1,u) = k_\lambda(u,1)\in C^1(0,1], k_\lambda'(1,u)\le 0,$$

$k_\lambda'(u,1)\le 0$ and $k_\lambda = \int_0^\infty k_\lambda(u,1)u^{\frac{\lambda}{2}-1}du > 0$,

then we have the following equivalent inequalities:

$$\int_a^b\int_a^b k_\lambda(x,y)f(x)g(y)dxdy$$
$$< k_\lambda[1-(\frac{a}{b})^{\frac{\lambda}{4}}]\{\int_a^b x^{p(1-\frac{\lambda}{2})-1}f^p(x)dx\}^{\frac{1}{p}}$$
$$\times\{\int_a^b x^{q(1-\frac{\lambda}{2})-1}g^q(x)dx\}^{\frac{1}{q}}; \quad (3.4.11)$$

$$\int_a^b y^{\frac{p\lambda}{2}-1}(\int_a^b k_\lambda(x,y)f(x)dx)^p dy$$
$$< \{k_\lambda[1-(\frac{a}{b})^{\frac{\lambda}{4}}]\}^p\int_a^b x^{p(1-\frac{\lambda}{2})-1}f^p(x)dx. \quad (3.4.12)$$

Note 3.4.5 (1) In view of (3.4.5) and (3.4.6), it follows that the constant factors in (3.4.1) and (3.4.2) are not the best possible; (2) for $a\to 0^+, b\to\infty$, inequalities (3.4.1) and (3.4.2) reduce to (2.2.1) and (2.2.2). It follows that (3.4.1) and (3.4.2) are extensions of (2.2.1) and (2.2.2); (3) for $r = s = 2$, (3.4.5) is weaker than (3.4.11).

3.4.2. SOME PARTICULAR EXAMPLES

In the following examples, we omit to write the assumption of f and g satisfying Theorem 3.4.1. Notice that the constant factors in the following inequalities are not the best possible

Example 3.4.6 If $k_\lambda(x,y) = \frac{1}{(x^\alpha+y^\alpha)^{\lambda/\alpha}}$ $(\lambda,\alpha > 0)$, $r,s > 1$, then we find $k_\lambda(u,1) = k_\lambda(1,u) = \frac{1}{(1+u^\alpha)^{\lambda/\alpha}}$, $l_\lambda = \frac{1}{2^{\lambda/\alpha}}$ and $k_\lambda(r) = \frac{1}{\alpha}B(\frac{\lambda}{r\alpha},\frac{\lambda}{s\alpha})$. By (3.4.5), (3.4.6), (3.4.11) and (3.4.12), we have the following two pairs of equivalent inequalities:

$$\int_a^b\int_a^b \frac{f(x)g(y)}{(x^\alpha+y^\alpha)^{\lambda/\alpha}}dxdy < [\frac{1}{\alpha}B(\frac{\lambda}{r\alpha},\frac{\lambda}{s\alpha}) - \frac{rs}{2^{\lambda/\alpha}\lambda}(\frac{a}{b})^{\frac{\lambda}{rs}}]$$
$$\times\{\int_a^b x^{p(1-\frac{\lambda}{r})-1}f^p(x)dx\}^{\frac{1}{p}}\{\int_a^b x^{q(1-\frac{\lambda}{s})-1}g^q(x)dx\}^{\frac{1}{q}}, \quad (3.4.13)$$

$$\int_a^b y^{\frac{p\lambda}{s}-1}[\int_a^b \frac{f(x)}{(x^\alpha+y^\alpha)^{\lambda/\alpha}}dx]^p dy$$
$$< [\frac{1}{\alpha}B(\frac{\lambda}{r\alpha},\frac{\lambda}{s\alpha}) - \frac{rs}{2^{\lambda/\alpha}\lambda}(\frac{a}{b})^{\frac{\lambda}{rs}}]^p\int_a^b x^{p(1-\frac{\lambda}{r})-1}f^p(x)dx; \quad (3.4.14)$$

$$\int_a^b\int_a^b \frac{f(x)g(y)}{(x^\alpha+y^\alpha)^{\lambda/\alpha}}dxdy < \frac{1}{\alpha}B(\frac{\lambda}{2\alpha},\frac{\lambda}{2\alpha})[1-(\frac{a}{b})^{\frac{\lambda}{4}}]$$
$$\times\{\int_a^b x^{p(1-\frac{\lambda}{2})-1}f^p(x)dx\}^{\frac{1}{p}}\{\int_a^b x^{q(1-\frac{\lambda}{2})-1}g^q(x)dx\}^{\frac{1}{q}}, \quad (3.4.15)$$

$$\int_a^b y^{\frac{p\lambda}{2}-1}[\int_a^b \frac{f(x)}{(x^\alpha+y^\alpha)^{\lambda/\alpha}}dx]^p dy$$
$$< \{\frac{1}{\alpha}B(\frac{\lambda}{2\alpha},\frac{\lambda}{2\alpha})[1-(\frac{a}{b})^{\frac{\lambda}{4}}]\}^p\int_a^b x^{p(1-\frac{\lambda}{2})-1}f^p(x)dx. \quad (3.4.16)$$

Note 3.4.7 For $\alpha = 1$, taking some particular values of (r,s) in (3.4.13) and (3.4.14), we deduce some results of (Yang NMJ 2003)[10], (Yang MIA 2003) [11].

Example 3.4.8 If $\lambda > 0, k_\lambda(x,y) = \frac{\ln(x/y)}{x^\lambda - y^\lambda}, r,s > 1$, then we find $k_\lambda(u,1) = k_\lambda(1,u) = \frac{\ln u}{u^\lambda - 1}, l_\lambda = \frac{1}{\lambda}$ and $k_\lambda(r) = [\frac{\pi}{\lambda \sin(\pi/r)}]^2$. By (3.4.5), (3.4.6), (3.4.11) and (3.4.12), we have the following two pairs of equivalent inequalities:

$$\int_a^b \int_a^b \frac{\ln(x/y)}{x^\lambda - y^\lambda} f(x)g(y)dxdy < \frac{1}{\lambda^2}\{[\frac{\pi}{\sin(\pi/r)}]^2 - rs(\frac{a}{b})^{\frac{\lambda}{rs}}\} \times \{\int_a^b x^{p(1-\frac{\lambda}{r})-1} f^p(x)dx\}^{\frac{1}{p}} \{\int_a^b x^{q(1-\frac{\lambda}{s})-1} g^q(x)dx\}^{\frac{1}{q}}, \tag{3.4.17}$$

$$\int_a^b y^{\frac{p\lambda}{s}-1}[\int_a^b \frac{\ln(x/y)}{x^\lambda - y^\lambda} f(x)dx]^p dy < \frac{1}{\lambda^{2p}}\{[\frac{\pi}{\sin(\pi/r)}]^2 - rs(\frac{a}{b})^{\frac{\lambda}{rs}}\}^p \int_a^b x^{p(1-\frac{\lambda}{r})-1} f^p(x)dx; \tag{3.4.18}$$

$$\int_a^b \int_a^b \frac{\ln(x/y)}{x^\lambda - y^\lambda} f(x)g(y)dxdy < (\frac{\pi}{\lambda})^2[1-(\frac{a}{b})^{\frac{\lambda}{4}}]\{\int_a^b x^{p(1-\frac{\lambda}{2})-1} f^p(x)dx\}^{\frac{1}{p}} \times\{\int_a^b x^{q(1-\frac{\lambda}{2})-1} g^q(x)dx\}^{\frac{1}{q}}, \tag{3.4.19}$$

$$\int_a^b y^{\frac{p\lambda}{2}-1}[\int_a^b \frac{\ln(x/y)}{x^\lambda - y^\lambda} f(x)dx]^p dy < (\frac{\pi}{\lambda})^{2p}[1-(\frac{a}{b})^{\frac{\lambda}{4}}]^p \int_a^b x^{p(1-\frac{\lambda}{2})-1} f^p(x)dx. \tag{3.4.20}$$

Example 3.4.9 If $k_\lambda(x,y) = \frac{1}{(\max\{x,y\})^\lambda} (\lambda > 0)$, $r,s > 1$, then we obtain

$$k_\lambda(u,1) = k_\lambda(1,u) = \frac{1}{(\max\{1,u\})^\lambda},$$

$l_\lambda = 1$ and $k_\lambda(r) = \frac{rs}{\lambda}$. By (3.4.5) and (3.4.6), we have the following equivalent inequalities:

$$\int_a^b \int_a^b \frac{1}{(\max\{x,y\})^\lambda} f(x)g(y)dxdy < \frac{rs}{\lambda}[1-(\frac{a}{b})^{\frac{\lambda}{rs}}]\{\int_a^b x^{p(1-\frac{\lambda}{r})-1} f^p(x)dx\}^{\frac{1}{p}} \times\{\int_a^b x^{q(1-\frac{\lambda}{s})-1} g^q(x)dx\}^{\frac{1}{q}}, \tag{3.4.21}$$

$$\int_a^b y^{\frac{p\lambda}{s}-1}[\int_a^b \frac{f(x)}{(\max\{x,y\})^\lambda} dx]^p dy < \{\frac{rs}{\lambda}[1-(\frac{a}{b})^{\frac{\lambda}{rs}}]\}^p \int_a^b x^{p(1-\frac{\lambda}{r})-1} f^p(x)dx. \tag{3.4.22}$$

Example 3.4.10 If $k_\lambda(x,y) = \frac{|\ln(x/y)|}{(\max\{x,y\})^\lambda} (\lambda > 0)$, $r,s > 1$, then we find $k_\lambda(r) = \frac{r^2+s^2}{\lambda^2}$ and

$$\varpi_\lambda(s,x,a,b) = \frac{r^2+s^2}{\lambda^2} - \int_0^{\frac{x}{b}}(-\ln u)u^{\frac{\lambda}{r}-1}du - \int_0^{\frac{a}{x}}(-\ln u)u^{\frac{\lambda}{s}-1}du$$

$$= \frac{r^2+s^2}{\lambda^2} + \frac{r}{\lambda}\int_0^{\frac{x}{b}}(\ln u)du^{\frac{\lambda}{r}} + \frac{s}{\lambda}\int_0^{\frac{a}{x}}(\ln u)du^{\frac{\lambda}{s}}$$

$$\le \frac{r^2+s^2}{\lambda^2} - \frac{rs}{\lambda^2}[\frac{1}{s}(\frac{x}{b})^{\frac{\lambda}{r}} + \frac{1}{r}(\frac{a}{x})^{\frac{\lambda}{s}}] \le \frac{r^2+s^2}{\lambda^2} - \frac{rs}{\lambda^2}(\frac{a}{b})^{\frac{\lambda}{rs}},$$

$$\omega_\lambda(r,y,a,b) \le \frac{r^2+s^2}{\lambda^2} - \frac{rs}{\lambda^2}(\frac{a}{b})^{\frac{\lambda}{rs}}.$$

By (3.4.5) and (3.4.6), we have the following equivalent inequalities:

$$\int_a^b \int_a^b \frac{|\ln(x/y)|}{(\max\{x,y\})^\lambda} f(x)g(y)dxdy < [\frac{r^2+s^2}{\lambda^2} - \frac{rs}{\lambda^2}(\frac{a}{b})^{\frac{\lambda}{rs}}]\{\int_a^b x^{p(1-\frac{\lambda}{r})-1} f^p(x)dx\}^{\frac{1}{p}} \times\{\int_a^b y^{q(1-\frac{\lambda}{s})-1} g^q(y)dy\}^{\frac{1}{q}}, \tag{3.4.23}$$

$$\int_a^b y^{\frac{p\lambda}{s}-1}[\int_a^b \frac{|\ln(x/y)|}{(\max\{x,y\})^\lambda} f(x)dx]^p dy < [\frac{r^2+s^2}{\lambda^2} - \frac{rs}{\lambda^2}(\frac{a}{b})^{\frac{\lambda}{rs}}]^p \int_a^b x^{p(1-\frac{\lambda}{r})-1} f^p(x)dx. \tag{3.4.24}$$

Example 3.4.11 If $r,s > 1$,

$$k_\lambda(x,y) = \frac{1}{|x^\alpha - y^\alpha|^{\lambda/\alpha}} (\alpha > 0, 0 < \lambda < \alpha),$$

then we find

$$k_\lambda(u,1) = k_\lambda(1,u) \ge l_\lambda = 1, u \in (0,1]$$

and

$$\tilde{k}_\lambda(r) = \frac{1}{\alpha}[B(1-\frac{\lambda}{\alpha}, \frac{\lambda}{\alpha r}) + B(1-\frac{\lambda}{\alpha}, \frac{\lambda}{\alpha s})].$$

By (3.4.5) and (3.4.6), we have the following equivalent inequalities:

$$\int_a^b \int_a^b \frac{1}{|x^\alpha - y^\alpha|^{\lambda/\alpha}} f(x)g(y)dxdy < [\tilde{k}_\lambda(r) - \frac{rs}{\lambda}(\frac{a}{b})^{\frac{\lambda}{rs}}]\{\int_a^b x^{p(1-\frac{\lambda}{r})-1} f^p(x)dx\}^{\frac{1}{p}} \times\{\int_a^b y^{q(1-\frac{\lambda}{s})-1} g^q(y)dy\}^{\frac{1}{q}}, \tag{3.4.25}$$

$$\int_a^b y^{\frac{p\lambda}{s}-1}(\int_a^b \frac{f(x)}{|x^\alpha - y^\alpha|^{\lambda/\alpha}} dx)^p dy < [\tilde{k}_\lambda(r) - \frac{rs}{\lambda}(\frac{a}{b})^{\frac{\lambda}{rs}}]^p \int_a^b x^{p(1-\frac{\lambda}{r})-1} f^p(x)dx. \tag{3.4.26}$$

3.5. SOME REVERSE HILBERT-TYPE INTEGRAL INEQUALITIES RESTRICTED IN THE SUBINTERVALS

3.5.1. TWO EQUIVALENT INEQUALITIES

Theorem 3.5.1 Suppose that $\lambda \in \mathbf{R}$, $r,s \neq 0$, $0 < p < 1$, $k_\lambda(x,y)(\ge 0)$ is a homogeneous function of $-\lambda$-degree in $(0,\infty)\times(0,\infty)$,

$k_\lambda(r)=\int_0^\infty k_\lambda(u,1)u^{\frac{\lambda}{r}-1}du$ is a positive number and $k_\lambda(u,1),k_\lambda(1,u)>0$ a.e. in $(0,1]$. For $a=0$ or $b=\infty$, $(a,b)\subset(0,\infty)$, define the weight functions $\varpi_\lambda(s,x,a,b)$ and $\omega_\lambda(r,y,a,b)$ as

$$\varpi_\lambda(s,x,a,b):$$
$$=x^{\frac{\lambda}{r}}\int_a^b k_\lambda(x,y)y^{\frac{\lambda}{s}-1}dy, x\in(a,b), \quad (3.5.1)$$
$$\omega_\lambda(r,y,a,b):$$
$$=y^{\frac{\lambda}{s}}\int_a^b k_\lambda(x,y)x^{\frac{\lambda}{r}-1}dx, y\in(a,b). \quad (3.5.2)$$

There exist two bounded measurable functions $\tilde{\mu}(x)$ and $\kappa(y)$ in (a,b), satisfying $0<\tilde{\mu}(x),\kappa(y)\le 1$, and

$$0<m_\lambda(r)\le\tilde{\mu}(x)k_\lambda(r)$$
$$\le\varpi_\lambda(s,x,a,b), x\in(a,b) \quad , \quad (3.5.3)$$
$$\omega_\lambda(r,y,a,b)\le\kappa(y)k_\lambda(r), y\in(a,b) \quad . \quad (3.5.4)$$

If $f,g\ge 0$, $0<\int_a^b x^{p(1-\frac{\lambda}{r})-1}f^p(x)dx<\infty$,

$$0<\int_a^b y^{q(1-\frac{\lambda}{s})-1}g^q(y)dy<\infty,$$

then we have the following equivalent inequalities:

$$I_\lambda(a,b):=\int_a^b\int_a^b k_\lambda(x,y)f(x)g(y)dxdy$$
$$>k_\lambda(r)\{\int_a^b\tilde{\mu}(x)x^{p(1-\frac{\lambda}{r})-1}f^p(x)dx\}^{\frac{1}{p}}$$
$$\times\{\int_a^b\kappa(y)y^{q(1-\frac{\lambda}{s})-1}g^q(y)dy\}^{\frac{1}{q}}, \quad (3.5.5)$$
$$J_\lambda(a,b):=\int_a^b\frac{y^{p\lambda/s-1}}{\kappa^{p-1}(y)}(\int_a^b k_\lambda(x,y)f(x)dx)^p dy$$
$$>k_\lambda^p(r)\int_a^b\tilde{\mu}(x)x^{p(1-\frac{\lambda}{r})-1}f^p(x)dx, \quad (3.5.6)$$
$$L_\lambda(a,b):=\int_a^b\frac{x^{q\lambda/r-1}}{\tilde{\mu}^{q-1}(x)}(\int_a^b k_\lambda(x,y)g(y)dy)^q dx$$
$$<k_\lambda^q(r)\int_a^b\kappa(y)y^{q(1-\frac{\lambda}{s})-1}g^q(y)dy. \quad (3.5.7)$$

Proof By the assumption, we have

$$0<\int_a^b\tilde{\mu}(x)x^{p(1-\frac{\lambda}{r})-1}f^p(x)dx<\infty$$

and $0<\int_a^b\kappa(y)y^{q(1-\frac{\lambda}{s})-1}g^q(y)dy<\infty$. By using the reverse Hölder's inequality (Kuang SSTP 2004) [1] and the same way of showing Theorem 3.1.3, in view of (3.5.3) and (3.5.4), we have (3.5.5). Setting

$$g(y):=\frac{y^{p\lambda/s-1}}{\kappa^{p-1}(y)}(\int_a^b k_\lambda(x,y)f(x)dx)^{p-1}$$

$(y\in(a,b))$. It is obvious that $J_\lambda(a,b)>0$. If $J_\lambda(a,b)=\infty$, then (3.5.6) is naturally valid; if

$$0<J_\lambda(a,b)=\int_a^b\kappa(y)y^{q(1-\frac{\lambda}{s})-1}g^q(y)dy<\infty,$$

then $0<\int_a^b y^{q(1-\frac{\lambda}{s})-1}g^q(y)dy<\infty$, by (3.5.5), it follows

$$\int_a^b\kappa(y)y^{q(1-\frac{\lambda}{s})-1}g^q(y)dy$$
$$=J_\lambda(a,b)=I_\lambda(a,b)$$
$$>k_\lambda(r)\{\int_a^b\tilde{\mu}(x)x^{p(1-\frac{\lambda}{r})-1}f^p(x)dx\}^{\frac{1}{p}}$$
$$\times\{\int_a^b\kappa(y)y^{q(1-\frac{\lambda}{s})-1}g^q(y)dy\}^{\frac{1}{q}}>0,$$
$$J_\lambda(a,b)=\int_a^b\kappa(y)y^{q(1-\frac{\lambda}{s})-1}g^q(y)dy$$
$$>k_\lambda^p(r)\int_a^b\tilde{\mu}(x)x^{p(1-\frac{\lambda}{r})-1}f^p(x)dx.$$

Then we have (3.5.6). On the other hand, suppose that (3.5.6) is valid. By the reverse Hölder's inequality, we have

$$I_\lambda(a,b)=\int_a^b[\frac{y^{\frac{-1}{p}+\frac{\lambda}{s}}}{\kappa^{1/q}(y)}\int_a^b k_\lambda(x,y)f(x)dx]$$
$$\times[\kappa^{\frac{1}{q}}(y)y^{\frac{1}{p}-\frac{\lambda}{s}}g(y)]dy$$
$$\ge[J_\lambda(a,b)]^{\frac{1}{p}}\{\int_a^b\kappa(y)y^{q(1-\frac{\lambda}{s})-1}g^q(y)dy\}^{\frac{1}{q}}. \quad (3.5.8)$$

Then by (3.5.6), since

$$0<\int_a^b\kappa(y)y^{q(1-\frac{\lambda}{s})-1}g^q(y)dy<\infty,$$

we have (3.5.5), which is equivalent to (3.5.6).

For $n\in\mathbf{N}, y\in(a,b)$, we set $[g(y)]_n$ as

$$[g(y)]_n:=\begin{cases} n, & g(y)>n,\\ g(y), & \frac{1}{n}\le g(y)\le n,\\ \frac{1}{n}, & g(y)<\frac{1}{n}.\end{cases}$$

Since $\int_a^b\kappa(y)y^{q(1-\frac{\lambda}{s})-1}g^q(y)dy>0$, there exist $[a_n,b_n]\subset(a,b), \lim_{n\to\infty}[a_n,b_n]=(a,b)$ and $n_0\in\mathbf{N}$,

$$0<\int_{a_n}^{b_n}\kappa(y)y^{q(1-\frac{\lambda}{s})-1}[g(y)]_n^q dy<\infty \ (n\ge n_0).$$

Setting

$$f_n(x):=\frac{x^{q\lambda/r-1}}{\tilde{\mu}^{q-1}(x)}(\int_{a_n}^{b_n}k_\lambda(x,y)[g(y)]_n dy)^{q-1},$$
$$x\in(a_n,b_n), n\ge n_0.$$

Then we find

$$0<\int_{a_n}^{b_n}\tilde{\mu}(x)x^{p(1-\frac{\lambda}{r})-1}f_n^p(x)dx<\infty \ (n\geq n_0).$$

By (3.5.5), it follows

$$\infty>\int_{a_n}^{b_n}\tilde{\mu}(x)x^{p(1-\frac{\lambda}{r})-1}f_n^p(x)dx$$
$$=\int_{a_n}^{b_n}\frac{x^{q\lambda/r-1}}{\tilde{\mu}^{q-1}(x)}(\int_{a_n}^{b_n}k_\lambda(x,y)[g(y)]_n dy)^q dx$$
$$=\int_{a_n}^{b_n}\int_{a_n}^{b_n}k_\lambda(x,y)f_n(x)[g(y)]_n dxdy$$
$$>k_\lambda(r)\{\int_{a_n}^{b_n}\tilde{\mu}(x)x^{p(1-\frac{\lambda}{r})-1}f_n^p(x)dx\}^{\frac{1}{p}}$$
$$\times\{\int_{a_n}^{b_n}\kappa(y)y^{q(1-\frac{\lambda}{s})-1}[g(y)]_n^q dy\}^{\frac{1}{q}}>0,$$
$$0<\int_{a_n}^{b_n}\tilde{\mu}(x)x^{p(1-\frac{\lambda}{r})-1}f_n^p(x)dx$$
$$<k_\lambda^q(r)\int_a^b\kappa(y)y^{q(1-\frac{\lambda}{s})-1}g^q(y)dy<\infty.$$

It follows $0<\int_a^b\tilde{\mu}(x)x^{p(1-\frac{\lambda}{r})-1}f_\infty^p(x)dx<\infty$, and then $0<\int_a^b x^{p(1-\frac{\lambda}{r})-1}f_\infty^p(x)dx<\infty$. For $n\to\infty$, by (3.5.5), the above two inequalities still keep the strict sign-inequalities and (3.5.7) is valid.

On the other hand, suppose that (3.5.7) is valid. By the reverse Hölder's inequality, we find

$$I_\lambda(a,b)=\int_a^b[\tilde{\mu}^{\frac{1}{p}}(x)x^{\frac{1}{q}-\frac{\lambda}{r}}f(x)]$$
$$\times[\tilde{\mu}^{\frac{-1}{p}}(x)x^{\frac{-1}{q}+\frac{\lambda}{r}}\int_a^b k_\lambda(x,y)g(y)dy]dx$$
$$\geq\{\int_a^b\tilde{\mu}(x)x^{p(1-\frac{\lambda}{r})-1}f^p(x)dx\}^{\frac{1}{p}}L_\lambda^{\frac{1}{q}}(a,b).\qquad(3.5.9)$$

Then by (3.5.7), we have (3.5.5), which is equivalent to (3.5.7). Hence inequalities (3.5.5), (3.5.6) and (3.5.7) are equivalent.

Note 3.5.2 For $0<a<b<\infty$, under the assumption of Theorem 3.5.1, inequalities (3.5.5), (3.5.6) and (3.5.7) may possibly keep the forms of equality. It is obvious that if at least one of the right hand sides of (3.5.3) and (3.5.4) keeps the strict sign-inequality $a.e.$ in (a,b), then we still have (3.5.5), (3.5.6) and (3.5.7).

3.5.2. SOME REVERSE HILBERT-TYPE INTEGRAL INEQUALITIES IN THE SUBINTERVAL (a,∞) $(a>0)$

Theorem 3.5.3 For $a>0,b=\infty$, as the assumption of Theorem 3.5.1, if $\lambda\in\mathbf{R}$, $r,s\neq 0$, there exists $\eta<\frac{\lambda}{s}$, such that

$$0<\theta_\lambda(s,x,a)=O((\tfrac{a}{x})^{\frac{\lambda}{s}-\eta})\leq\tilde{l}_\lambda<1(x\to\infty)$$

and

$$\tilde{\mu}(x)=1-\theta_\lambda(s,x,a)$$
$$\leq\tfrac{1}{k_\lambda(r)}\varpi_\lambda(s,x,a,\infty),x\in(a,\infty),$$

Then we have the following equivalent inequalities:

$$\int_a^\infty\int_a^\infty k_\lambda(x,y)f(x)g(y)dxdy$$
$$>k_\lambda(r)\ \{\int_a^\infty[1-\theta_\lambda(s,x,a)]x^{p(1-\frac{\lambda}{r})-1}f^p(x)dx\}^{\frac{1}{p}}$$
$$\times\{\int_a^\infty y^{q(1-\frac{\lambda}{s})-1}g^q(y)dy\}^{\frac{1}{q}},\qquad(3.5.10)$$
$$\int_a^\infty y^{\frac{p\lambda}{s}-1}(\int_a^\infty k_\lambda(x,y)f(x)dx)^p dy$$
$$>k_\lambda^p(r)\int_a^\infty[1-\theta_\lambda(s,x,a)]x^{p(1-\frac{\lambda}{r})-1}f^p(x)dx,\qquad(3.5.11)$$
$$\int_a^\infty\frac{x^{q\lambda/r-1}}{[1-\theta_\lambda(s,x,a)]^{q-1}}(\int_a^\infty k_\lambda(x,y)g(y)dy)^q dx$$
$$<k_\lambda^q(r)\int_a^\infty y^{q(1-\frac{\lambda}{s})-1}g^q(y)dy,\qquad(3.5.12)$$

where the constant factors $k_\lambda(r)$ and $k_\lambda^\rho(r)$ $(\rho=p,q)$ are the best possible.

Proof By Theorem 3.5.1, for $a>0,b=\infty$, it is obvious that inequalities (3.5.10), (3.5.11) and (3.5.12) are valid and equivalent. In the following, we prove that the constant factor $k_\lambda(r)$ and $k_\lambda^\rho(r)$ $(\rho=p,q)$ are the best possible. For $n\in\mathbf{N}$, we set

$$f_n(x)=x^{\frac{\lambda}{r}-1-\frac{1}{np}},g_n(x)=x^{\frac{\lambda}{s}-1-\frac{1}{nq}},\ x\in(a,\infty).$$

If there exists a constant $K\geq k_\lambda(r)$, such that (3.5.10) is still valid as we replace $k_\lambda(r)$ by K, then in particular, we find

$$I_n:=\int_a^\infty\int_a^\infty k_\lambda(x,y)f_n(x)g_n(y)dxdy$$
$$\geq K\{\int_a^\infty[1-\theta_\lambda(s,x,a)]x^{p(1-\frac{\lambda}{r})-1}f_n{}^p(x)dx\}^{\frac{1}{p}}$$
$$\times\{\int_a^\infty y^{q(1-\frac{\lambda}{s})-1}g_n{}^q(y)dy\}^{\frac{1}{q}}$$
$$=K\{\int_a^\infty x^{-1-\frac{1}{n}}dx-\int_a^\infty O((\tfrac{a}{x})^{\frac{\lambda}{s}-\eta})x^{-1-\frac{1}{n}}dx\}^{\frac{1}{p}}$$
$$\times\{\int_a^\infty y^{-1-\frac{1}{n}}dy\}^{\frac{1}{q}}=\tfrac{Kn}{a^{1/n}}(1-\tfrac{O(n)}{n})^{\frac{1}{p}}.\qquad(3.5.13)$$

For fixed y, setting $u=y/x$, we obtain

$$I_n\leq\int_a^\infty y^{\frac{\lambda}{s}-1-\frac{1}{nq}}[\int_0^\infty k_\lambda(x,y)x^{\frac{\lambda}{r}-1-\frac{1}{np}}dx]dy$$
$$=\tfrac{n}{a^{1/n}}\int_0^\infty k_\lambda(1,u)u^{\frac{\lambda}{s}-1+\frac{1}{np}}du.\qquad(3.5.14)$$

Then in view of (3.5.13) and (3.5.14), we find

$$\int_0^\infty k_\lambda(1,u)u^{\frac{\lambda}{s}-1+\frac{1}{np}}du \ge K\,[1-\tfrac{O(n)}{n}]^{\frac{1}{p}}.$$

For $n \to \infty$, referring to the proof of Lemma 2.2.5, by Lebesgue control theorem, we have

$$k_\lambda(r)=\int_0^\infty k_\lambda(1,u)u^{\frac{\lambda}{s}-1}du \ge K$$

and $K=k_\lambda(r)$ is the best value of (3.5.10). We conform that the constant factor in (3.5.11) (in (3.5.12)) is the best possible, otherwise we can get a contradiction by (3.5.8) for $b=\infty$ (by (3.5.9) for $b=\infty$) that the constant factor in (3.5.10) is not the best possible. □

Note 3.5.4 For $a \to 0^+$ in (3.5.10), it deduces to (2.2.1). It follows that (3.5.10) is an extension of (2.2.1).

Corollary 3.5.5 For $\lambda \in \mathbf{R}, r,s \ne 0, a>0, b=\infty$, as the assumption of Theorem 3.5.1, if there exists $0<\tilde{\rho}_\lambda(s)<1$, such that

$$\tilde{\mu}(x)=1-\tilde{\rho}_\lambda(s)(\tfrac{a}{x})^{\frac{\lambda}{s}} \le \tfrac{1}{k_\lambda(r)}\varpi_\lambda(s,x,a,b), x\in(a,\infty),$$

then we have the following equivalent inequalities:

$$\int_a^\infty\int_a^\infty k_\lambda(x,y)f(x)g(y)dxdy > k_\lambda(r)\,\{\int_a^\infty[1-\tilde{\rho}_\lambda(s)(\tfrac{a}{x})^{\frac{\lambda}{s}}]x^{p(1-\frac{\lambda}{r})-1}f^p(x)dx\}^{\frac{1}{p}} \times\{\int_a^\infty y^{q(1-\frac{\lambda}{s})-1}g^q(y)dy\}^{\frac{1}{q}}, \quad (3.5.15)$$

$$\int_a^\infty y^{\frac{p\lambda}{s}-1}(\int_a^\infty k_\lambda(x,y)f(x)dx)^p dy > k_\lambda^p(r)\int_a^\infty[1-\tilde{\rho}_\lambda(s)(\tfrac{a}{x})^{\frac{\lambda}{s}}]x^{p(1-\frac{\lambda}{r})-1}f^p(x)dx, \quad (3.5.16)$$

$$\int_a^\infty \frac{x^{q\lambda/r-1}}{[1-\tilde{\rho}_\lambda(s)(\frac{a}{x})^{\lambda/s}]^{q-1}}(\int_a^\infty k_\lambda(x,y)g(y)dy)^q dx < k_\lambda^q(r)\int_a^\infty y^{q(1-\frac{\lambda}{s})-1}g^q(y)dy, \quad (3.5.17)$$

where the constant factors $k_\lambda(r)$ and $k_\lambda^\rho(r)$ $(\rho=p,q)$ are the best possible.

Proof Setting $\theta_\lambda(s,x,a)=\tilde{\rho}_\lambda(s)(\frac{a}{x})^{\frac{\lambda}{s}}$ $(\eta=0)$ in Theorem 3.5.3, we may prove the corollary. □

Example 3.5.6 If $r,s>1, a>0, \alpha>0, 0<\lambda<\alpha\min\{r,s\}$, $k_\lambda(x,y)=\frac{1}{(x^\alpha+y^\alpha)^{\lambda/\alpha}}$, then we find

$$\varpi_\lambda(s,x,a,\infty) =\tfrac{1}{\alpha}B(\tfrac{\lambda}{\alpha r},\tfrac{\lambda}{\alpha s})-\int_0^{\frac{a}{x}}\frac{1}{(u^\alpha+1)^{\lambda/\alpha}}u^{\frac{\lambda}{s}-1}du >\tfrac{1}{\alpha}B(\tfrac{\lambda}{\alpha r},\tfrac{\lambda}{\alpha s})-\int_0^{\frac{a}{x}}u^{\frac{\lambda}{s}-1}du =\tfrac{1}{\alpha}B(\tfrac{\lambda}{\alpha r},\tfrac{\lambda}{\alpha s})[1-\tfrac{s\alpha}{\lambda B(\frac{\lambda}{\alpha r},\frac{\lambda}{\alpha s})}(\tfrac{a}{x})^{\frac{\lambda}{s}}],$$

where setting $\tilde{\rho}(s)=1-\frac{s\alpha}{\lambda B(\frac{\lambda}{\alpha r},\frac{\lambda}{\alpha s})}<1$. Since $\alpha>0, 0<\lambda<\alpha r$, then $\frac{\lambda}{\alpha r}-1<0$ and

$$\tfrac{1}{\alpha}B(\tfrac{\lambda}{\alpha r},\tfrac{\lambda}{\alpha s}) =\tfrac{1}{\alpha}\int_0^1(1-u)^{\frac{\lambda}{\alpha r}-1}u^{\frac{\lambda}{\alpha s}-1}du>\tfrac{1}{\alpha}\int_0^1 u^{\frac{\lambda}{\alpha s}-1}du=\tfrac{s}{\lambda}.$$

It follows $\tilde{\rho}(s)>0$. By Corollary 3.5.5, we have the following equivalent inequalities:

$$\int_a^\infty\int_a^\infty\frac{1}{(x^\alpha+y^\alpha)^{\lambda/\alpha}}f(x)g(y)dxdy>\tfrac{1}{\alpha}B(\tfrac{\lambda}{\alpha r},\tfrac{\lambda}{\alpha s}) \times\{\int_a^\infty[1-\tfrac{\alpha s}{\lambda B(\frac{\lambda}{\alpha r},\frac{\lambda}{\alpha s})}(\tfrac{a}{x})^{\frac{\lambda}{s}}]x^{p(1-\frac{\lambda}{r})-1}f^p(x)dx\}^{\frac{1}{p}} \times\{\int_a^\infty y^{q(1-\frac{\lambda}{s})-1}g^q(y)dy\}^{\frac{1}{q}}, \quad (3.5.18)$$

$$\int_a^\infty y^{\frac{p\lambda}{s}-1}(\int_a^\infty\frac{f(x)}{(x^\alpha+y^\alpha)^{\lambda/\alpha}}dx)^p dy>(\tfrac{1}{\alpha}B(\tfrac{\lambda}{\alpha r},\tfrac{\lambda}{\alpha s}))^p \times\int_a^\infty[1-\tfrac{\alpha s}{\lambda B(\frac{\lambda}{\alpha r},\frac{\lambda}{\alpha s})}(\tfrac{a}{x})^{\frac{\lambda}{s}}]x^{p(1-\frac{\lambda}{r})-1}f^p(x)dx, \quad (3.5.19)$$

$$\int_a^\infty\frac{x^{q\lambda/r-1}}{[1-\frac{\alpha s}{\lambda B(\frac{\lambda}{\alpha r},\frac{\lambda}{\alpha s})}(\frac{a}{x})^{\lambda/s}]^{q-1}}[\int_a^\infty\frac{g(y)}{(x^\alpha+y^\alpha)^{\lambda/\alpha}}dy]^q dx <(\tfrac{1}{\alpha}B(\tfrac{\lambda}{\alpha r},\tfrac{\lambda}{\alpha s}))^q\int_a^\infty y^{q(1-\frac{\lambda}{s})-1}g^q(y)dy, \quad (3.5.20)$$

where the constant factors are the best possible.

Example 3.5.7 If $r,s>1, a>0, \lambda>0$,

$$k_\lambda(x,y)=\frac{1}{(\max\{x,y\})^\lambda},$$

then we find $k_\lambda(u,1)=k_\lambda(1,u)=\frac{1}{(\max\{1,u\})^\lambda}$, and

$$\varpi_\lambda(s,x,a,\infty) =\tfrac{rs}{\lambda}-\int_0^{\frac{a}{x}}u^{\frac{\lambda}{s}-1}du=\tfrac{rs}{\lambda}[1-\tfrac{1}{r}(\tfrac{a}{x})^{\frac{\lambda}{s}}],$$

$$\omega_\lambda(r,y,a,\infty)\le\tfrac{rs}{\lambda}(x,y\in(a,\infty)).$$

By Corollary 3.5.5, we have the following equivalent inequalities:

$$\int_a^\infty\int_a^\infty\frac{1}{(\max\{x,y\})^\lambda}f(x)g(y)dxdy >\tfrac{rs}{\lambda}\{\int_a^\infty[1-\tfrac{1}{r}(\tfrac{a}{x})^{\frac{\lambda}{s}}]x^{p(1-\frac{\lambda}{r})-1}f^p(x)dx\}^{\frac{1}{p}} \times\{\int_a^\infty x^{q(1-\frac{\lambda}{s})-1}g^q(x)dx\}^{\frac{1}{q}}, \quad (3.5.21)$$

$$\int_a^\infty y^{\frac{p\lambda}{s}-1}\Big[\int_a^\infty \frac{f(x)}{(\max\{x,y\})^\lambda}dx\Big]^p dy$$
$$>(\tfrac{rs}{\lambda})^p\int_a^\infty [1-\tfrac{1}{r}(\tfrac{a}{x})^{\frac{\lambda}{s}}]x^{p(1-\frac{\lambda}{r})-1}f^p(x)dx \qquad (3.5.22)$$
$$\int_a^\infty \frac{x^{q\lambda/r-1}}{[1-\frac{1}{r}(\frac{a}{x})^{\lambda/s}]^{q-1}}\Big[\int_a^\infty \frac{g(y)}{(\max\{x,y\})^\lambda}dy\Big]^q dx$$
$$<(\tfrac{rs}{\lambda})^q\int_a^\infty y^{q(1-\frac{\lambda}{s})-1}g^q(y)dy. \qquad (3.5.23)$$

Example 3.5.8 If $k_\lambda(x,y)=\frac{|\ln(x/y)|}{(\max\{x,y\})^\lambda}(\lambda>0)$, $r,s>1$, then we find $k_\lambda(r)=\frac{r^2+s^2}{\lambda^2}$ and

$$\varpi_\lambda(s,x,a,\infty)=\tfrac{r^2+s^2}{\lambda^2}-\int_0^{\frac{a}{x}}(-\ln u)u^{\frac{\lambda}{s}-1}du$$
$$=\tfrac{r^2+s^2}{\lambda^2}-\tfrac{s}{\lambda}\int_0^{\frac{a}{x}}(-\ln u)du^{\frac{\lambda}{s}}$$
$$=\tfrac{r^2+s^2}{\lambda^2}\{1-\tfrac{s^2}{r^2+s^2}[1-\tfrac{\lambda}{s}\ln(\tfrac{a}{x})](\tfrac{a}{x})^{\frac{\lambda}{s}}\},$$
$$\theta_\lambda(s,x,a):=\tfrac{s^2}{r^2+s^2}[1-\tfrac{\lambda}{s}\ln(\tfrac{a}{x})](\tfrac{a}{x})^{\frac{\lambda}{s}}$$
$$=O((\tfrac{a}{x})^{\frac{\lambda}{s}-\eta})\le\tfrac{s^2}{r^2+s^2}<1,0<\eta<\tfrac{\lambda}{s}.$$

By Theorem 3.5.3, we have the following equivalent inequalities:

$$\int_a^\infty\int_a^\infty \frac{|\ln(x/y)|}{(\max\{x,y\})^\lambda}f(x)g(y)dxdy$$
$$>\tfrac{r^2+s^2}{\lambda^2}\{\int_a^\infty[1-\theta_\lambda(s,x,a)]x^{p(1-\frac{\lambda}{r})-1}f^p(x)dx\}^{\frac{1}{p}}$$
$$\times\{\int_a^\infty y^{q(1-\frac{\lambda}{s})-1}g^q(y)dy\}^{\frac{1}{q}}, \qquad (3.5.24)$$
$$\int_a^\infty y^{\frac{p\lambda}{s}-1}\Big[\int_a^\infty \frac{|\ln(x/y)|}{(\max\{x,y\})^\lambda}f(x)dx\Big]^p dy$$
$$>(\tfrac{r^2+s^2}{\lambda^2})^p\int_a^\infty[1-\theta_\lambda(s,x,a)]x^{p(1-\frac{\lambda}{r})-1}f^p(x)dx, \qquad (3.5.25)$$
$$\int_a^\infty \frac{x^{q\lambda/r-1}}{[1-\theta_\lambda(s,x,a)]^{q-1}}\Big[\int_a^\infty \frac{|\ln(x/y)|}{(\max\{x,y\})^\lambda}g(y)dy\Big]^q dx$$
$$<(\tfrac{r^2+s^2}{\lambda^2})^q\int_a^\infty y^{q(1-\frac{\lambda}{s})-1}g^q(y)dy. \qquad (3.5.26)$$

Example 3.5.9 If $0<a<\infty,\lambda>0$, $k_\lambda(x,y)=\frac{1}{x^\lambda+y^\lambda}$, then we find

$$\varpi_\lambda(2,x,a,\infty)=\tfrac{\pi}{\lambda}-\int_0^{\frac{a}{x}}\frac{u^{\frac{\lambda}{2}-1}}{1+u^\lambda}du$$
$$=\tfrac{\pi}{\lambda}[1-\tfrac{2}{\pi}\arctan(\tfrac{a}{x})^{\frac{\lambda}{2}}]\le\tfrac{\pi}{\lambda},\ x\in(a,\infty),$$
$$\theta_\lambda(2,y,a)=\tfrac{2}{\pi}\arctan(\tfrac{a}{x})^{\frac{\lambda}{2}}$$
$$=O((\tfrac{a}{x})^{\frac{\lambda}{2}})\le\tfrac{1}{2}(x\to\infty).$$

By Theorem 3.5.3, setting $r=s=2$, we have the following equivalent inequalities:

$$\int_a^\infty\int_a^\infty \frac{1}{x^\lambda+y^\lambda}f(x)g(y)dxdy>\tfrac{\pi}{\lambda}$$
$$\times\{\int_a^\infty[1-\tfrac{2}{\pi}\arctan(\tfrac{a}{x})^{\frac{\lambda}{2}}]x^{p(1-\frac{\lambda}{2})-1}f^p(x)dx\}^{\frac{1}{p}}$$
$$\times\{\int_a^\infty y^{q(1-\frac{\lambda}{2})-1}g^q(y)dy\}^{\frac{1}{q}}, \qquad (3.5.27)$$
$$\int_a^\infty y^{\frac{p\lambda}{2}-1}(\int_a^\infty \frac{f(x)}{x^\lambda+y^\lambda}dx)^p dy$$
$$>(\tfrac{\pi}{\lambda})^p\int_a^\infty[1-\tfrac{2}{\pi}\arctan(\tfrac{a}{x})^{\frac{\lambda}{2}}]x^{p(1-\frac{\lambda}{2})-1}f^p(x)dx, \qquad (3.5.28)$$
$$\int_a^\infty \frac{x^{q\lambda/2-1}}{[1-\frac{2}{\pi}\arctan(\frac{a}{x})^{\lambda/2}]^{q-1}}(\int_a^\infty \frac{g(y)}{x^\lambda+y^\lambda}dy)^q dx$$
$$<(\tfrac{\pi}{\lambda})^q\int_a^\infty y^{q(1-\frac{\lambda}{2})-1}g^q(y)dy. \qquad (3.5.29)$$

Note 3.5.10 The constant factors in the above inequalities are all the best possible. (Zhong JWSU 2007)[12] studied the reverse forms in $k_\lambda(x,y)=\frac{1}{(x+y)^\lambda}$ $(\lambda>0)$ and (Yang MJ 2009) [13] considered some reverse forms in the general kernel.

3.5.3. SOME REVERSE HILBERT-TYPE INTEGRAL INEQUALITIES IN THE SUBINTERVAL $(0,b)$ $(0<b<\infty)$

Theorem 3.5.11 For $a=0,0<b<\infty$, as the assumption of Theorem 3.5.1, if $\lambda\in\mathbf{R},r,s\neq0$, there exists a $\eta<\frac{\lambda}{r}$, such that

$$0<\theta_\lambda(r,x,b)=O((\tfrac{x}{b})^{\frac{\lambda}{r}-\eta})\le\tilde{l}_\lambda<1(x\to0^+)$$

and

$$\tilde{\mu}(x)=1-\theta_\lambda(r,x,b)$$
$$\le\tfrac{1}{k_\lambda(r)}\varpi_\lambda(s,x,0,b),\ x\in(0,b),$$

then we have the following equivalent inequalities:

$$\int_0^b\int_0^b k_\lambda(x,y)f(x)g(y)dxdy$$
$$>k_\lambda(r)\{\int_0^b[1-\theta_\lambda(r,x,b)]x^{p(1-\frac{\lambda}{r})-1}f^p(x)dx\}^{\frac{1}{p}}$$
$$\times\{\int_0^b y^{q(1-\frac{\lambda}{s})-1}g^q(y)dy\}^{\frac{1}{q}}, \qquad (3.5.30)$$
$$\int_0^b y^{\frac{p\lambda}{s}-1}(\int_0^b k_\lambda(x,y)f(x)dx)^p dy$$
$$>k_\lambda^p(r)\int_0^b[1-\theta_\lambda(r,x,b)]x^{p(1-\frac{\lambda}{r})-1}f^p(x)dx, \qquad (3.5.31)$$
$$\int_0^b \frac{x^{q\lambda/r-1}}{[1-\theta_\lambda(r,x,b)]^{q-1}}(\int_0^b k_\lambda(x,y)g(y)dy)^q dx$$

$$< k_\lambda^q(r)\int_0^b y^{q(1-\frac{\lambda}{s})-1} g^q(y)dy, \quad (3.5.32)$$

where the constant factors $k_\lambda(r)$ and $k_\lambda^\rho(r)$ $(\rho = p,q)$ are the best possible.

Proof It is obvious that by Theorem 3.5.1 and the assumption, we have (3.5.30), (3.5.31) and (3.5.32). For $n\in \mathbf{N}$, we set

$$f_n(x) = x^{\frac{\lambda}{r}-1+\frac{1}{np}}, g_n(x) = x^{\frac{\lambda}{s}-1+\frac{1}{nq}} \quad (x\in (0,b)).$$

If there exists a constant $K \ge k_\lambda(r)$, such that (3.5.30) is still valid as we replace $k_\lambda(r)$ by K, then in particular, we find

$$\tilde{I}_n := \int_0^b\int_0^b k_\lambda(x,y)f_n(x)g_n(y)dxdy$$

$$\ge K\{\int_0^b [1-\theta_\lambda(r,x,b)]x^{p(1-\frac{\lambda}{r})-1} f_n^{\ p}(x)dx\}^{\frac{1}{p}}$$

$$\times\{\int_0^b y^{q(1-\frac{\lambda}{s})-1} g_n^{\ q}(y)dy\}^{\frac{1}{q}}$$

$$= K\{\int_0^b x^{-1+\frac{1}{n}}dx - \int_0^b O((\tfrac{x}{b})^{\frac{\lambda}{r}-\eta})x^{-1+\frac{1}{n}}dx\}^{\frac{1}{p}}$$

$$\times\{\int_0^b y^{-1+\frac{1}{n}}dy\}^{\frac{1}{q}} = Knb^{\frac{1}{n}}[1-\tfrac{O(n)}{n}]^{\frac{1}{p}}; \quad (3.5.33)$$

For fixed $y\in (0,b]$, setting $u = \frac{y}{x}$, we find

$$\tilde{I}_n \le \int_0^b y^{\frac{\lambda}{s}-1+\frac{1}{nq}}[\int_0^\infty k_\lambda(x,y)x^{\frac{\lambda}{r}-1+\frac{1}{np}}dx]dy$$

$$= nb^{\frac{1}{n}}\int_0^\infty k_\lambda(1,u)u^{\frac{\lambda}{s}-1-\frac{1}{np}}du. \quad (3.5.34)$$

Then by (3.5.33), we obtain

$$\int_0^\infty k_\lambda(1,u)u^{\frac{\lambda}{s}-1-\frac{1}{np}}du \ge K\,[1-\tfrac{O(n)}{n}]^{\frac{1}{p}}.$$

For $n \to \infty$, referring to the proof of Lemma 2.2.5, by Lebesgue control theorem, we have

$$k_\lambda(r) = \int_0^\infty k_\lambda(1,u)u^{\frac{\lambda}{s}-1}du \ge K$$

and $K = k_\lambda(r)$ is the best value of (3.5.30). We conform that the constant factor in (3.5.31) (in (3.5.32)) is the best possible, otherwise we can get a contradiction by (3.5.8) (by (3.5.9) for $a = 0$ that the constant factor in (3.5.30) is not the best possible. □

Note 3.5.12 For $b \to \infty$, (3.5.30) deduces to (2.2.1). It follows that (3.5.30) is an extension of (2.2.1).

Corollary 3.5.13 For $\lambda\in \mathbf{R}$,$r,s \ne 0, 0<b<\infty$, as the assumption of Theorem 3.5.1, if there exists $0<\vartheta_\lambda(r)<1$ such that

$$\tilde{\mu}(x) := 1-\vartheta(r)(\tfrac{x}{b})^{\frac{\lambda}{r}}$$

$$\le \tfrac{1}{k_\lambda(r)}\varpi_\lambda(s,x,0,b), \quad x\in (0,b),$$

then we have the following equivalent inequalities:

$$\int_0^b\int_0^b k_\lambda(x,y)f(x)g(y)dxdy$$

$$> k_\lambda(r)\{\int_0^b [1-\vartheta_\lambda(r)(\tfrac{x}{b})^{\frac{\lambda}{r}}]x^{p(1-\frac{\lambda}{r})-1} f^p(x)dx\}^{\frac{1}{p}}$$

$$\times\{\int_0^b y^{q(1-\frac{\lambda}{s})-1} g^q(y)dy\}^{\frac{1}{q}}, \quad (3.5.35)$$

$$\int_0^b y^{\frac{p\lambda}{s}-1}(\int_0^b k_\lambda(x,y)f(x)dx)^p dy$$

$$> k_\lambda^p(r)\int_0^b [1-\vartheta_\lambda(r)(\tfrac{x}{b})^{\frac{\lambda}{r}}]x^{p(1-\frac{\lambda}{r})-1} f^p(x)dx, \quad (3.5.36)$$

$$\int_0^b \frac{x^{q\lambda/r-1}}{[1-\vartheta_\lambda(r)(\frac{x}{b})^{\lambda/r}]^{q-1}}(\int_0^b k_\lambda(x,y)g(y)dy)^q dx$$

$$< k_\lambda^q(r)\int_0^b y^{q(1-\frac{\lambda}{s})-1} g^q(y)dy, \quad (3.5.37)$$

where the constant factors $k_\lambda(r)$ and $k_\lambda^\rho(r)$ $(\rho = p,q)$ are the best possible.

Proof Setting $\theta_\lambda(r,x,b) = \vartheta_\lambda(r)(\frac{x}{b})^{\frac{\lambda}{r}}$ $(\eta = 0)$ in Theorem 3.5.10, we have (3.5.35)- (3.5.37). □

Example 3.5.14 If $r,s>1, b>0, \alpha>0$,

$$0<\lambda<\alpha\min\{r,s\}, k_\lambda(x,y) = \tfrac{1}{(x^\alpha+y^\alpha)^{\lambda/\alpha}},$$

then we find

$$\varpi_\lambda(s,x,0,b)$$

$$= \tfrac{1}{\alpha}B(\tfrac{\lambda}{\alpha r},\tfrac{\lambda}{\alpha s}) - \int_0^{\frac{x}{b}} \tfrac{1}{(u^\alpha+1)^{\lambda/\alpha}}u^{\frac{\lambda}{r}-1}du$$

$$> \tfrac{1}{\alpha}B(\tfrac{\lambda}{\alpha r},\tfrac{\lambda}{\alpha s}) - \int_0^{\frac{x}{b}} u^{\frac{\lambda}{r}-1}du$$

$$= \tilde{\mu}(x) := \tfrac{1}{\alpha}B(\tfrac{\lambda}{\alpha r},\tfrac{\lambda}{\alpha s})[1-\tfrac{r\alpha}{\lambda B(\frac{\lambda}{\alpha r},\frac{\lambda}{\alpha s})}(\tfrac{x}{b})^{\frac{\lambda}{r}}].$$

where taking $\vartheta(r) = 1-\frac{r\alpha}{\lambda B(\frac{\lambda}{\alpha r},\frac{\lambda}{\alpha s})}<1$. Since $\alpha>0$, $0<\lambda<\alpha s$, then $\frac{\lambda}{\alpha s}-1<0$ and

$$\tfrac{1}{\alpha}B(\tfrac{\lambda}{\alpha r},\tfrac{\lambda}{\alpha s}) = \tfrac{1}{\alpha}\int_0^1 (1-u)^{\frac{\lambda}{\alpha s}-1}u^{\frac{\lambda}{\alpha r}-1}du$$

$$> \tfrac{1}{\alpha}\int_0^1 u^{\frac{\lambda}{\alpha r}-1}du = \tfrac{r}{\lambda}.$$

It follows $\vartheta(r)>0$. By Corollary 3.5.12, we have the following equivalent inequalities:

$$\int_0^b\int_0^b \tfrac{f(x)g(y)}{(x^\alpha+y^\alpha)^{\lambda/\alpha}}dxdy > \tfrac{1}{\alpha}B(\tfrac{\lambda}{\alpha r},\tfrac{\lambda}{\alpha s})$$

$$\times\{\int_0^b [1-\tfrac{\alpha r}{\lambda B(\frac{\lambda}{\alpha r},\frac{\lambda}{\alpha s})}(\tfrac{x}{b})^{\frac{\lambda}{r}}]x^{p(1-\frac{\lambda}{r})-1} f^p(x)dx\}^{\frac{1}{p}}$$

$$\times\{\int_a^\infty y^{q(1-\frac{\lambda}{s})-1}g^q(y)dy\}^{\frac{1}{q}}, \quad (3.5.38)$$

$$\int_0^b y^{\frac{p\lambda}{s}-1}(\int_0^b \frac{f(x)}{(x^\alpha+y^\alpha)^{\lambda/\alpha}}dx)^p dy > (\frac{1}{\alpha}B(\frac{\lambda}{\alpha r},\frac{\lambda}{\alpha s}))^p$$
$$\times\int_0^b [1-\frac{\alpha r}{\lambda B(\frac{\lambda}{\alpha r},\frac{\lambda}{\alpha s})}(\frac{x}{b})^{\frac{\lambda}{r}}]x^{p(1-\frac{\lambda}{r})-1}f^p(x)dx, \quad (3.5.39)$$

$$\int_0^b \frac{x^{q\lambda/r-1}}{[1-\frac{\alpha r}{\lambda B(\frac{\lambda}{\alpha r},\frac{\lambda}{\alpha s})}(\frac{x}{b})^{\lambda/r}]^{q-1}}(\int_0^b \frac{g(y)}{(x^\alpha+y^\alpha)^{\lambda/\alpha}}dy)^q dx$$
$$< (\frac{1}{\alpha}B(\frac{\lambda}{\alpha r},\frac{\lambda}{\alpha s}))^q\int_0^b y^{q(1-\frac{\lambda}{s})-1}g^q(y)dy, \quad (3.5.40)$$

where the constant factors $\frac{1}{\alpha}B(\frac{\lambda}{\alpha r},\frac{\lambda}{\alpha s})$ and $(\frac{1}{\alpha}B(\frac{\lambda}{\alpha r},\frac{\lambda}{\alpha s}))^\rho$ $(\rho = p,q)$ are the best possible.

Example 3.5.15 If $r,s>1, 0<b<\infty, \lambda>0$, $k_\lambda(x,y)=\frac{1}{(\max\{x,y\})^\lambda}$, then we find

$$\varpi_\lambda(s,x,0,b)$$
$$=\frac{rs}{\lambda}-\int_0^{\frac{x}{b}} u^{\frac{\lambda}{r}-1}du=\frac{rs}{\lambda}[1-\frac{1}{s}(\frac{x}{b})^{\frac{\lambda}{r}}].$$

By Corollary 3.5.12, we have the following equivalent inequalities:

$$\int_0^b\int_0^b \frac{1}{(\max\{x,y\})^\lambda}f(x)g(y)dxdy$$
$$>\frac{rs}{\lambda}\{\int_0^b [1-\frac{1}{s}(\frac{x}{b})^{\frac{\lambda}{r}}]x^{p(1-\frac{\lambda}{r})-1}f^p(x)dx\}^{\frac{1}{p}}$$
$$\times\{\int_0^b x^{q(1-\frac{\lambda}{s})-1}g^q(x)dx\}^{\frac{1}{q}}, \quad (3.5.41)$$

$$\int_0^b y^{\frac{p\lambda}{s}-1}[\int_0^b \frac{f(x)}{(\max\{x,y\})^\lambda}dx]^p dy$$
$$>(\frac{rs}{\lambda})^p\int_0^b [1-\frac{1}{s}(\frac{x}{b})^{\frac{\lambda}{r}}]x^{p(1-\frac{\lambda}{r})-1}f^p(x)dx, \quad (3.5.42)$$

$$\int_0^b \frac{x^{q\lambda/r-1}}{[1-\frac{1}{s}(\frac{x}{b})^{\lambda/r}]^{q-1}}[\int_0^b \frac{g(y)}{(\max\{x,y\})^\lambda}dy]^q dx$$
$$<(\frac{rs}{\lambda})^q\int_0^b y^{q(1-\frac{\lambda}{s})-1}g^q(y)dy, \quad (3.5.43)$$

where the constant factors $\frac{rs}{\lambda}$, $(\frac{rs}{\lambda})^p$ and $(\frac{rs}{\lambda})^q$ are the best possible.

Example 3.5.16 If $r,s>1, \lambda>0$, $k_\lambda(x,y)=\frac{|\ln(x/y)|}{(\max\{x,y\})^\lambda}$, then we find $k_\lambda(r)=\frac{r^2+s^2}{\lambda^2}$,

$$\varpi_\lambda(s,x,0,b)=\frac{r^2+s^2}{\lambda^2}-\int_0^{\frac{x}{b}}(-\ln u)u^{\frac{\lambda}{r}-1}du$$
$$=\frac{r^2+s^2}{\lambda^2}-\frac{r}{\lambda}\int_0^{\frac{x}{b}}(-\ln u)du^{\frac{\lambda}{r}}$$
$$=\frac{r^2+s^2}{\lambda^2}\{1-\frac{r^2}{r^2+s^2}[1-\frac{\lambda}{r}\ln(\frac{x}{b})](\frac{x}{b})^{\frac{\lambda}{r}}\},$$

$$\theta_\lambda(r,x,b):=\frac{r^2}{r^2+s^2}[1-\frac{\lambda}{r}\ln(\frac{x}{b})](\frac{x}{b})^{\frac{\lambda}{r}}$$
$$=O((\frac{x}{b})^{\frac{\lambda}{r}-\eta})\le\frac{r^2}{r^2+s^2}<1,\ 0<\eta<\frac{\lambda}{r}.$$

By Theorem 3.5.10, we have the following equivalent inequalities:

$$\int_0^b\int_0^b \frac{|\ln(x/y)|}{(\max\{x,y\})^\lambda}f(x)g(y)dxdy$$
$$>\frac{r^2+s^2}{\lambda^2}\{\int_0^b [1-\theta_\lambda(r,x,b)]x^{p(1-\frac{\lambda}{r})-1}f^p(x)dx\}^{\frac{1}{p}}$$
$$\times\{\int_0^b y^{q(1-\frac{\lambda}{s})-1}g^q(y)dy\}^{\frac{1}{q}}, \quad (3.5.44)$$

$$\int_0^b y^{\frac{p\lambda}{s}-1}[\int_0^b \frac{|\ln(x/y)|}{(\max\{x,y\})^\lambda}f(x)dx]^p dy$$
$$>(\frac{r^2+s^2}{\lambda^2})^p\int_0^b [1-\theta_\lambda(r,x,b)]x^{p(1-\frac{\lambda}{r})-1}f^p(x)dx, \quad (3.5.45)$$

$$\int_0^b \frac{x^{q\lambda/r-1}}{[1-\theta_\lambda(r,x,b)]^{q-1}}[\int_0^b \frac{|\ln(x/y)|}{(\max\{x,y\})^\lambda}g(y)dy]^q dx$$
$$<(\frac{r^2+s^2}{\lambda^2})^q\int_0^b y^{q(1-\frac{\lambda}{s})-1}g^q(y)dy. \quad (3.5.46)$$

Example 3.5.17 If $r,s>1, 0<b<\infty, \lambda>0$, $k_\lambda(x,y)=\frac{1}{x^\lambda+y^\lambda}$, then we find

$$\varpi_\lambda(2,x,0,b)=\frac{\pi}{\lambda}-\int_0^{\frac{x}{b}}\frac{1}{1+u^\lambda}u^{\frac{\lambda}{2}-1}du$$
$$=\frac{\pi}{\lambda}[1-\frac{2}{\pi}\arctan(\frac{x}{b})^{\frac{\lambda}{2}}],\ x\in(0,b).$$

By Theorem 3.5.10, setting $r=s=2$, we have the following equivalent inequalities:

$$\int_0^b\int_0^b \frac{1}{x^\lambda+y^\lambda}f(x)g(y)dxdy>\frac{\pi}{\lambda}$$
$$\times\{\int_0^b [1-\frac{2}{\pi}\arctan(\frac{x}{b})^{\frac{\lambda}{2}}]x^{p(1-\frac{\lambda}{2})-1}f^p(x)dx\}^{\frac{1}{p}}$$
$$\times\{\int_0^b y^{q(1-\frac{\lambda}{2})-1}g^q(y)dy\}^{\frac{1}{q}}, \quad (3.5.47)$$

$$\int_0^b y^{\frac{p\lambda}{2}-1}(\int_0^b \frac{f(x)}{x^\lambda+y^\lambda}dx)^p dy$$
$$>(\frac{\pi}{\lambda})^p\int_0^b [1-\frac{2}{\pi}\arctan(\frac{x}{b})^{\frac{\lambda}{2}}]x^{p(1-\frac{\lambda}{2})-1}f^p(x)dx, \quad (3.5.48)$$

$$\int_0^b \frac{x^{q\lambda/2-1}}{[1-\frac{2}{\pi}\arctan(\frac{x}{b})^{\lambda/2}]^{q-1}}(\int_0^b \frac{g(y)}{x^\lambda+y^\lambda}dy)^q dx$$
$$<(\frac{\pi}{\lambda})^q\int_0^b y^{q(1-\frac{\lambda}{2})-1}g^q(y)dy. \quad (3.5.49)$$

3.5.4. SOME REVERSE HILBERT-TYPE INTEGRAL INEQUALITIES IN THE SUBINTERVAL (a,b) $(0<a<b<\infty)$

Lemma 3.5.18 For $0<a<b<\infty, \lambda>0, r,s>1$, if $k_\lambda(u,1)$, $k_\lambda(1,u)\ge l_\lambda>0$, $u\in(0,1]$, define

$$d_\lambda(a,b) := \sup_{a<x<b}[\int_0^{\frac{a}{x}} k_\lambda(1,u)u^{\frac{\lambda}{s}-1}du + \int_{\frac{b}{x}}^{\infty} k_\lambda(1,u)u^{\frac{\lambda}{s}-1}du],$$

$$c_\lambda := \sup_{a<x<b}[\tfrac{1}{r}(\tfrac{a}{x})^{\frac{\lambda}{s}} + \tfrac{1}{s}(\tfrac{x}{b})^{\frac{\lambda}{r}}].$$

Then we have the following inequalities:

$$0 < m_\lambda(r) := \tfrac{rsl_\lambda}{\lambda}(1-c_\lambda) \le k_\lambda(r) - d_\lambda(a,b) \le \varpi_\lambda(s,x,a,b) \le k_\lambda(r) - \tfrac{rsl_\lambda}{\lambda}(\tfrac{a}{b})^{\frac{\lambda}{rs}} \le k_\lambda(r), x \in (a,b), \quad (3.5.50)$$

$$0 < m_\lambda(r) \le \omega_\lambda(r,y,a,b) \le k_\lambda(r) - \tfrac{rsl_\lambda}{\lambda}(\tfrac{a}{b})^{\frac{\lambda}{rs}} \le k_\lambda(r), y \in (a,b). \quad (3.5.51)$$

Proof Since $k_\lambda(u,1) \ge l_\lambda > 0$, $u \in (0,1]$, then for $u \in (1,\infty)$, we have $k_\lambda(1,u) = u^{-\lambda}k_\lambda(\frac{1}{u},1) \ge l_\lambda u^{-\lambda}$. In (3.1.1), for fixed $x \in (a,b)$, setting $u = \frac{y}{x}$, due to the mean value inequality, it follows

$$\varpi_\lambda(s,x,a,b) = \int_{\frac{a}{x}}^{\frac{b}{x}} k_\lambda(1,u)u^{\frac{\lambda}{s}-1}du \ge k_\lambda(r) - d_\lambda(a,b)$$

$$= \inf_{a<x<b}[\int_{\frac{a}{x}}^{1} k_\lambda(1,u)u^{\frac{\lambda}{s}-1}du + \int_1^{\frac{b}{x}} k_\lambda(1,u)u^{\frac{\lambda}{s}-1}du]$$

$$\ge l_\lambda \inf_{a<x<b}[\int_{\frac{a}{x}}^{1} u^{\frac{\lambda}{s}-1}du + \int_1^{\frac{b}{x}} u^{\frac{-\lambda}{r}-1}du]$$

$$= \tfrac{rsl_\lambda}{\lambda}(1-c_\lambda) > 0,$$

$$\varpi_\lambda(s,x,a,b) = k_\lambda(r) - [\int_0^{\frac{a}{x}} k_\lambda(1,u)u^{\frac{\lambda}{s}-1}du + \int_{\frac{b}{x}}^{\infty} k_\lambda(1,u)u^{\frac{\lambda}{s}-1}du]$$

$$\le k_\lambda(r) - l_\lambda[\int_0^{\frac{a}{x}} u^{\frac{\lambda}{s}-1}du + \int_{\frac{b}{x}}^{\infty} u^{-\lambda}u^{\frac{\lambda}{s}-1}du]$$

$$= k_\lambda(r) - \tfrac{rsl_\lambda}{\lambda}[\tfrac{1}{r}(\tfrac{a}{x})^{\frac{\lambda}{s}} + \tfrac{1}{s}(\tfrac{x}{b})^{\frac{\lambda}{r}}]$$

$$\le k_\lambda(r) - \tfrac{rsl_\lambda}{\lambda}(\tfrac{a}{b})^{\frac{\lambda}{rs}} \le k_\lambda(r).$$

Then (3.5.50) is valid. By the same way, we have (3.5.51).

By Note 3.5.2, setting $\tilde{\mu}(x) = 1 - \frac{d_\lambda(a,b)}{k_\lambda(r)}$ and $\kappa(y) = 1 - \frac{rsl_\lambda}{\lambda k_\lambda(r)}(\frac{a}{b})^{\frac{\lambda}{rs}}$, we have

Theorem 3.5.19 For $0 < a < b < \infty, \lambda > 0, r,s > 1$, as the assumption of Theorem 3.5.1, if

$$k_\lambda(u,1), k_\lambda(1,u) \ge l_\lambda > 0,\ u \in (0,1],$$

the we have the following equivalent inequalities:

$$I_\lambda(a,b) > (k_\lambda(r) - d_\lambda(a,b))^{\frac{1}{p}}[k_\lambda(r) - \tfrac{rsl_\lambda}{\lambda}(\tfrac{a}{b})^{\frac{\lambda}{sr}}]^{\frac{1}{q}} \times \{\int_a^b x^{p(1-\frac{\lambda}{r})-1} f^p(x)dx\}^{\frac{1}{p}} \{\int_a^b y^{q(1-\frac{\lambda}{s})-1} g^q(y)dy\}^{\frac{1}{q}}, \quad (3.5.52)$$

$$\int_a^b y^{\frac{p\lambda}{s}-1}(\int_a^b k_\lambda(x,y)f(x)dx)^p dy > [k_\lambda(r) - \tfrac{rsl_\lambda}{\lambda}(\tfrac{a}{b})^{\frac{\lambda}{rs}}]^{p-1}(k_\lambda(r) - d_\lambda(a,b)) \times \int_a^b x^{p(1-\frac{\lambda}{r})-1} f^p(x)dx, \quad (3.5.53)$$

$$\int_a^b x^{\frac{q\lambda}{r}-1}(\int_a^b k_\lambda(x,y)g(y)dy)^q dx < (k_\lambda(r) - d_\lambda(a,b))^{q-1}[k_\lambda(r) - \tfrac{rsl_\lambda}{\lambda}(\tfrac{a}{b})^{\frac{\lambda}{rs}}] \times \int_a^b y^{q(1-\frac{\lambda}{s})-1} g^q(y)dy, \quad (3.5.54)$$

where,

$$d_\lambda(a,b) = \sup_{a<x<b}[\int_0^{\frac{a}{x}} k_\lambda(1,u)u^{\frac{\lambda}{s}-1}du + \int_{\frac{b}{x}}^{\infty} k_\lambda(1,u)u^{\frac{\lambda}{s}-1}du] < k_\lambda(r).$$

Note 3.5.20 (1) If $k_\lambda(u,1), k_\lambda(1,u)$ are decreasing continuous in $(0,1]$, we may get $l_\lambda = k_\lambda(1,1) > 0$ in Theorem 3.5.18. (2) For $a \to 0^+, b \to \infty$, $d_\lambda(a,b) \to 0$, inequality (3.5.52) reduces to (2.2.1). It follows that (3.5.52) is an extension of (2.2.1). (3) For particular kernel $k_\lambda(x,y)$, if we obtain $k_\lambda(r)$ and l_λ, then by (3.5.52)-(3.5.54), we may build some new inequalities. For example, taking

$$k_\lambda(x,y) = \frac{1}{(x^\alpha + y^\alpha)^{\lambda/\alpha}} (\lambda, \alpha > 0),$$

we get $k_\lambda(r) = \frac{1}{\alpha}B(\frac{\lambda}{r\alpha}, \frac{\lambda}{s\alpha})$ and $l_\lambda = \frac{1}{2^{\lambda/\alpha}}$; taking

$$k_\lambda(x,y) = \frac{\ln(x/y)}{x^\lambda - y^\lambda} (\lambda > 0),$$

we get $k_\lambda(r) = [\frac{\pi}{\lambda \sin(\pi/r)}]^2$ and $l_\lambda = \frac{1}{\lambda}$.

Example 3.5.21 If $0 < a < b < \infty, r,s > 1$, $k_\lambda(x,y) = \frac{1}{(\max\{x,y\})^\lambda}$ $(\lambda > 0)$, then we find $k_\lambda(u,1) = k_\lambda(1,u) = \frac{1}{(\max\{1,u\})^\lambda}$, $k_\lambda(r) = \frac{rs}{\lambda}$ and

$$\varpi_\lambda(s,x,a,b) = \tfrac{rs}{\lambda}[1 - \tfrac{1}{r}(\tfrac{a}{x})^{\frac{\lambda}{s}} - \tfrac{1}{s}(\tfrac{x}{b})^{\frac{\lambda}{r}}] \le \tfrac{rs}{\lambda},$$

$$\omega_\lambda(r,y,a,b) = \tfrac{rs}{\lambda}[1 - \tfrac{1}{s}(\tfrac{a}{y})^{\frac{\lambda}{r}} - \tfrac{1}{r}(\tfrac{y}{b})^{\frac{\lambda}{s}}] \le \tfrac{rs}{\lambda}[1 - (\tfrac{a}{b})^{\frac{\lambda}{rs}}].$$

By Note 3.5.2, we have the following equivalent inequalities:

$$\int_a^b\int_a^b \frac{f(x)g(y)}{(\max\{x,y\})^\lambda}dxdy > \frac{rs}{\lambda}[1-(\frac{a}{b})^{\frac{\lambda}{rs}}]^{\frac{1}{q}}$$

$$\times\{\int_a^b [1-\frac{1}{r}(\frac{a}{x})^{\frac{\lambda}{s}}-\frac{1}{s}(\frac{x}{b})^{\frac{\lambda}{r}}]x^{p(1-\frac{\lambda}{r})-1}f^p(x)dx\}^{\frac{1}{p}}$$

$$\times\{\int_a^b x^{q(1-\frac{\lambda}{s})-1}g^q(x)dx\}^{\frac{1}{q}}, \qquad (3.5.55)$$

$$\int_a^b y^{\frac{p\lambda}{s}-1}(\int_a^b \frac{f(x)}{(\max\{x,y\})^\lambda}dx)^p dy$$

$$> (\frac{rs}{\lambda})^p[1-(\frac{a}{b})^{\frac{\lambda}{rs}}]^{p-1}$$

$$\times\int_a^b [1-\frac{1}{r}(\frac{a}{x})^{\frac{\lambda}{s}}-\frac{1}{s}(\frac{x}{b})^{\frac{\lambda}{r}}]x^{p(1-\frac{\lambda}{r})-1}f^p(x)dx, \qquad (3.5.56)$$

$$\int_a^b \frac{x^{q\lambda/r-1}}{[1-\frac{1}{r}(\frac{a}{x})^{\lambda/s}-\frac{1}{s}(\frac{x}{b})^{\lambda/r}]^{q-1}}(\int_a^b \frac{g(y)}{(\max\{x,y\})^\lambda}dy)^q dx$$

$$< (\frac{rs}{\lambda})^q[1-(\frac{a}{b})^{\frac{\lambda}{rs}}]\int_a^b y^{q(1-\frac{\lambda}{s})-1}g^q(y)dy. \qquad (3.5.57)$$

Example 3.5.22 If $0<a<b<\infty, r>1, k_\lambda(x,y) = \frac{1}{x^\lambda+y^\lambda} (\lambda>0)$, then we find $k_\lambda(u,1) = k_\lambda(1,u) = \frac{1}{1+u^\lambda}$. By (3.5.2) and (3.5.3), setting $v=u^\lambda$ and

$$\theta_\lambda(x)=\frac{2}{\pi}[\arctan(\frac{a}{x})^{\frac{\lambda}{2}}+\arctan(\frac{x}{b})^{\frac{\lambda}{2}}],$$

we obtain

$$\varpi_\lambda(2,x,a,b) = \frac{\pi}{\lambda}-\int_0^{\frac{a}{x}}\frac{u^{\frac{\lambda}{2}-1}}{1+u^\lambda}du-\int_0^{\frac{x}{b}}\frac{u^{\frac{\lambda}{2}-1}}{1+u^\lambda}du$$

$$=\frac{\pi}{\lambda}-\frac{2}{\lambda}[\int_0^{(\frac{a}{x})^\lambda}\frac{1}{1+(\sqrt{v})^2}d\sqrt{v}+\int_0^{(\frac{x}{b})^\lambda}\frac{1}{1+(\sqrt{v})^2}d\sqrt{v}]$$

$$=\frac{\pi}{\lambda}[1-\theta_\lambda(x)], \qquad (3.5.58)$$

$$\omega_\lambda(2,y,a,b) = \frac{\pi}{\lambda}-\int_0^{\frac{a}{y}}\frac{u^{\frac{\lambda}{2}-1}}{1+u^\lambda}du-\int_0^{\frac{y}{b}}\frac{u^{\frac{\lambda}{2}-1}}{1+u^\lambda}du$$

$$\le\frac{\pi}{\lambda}-\frac{1}{2}(\int_0^{\frac{a}{y}}u^{\frac{\lambda}{2}-1}du+\int_0^{\frac{y}{b}}u^{\frac{\lambda}{2}-1}du)$$

$$=\frac{\pi}{\lambda}-\frac{1}{\lambda}[(\frac{a}{y})^{\frac{\lambda}{2}}+(\frac{y}{b})^{\frac{\lambda}{2}}]$$

$$\le\frac{\pi}{\lambda}[1-\frac{2}{\pi}(\frac{a}{b})^{\frac{\lambda}{4}}],\ y\in(a,b). \qquad (3.5.59)$$

By Note 3.5.2, setting $r=s=2$, we obtain the following equivalent inequalities:

$$\int_a^b\int_a^b \frac{1}{x^\lambda+y^\lambda}f(x)g(y)dxdy > \frac{\pi}{\lambda}[1-\frac{2}{\pi}(\frac{a}{b})^{\frac{\lambda}{4}}]^{\frac{1}{q}}$$

$$\times\{\int_a^b [1-\theta_\lambda(x)]x^{p(1-\frac{\lambda}{2})-1}f^p(x)dx\}^{\frac{1}{p}}$$

$$\times\{\int_a^b y^{q(1-\frac{\lambda}{2})-1}g^q(y)dy\}^{\frac{1}{q}}, \qquad (3.5.60)$$

$$\int_a^b y^{\frac{p\lambda}{2}-1}(\int_a^b \frac{f(x)}{x^\lambda+y^\lambda}dx)^p dy > (\frac{\pi}{\lambda})^p[1-\frac{2}{\pi}(\frac{a}{b})^{\frac{\lambda}{4}}]^{p-1}$$

$$\times\int_a^b [1-\theta_\lambda(x)]x^{p(1-\frac{\lambda}{2})-1}f^p(x)dx, \qquad (3.5.61)$$

$$\int_a^b \frac{x^{q\lambda/2-1}}{(1-\theta_\lambda(x))^{q-1}}(\int_a^b \frac{g(y)}{x^\lambda+y^\lambda}dy)^q dx$$

$$< (\frac{\pi}{\lambda})^q[1-\frac{2}{\pi}(\frac{a}{b})^{\frac{\lambda}{4}}]\int_a^b y^{q(1-\frac{\lambda}{2})-1}g^q(y)dy. \qquad (3.5.62)$$

3.6 REFERENCES

1. Kuang JC. Applied inequalities. Jinan: Shandong Science Technic Press, 2004.
2. Yang BC. On generalizations of Hardy-Hilbert's integral inequalities, Acta Mathematica Sinica, 1998; 41(4): 839-844.
3. Yang BC. A general Hardy-Hilbert's integral inequality with a best constant factor, Chinese Mathematics Annals, 2000; 21A (4): 401-408.
4. Yang BC. On Hilbert's integral inequality , J. Math. Anal. Appl., 1998; 220: 778~785.
5. Yang BC. On generalization of Hilbert's theorem, Journal of Nanjing University Mathematical Biquarterly, 2000; 17(1): 152-156.
6. Yang BC. On Hardy-Littlewood-Polya's theorem, Journal of Changde Teachers University (Natural Science Edition), 2002; 14(3): 29-32.
7. Yang BC, Debnath L. On the extended Hardy-Hilbert's inequality, J. Math. Anal. Appl., 2002; 272: 187-199.
8. Yang BC, On Hardy-Hilbert's integral inequality, J. Math. Anal. Appl., 2001; 261: 295-306.
9. Zhong WY, An equivalent form of extended Hardy-Hilbert integral inequality, Journal of Shanghai University (Natural Edition), 2007; 13(1): 51-54.
10. Yang BC, On Hardy-Hilbert's integral inequality and its equivalent form, Northeast Math. J, 2003; 19(2): 139-148.
11. Yang BC, Rassias T M, On the way of weight coefficient and research for Hilbert-type inequalities, Math. Ineq. Appl., 2003; 6(4): 625-658.
12. Zhong WY, Yang BC, On the reverse Hardy-Hilbert's integral inequality, Journal of Southwest University (Natural Edition), 2007; 29(4): 44-48.
13. Yang BC, A reverse Hilbert-type integral inequality on a subinterval, Journal of Mathematics, 2009; 29.(to appear)

4. Some Innovative Hilbert-Type Integral Inequalities

Bicheng Yang

Department of Mathematics, Guangdong Education Institute, Guangzhou, Guangdong 510303, P. R. China; E-mail: bcyang@pub.guangzhou.gd.cn

Abstract: In this chapter, based on some theorems of Chapters 2, by using the technique of real analysis, we discuss how to use some particular parameters to deduce some new Hilbert-type integral inequalities and the reverses with the best constant factors.

4.1. HILBERT-TYPE INTEGRAL INEQUALITIES WITH THE HOMOGENEOUS KERNELS OF 0-DEGREE

4.1.1. A COROLLARY

For $\lambda = 0$ in Theorem 2.2.3 and Theorem 2.2.7, we have

Corollary 4.1.1 If (p,q) is one pair of conjugate exponents $p>0(p\neq 1)$, $k_0(x,y)\geq 0$ is a homogeneous function of 0 -degree in $(0,\infty)\times(0,\infty)$,

$$0<k_0:=\int_0^\infty k_0(u,1)u^{-1}du$$
$$=\int_0^1[k_0(u,1)+k_0(1,u)]u^{-1}du<\infty,$$
$$\tilde{\phi}_\rho(x)=x^{\rho-1}\ (x>0;\rho=p,q),$$

$f(x),g(x)\geq 0$, such that

$$0<\|f\|_{p,\tilde{\phi}_p}=\{\int_0^\infty x^{p-1}f^p(x)dx\}^{\frac{1}{p}}<\infty$$

and $0<\|g\|_{q,\tilde{\phi}_q}=\{\int_0^\infty x^{q-1}g^q(x)dx\}^{\frac{1}{q}}<\infty$, then (1) for $p>1$, we have the following equivalent inequalities:

$$\int_0^\infty\int_0^\infty k_0(x,y)f(x)g(y)dxdy$$
$$<k_0\|f\|_{p,\tilde{\phi}_p}\|g\|_{q,\tilde{\phi}_q}, \qquad (4.1.1)$$

$$\int_0^\infty \frac{1}{y}(\int_0^\infty k_0(x,y)f(x)dx)^p dy$$
$$<k_0^p\|f\|_{p,\tilde{\phi}_p}^p, \qquad (4.1.2)$$

where the constant factors k_0 and k_0^p are the best possible; (2) for $0<p<1$, if there exists a neighborhood I_0 of 0, such that

$$0<K_0(x)=\int_0^\infty k_0(u,1)u^{x-1}du$$
$$\leq\int_0^1[k_0(u,1)+k_0(1,u)]u^{-|x|-1}du<\infty, x\in I_0,$$

then we have the equivalent reverses of (4.1.1) and (4.1.2) with the same best constant factors.

4.1.2. SOME EXAMPLES WITH MULTI-PARAMETERS

Example 4.1.2 If $\alpha>0,\beta\in\mathbf{R},\gamma>-1$,

$$k_0(x,y)=(\tfrac{\min\{x,y\}}{x+y})^\alpha\,|\ln(\tfrac{x}{y})|^\gamma\arctan(\tfrac{x}{y})^\beta,$$

then for $|x|<\alpha$, we find

$$0<K_0(x)=\int_0^\infty k_0(u,1)u^{x-1}du$$
$$\leq\int_0^1\tfrac{(-\ln u)^\gamma}{(u+1)^\alpha}(\arctan u^\beta+\arctan u^{-\beta})u^{\alpha-|x|-1}du$$
$$=\tfrac{\pi}{2}\int_0^1(1+u)^{-\alpha}(-\ln u)^\gamma u^{\alpha-|x|-1}du$$
$$<\tfrac{\pi}{2}\int_0^1(-\ln u)^\gamma u^{\alpha-|x|-1}du=\tfrac{\pi\Gamma(\gamma+1)}{2(\alpha-|x|)^{\gamma+1}}<\infty,$$

$$k_0=\tfrac{\pi}{2}\int_0^1\sum_{k=0}^\infty\binom{-\alpha}{k}(-\ln u)^\gamma u^{k+\alpha-1}du$$
$$=\tfrac{\pi}{2}\sum_{k=0}^\infty\binom{-\alpha}{k}\int_0^1(-\ln u)^\gamma u^{k+\alpha-1}du$$
$$=\tfrac{\pi}{2}\Gamma(\gamma+1)\sum_{k=0}^\infty\binom{-\alpha}{k}\tfrac{1}{(k+\alpha)^{\gamma+1}}.$$

As the assumption of Corollary 4.1.1, in view of (4.1.1) and (4.1.2), (1) for $p>1$, we have the following equivalent inequalities with the best constant factors:

$$\int_0^\infty\int_0^\infty(\tfrac{\min\{x,y\}}{x+y})^\alpha\,|\ln\tfrac{x}{y}|^\gamma\arctan(\tfrac{x}{y})^\beta f(x)g(y)dxdy$$
$$<k_0\|f\|_{p,\tilde{\phi}_p}\|g\|_{q,\tilde{\phi}_q}, \qquad (4.1.3)$$

$$\int_0^\infty\tfrac{1}{y}(\int_0^\infty(\tfrac{\min\{x,y\}}{x+y})^\alpha\,|\ln\tfrac{x}{y}|^\gamma\arctan(\tfrac{x}{y})^\beta f(x)dx)^p dy$$
$$<k_0^p\|f\|_{p,\tilde{\phi}_p}^p; \qquad (4.1.4)$$

(2) for $0 < p < 1$, we have the equivalent reverses of (4.1.3) and (4.1.4) with the same best constant factors.

Example 4.1.3 If $\alpha > 0, \beta \in \mathbf{R}$,

$$k_0(x,y) = \frac{(\min\{x,y\})^\alpha}{x^\alpha + y^\alpha} \arctan(\tfrac{x}{y})^\beta,$$

then for $|x| < \alpha$, we find

$$0 < K_0(x) = \int_0^\infty k_0(u,1) u^{x-1} du$$

$$\leq \int_0^1 \frac{u^{\alpha-|x|-1}}{u^\alpha + 1} (\arctan u^\beta + \arctan u^{-\beta}) du$$

$$= \frac{\pi}{2} \int_0^1 \sum_{k=0}^\infty (-1)^k u^{(k+1)\alpha - |x| - 1} du$$

$$= \frac{\pi}{2} \sum_{k=0}^\infty (-1)^k \int_0^1 u^{(k+1)\alpha - |x| - 1} du$$

$$= \frac{\pi}{2} \sum_{k=1}^\infty (-1)^{k-1} \frac{1}{k\alpha - |x|} < \infty,$$

and

$$k_0 = \frac{\pi}{2\alpha} \sum_{k=1}^\infty (-1)^{k-1} \frac{1}{k} = \frac{\pi}{2\alpha} \ln 2.$$

As the assumption of Corollary 4.1.1, in view of (4.1.1) and (4.1.2), (1) for $p > 1$, we have the following equivalent inequalities with the best constant factors:

$$\int_0^\infty \int_0^\infty \frac{(\min\{x,y\})^\alpha}{x^\alpha + y^\alpha} \arctan(\tfrac{x}{y})^\beta f(x) g(y) dx dy$$
$$< \frac{\pi}{2\alpha} \ln 2 \| f \|_{p,\tilde{\phi}_p} \| g \|_{q,\tilde{\phi}_q}, \qquad (4.1.5)$$

$$\int_0^\infty \frac{1}{y} \left(\int_0^\infty \frac{(\min\{x,y\})^\alpha}{x^\alpha + y^\alpha} \arctan(\tfrac{x}{y})^\beta f(x) dx \right)^p dy$$
$$< \left(\frac{\pi}{2\alpha} \ln 2\right)^p \| f \|_{p,\tilde{\phi}_p}^p; \qquad (4.1.6)$$

(2) for $0 < p < 1$, we have the equivalent reverses of (4.1.5) and (4.1.6) with the same best constant factors.

Example 4.1.4 If $\alpha > 0, \beta \in \mathbf{R}$,

$$k_0(x,y) = \left(\frac{\min\{x,y\}}{\max\{x,y\}}\right)^\alpha \arctan(\tfrac{x}{y})^\beta,$$

then for $|x| < \alpha$, we find

$$0 < K_0(x) = \int_0^\infty k_0(u,1) u^{x-1} du$$

$$\leq \int_0^1 u^\alpha (\arctan u^\beta + \arctan u^{-\beta}) u^{-|x|-1} du$$

$$= \frac{\pi}{2} \int_0^1 u^{\alpha - |x| - 1} du = \frac{\pi}{2(\alpha - |x|)} < \infty,$$

$$k_0 = \frac{\pi}{2\alpha}.$$

As the assumption of Corollary 4.1.1, in view of (4.1.1) and (4.1.2), (1) for $p > 1$, we have the following equivalent inequalities with the best constant factors:

$$\int_0^\infty \int_0^\infty \left(\frac{\min\{x,y\}}{\max\{x,y\}}\right)^\alpha \arctan(\tfrac{x}{y})^\beta f(x) g(y) dx dy$$
$$< \frac{\pi}{2\alpha} \| f \|_{p,\tilde{\phi}_p} \| g \|_{q,\tilde{\phi}_q}, \qquad (4.1.7)$$

$$\int_0^\infty \frac{1}{y} \left(\int_0^\infty \left(\frac{\min\{x,y\}}{x+y}\right)^\alpha \arctan(\tfrac{x}{y})^\beta f(x) dx \right)^p dy$$
$$< \left(\frac{\pi}{2\alpha}\right)^p \| f \|_{p,\tilde{\phi}_p}^p; \qquad (4.1.8)$$

(2) for $0 < p < 1$, we have the equivalent reverses of (4.1.7) and (4.1.8) with the same best constant factors.

Example 4.1.5 If $\alpha > 0, \beta \in \mathbf{R}$,

$$k_0(x,y) = \frac{(\min\{x,y\})^\alpha}{x^\alpha - y^\alpha} \ln(\tfrac{x}{y}) \arctan(\tfrac{x}{y})^\beta,$$

then for $|x| < 1/2$, we find

$$0 < K_0(x) = \int_0^\infty k_0(u,1) u^{x-1} du$$

$$\leq \int_0^1 \frac{u^{\alpha-|x|-1} \ln u}{u^\alpha - 1} (\arctan u^\beta + \arctan u^{-\beta}) du$$

$$\leq \frac{\pi}{2\alpha} \int_0^\infty \frac{\ln v}{v-1} v^{-\frac{1}{2}} dv = \frac{\pi^3}{2\alpha} < \infty,$$

$$k_0 = \frac{\pi}{2\alpha} \int_0^1 \frac{\ln v}{v-1} dv = \frac{\pi}{2\alpha} \int_0^1 (-\ln v) \sum_{k=0}^\infty v^k dv$$

$$= \frac{\pi}{2\alpha} \sum_{k=0}^\infty \int_0^1 (-\ln v) v^k dv = \frac{\pi}{2\alpha} \sum_{k=1}^\infty \frac{1}{k^2} = \frac{\pi^3}{12\alpha}.$$

As the assumption of Corollary 4.1.1, in view of (4.1.1) and (4.1.2), (1) for $p > 1$, we have the following equivalent inequalities with the best constant factors:

$$\int_0^\infty \int_0^\infty \frac{(\min\{x,y\})^\alpha}{x^\alpha - y^\alpha} \ln(\tfrac{x}{y}) \arctan(\tfrac{x}{y})^\beta f(x) g(y) dx dy$$
$$< \frac{\pi^3}{12\alpha} \| f \|_{p,\tilde{\phi}_p} \| g \|_{q,\tilde{\phi}_q}, \qquad (4.1.9)$$

$$\int_0^\infty \frac{1}{y} \left[\int_0^\infty \frac{(\min\{x,y\})^\alpha}{x^\alpha - y^\alpha} \ln(\tfrac{x}{y}) \arctan(\tfrac{x}{y})^\beta f(x) dx \right]^p dy$$
$$< \left(\frac{\pi^3}{12\alpha}\right)^p \| f \|_{p,\tilde{\phi}_p}^p; \qquad (4.1.10)$$

(2) for $0 < p < 1$, we have the equivalent reverses of (4.1.7) and (4.1.8) with the same best constant factors.

Example 4.1.6 If $0 < \alpha < 1, \beta \in \mathbf{R}$,

$$k_0(x,y) = \left(\frac{\min\{x,y\}}{|x-y|}\right)^\alpha \arctan(\tfrac{x}{y})^\beta,$$

then for $|x| < \alpha$, we find

$$0 < K_0(x) = \int_0^\infty k_0(u,1) u^{x-1} du$$

$$\leq \int_0^1 \frac{u^{\alpha-|x|-1}}{(1-u)^\alpha} (\arctan u^\beta + \arctan u^{-\beta}) du$$

$$=\frac{\pi}{2}\int_0^1(1-u)^{(1-\alpha)-1}u^{\alpha-|x|-1}du$$
$$=\frac{\pi}{2}B(1-\alpha,\alpha-|x|)<\infty,$$

and

$$k_0=\frac{\pi}{2}B(1-\alpha,\alpha).$$

As the assumption of Corollary 4.1.1, in view of (4.1.1) and (4.1.2), (1) for $p>1$, we have the following equivalent inequalities with the best constant factors:

$$\int_0^\infty\int_0^\infty(\frac{\min\{x,y\}}{|x-y|})^\alpha\arctan(\frac{x}{y})^\beta f(x)g(y)dxdy$$
$$<\frac{\pi}{2}B(1-\alpha,\alpha)\|f\|_{p,\tilde{\phi}_p}\|g\|_{q,\tilde{\phi}_q},\qquad(4.1.11)$$

$$\int_0^\infty\frac{1}{y}(\int_0^\infty(\frac{\min\{x,y\}}{|x-y|})^\alpha\arctan(\frac{y}{x})^\beta f(x)dx)^p dy$$
$$<(\frac{\pi}{2}B(1-\alpha,\alpha))^p\|f\|_{p,\tilde{\phi}_p}^p;\qquad(4.1.12)$$

(2) for $0<p<1$, we have the equivalent reverses of (4.1.11) and (4.1.12) with the best constant factors.

4.2. HILBERT-TYPE INTEGRAL INEQUALITIES WITH THE HOMOGENEOUS KERNELS OF -1-DEGREE

4.2.1. A COROLLARY

For $r=s=2$ in Theorem 2.2.3 and Theorem 2.2.7, we have

Corollary 4.2.1 If $p>0(p\neq1)$, $\lambda>0$, $k_\lambda(x,y)\geq0$ is a homogeneous function of $-\lambda$-degree in $(0,\infty)\times(0,\infty)$,

$$0<k_\lambda:=\int_0^\infty k_\lambda(u,1)u^{\frac{\lambda}{2}-1}du<\infty,$$

$$\phi_\rho(x)=x^{\rho(1-\frac{\lambda}{2})-1}\ (x>0;\rho=p,q),$$

$f(x),g(x)\geq0$, such that

$$0<\|f\|_{p,\phi_p}=\{\int_0^\infty x^{p(1-\frac{\lambda}{2})-1}f^p(x)dx\}^{\frac{1}{p}}<\infty$$

and

$$0<\|g\|_{q,\phi_q}=\{\int_0^\infty x^{q(1-\frac{\lambda}{2})-1}g^q(x)dx\}^{\frac{1}{q}}<\infty,$$

then (1) for $p>1$, we have the following equivalent inequalities:

$$\int_0^\infty\int_0^\infty k_\lambda(x,y)f(x)g(y)dxdy$$
$$<k_\lambda\|f\|_{p,\phi_p}\|g\|_{q,\phi_q},\qquad(4.2.1)$$

$$\int_0^\infty y^{\frac{p\lambda}{2}-1}(\int_0^\infty k_\lambda(x,y)f(x)dx)^p dy$$
$$<k_\lambda^p\|f\|_{p,\phi_p}^p,\qquad(4.2.2)$$

where the constant factors k_λ and k_λ^p are the best possible; (2) for $0<p<1$, if there exists a neighborhood I_λ of $\frac{\lambda}{r}$, such that

$$0<K_\lambda(x)=\int_0^\infty k_\lambda(u,1)u^{x-1}du<\infty,x\in I_\lambda,$$

we have the equivalent reverses of (4.2.1) and (4.2.2) with the same best constant factors.

4.2.2. A RELATION OF HILBERT'S INTEGRAL INEQUALITY AND A H-L-P INTEGRAL INEQUALITY

In the following examples, we use the following inequalities: for $x,y>0$,

$$\min\{x,y\}\leq\frac{2}{x^{-1}+y^{-1}}\leq(xy)^{\frac{1}{2}}\leq\frac{x+y}{2}\leq\max\{x,y\},$$

and omit to write the assumption of f and g in Corollary 4.2.1 and the conclusions that the constant factors are the best possible.

Example 4.2.2 If $\lambda>0,B,C\geq0$, $A>-\min\{B,C\}$,

$$k_\lambda(x,y)=\frac{1}{A\max\{x^\lambda,y^\lambda\}+Bx^\lambda+Cy^\lambda},$$

then we find

$$k_\lambda=\int_0^\infty k_\lambda(u,1)u^{\frac{\lambda}{2}-1}du$$
$$=\int_0^\infty\frac{1}{A\max\{u^\lambda,1\}+Bu^\lambda+C}u^{\frac{\lambda}{2}-1}du$$
$$=\int_0^1\frac{1}{Bu^\lambda+(A+C)}u^{\frac{\lambda}{2}-1}du+\int_1^\infty\frac{1}{(A+B)u^\lambda+C}u^{\frac{\lambda}{2}-1}du.\qquad(4.2.3)$$

(1) For $B,C>0,A>-\min\{B,C\}$, setting $v=\sqrt{\frac{B}{A+C}}u^{\lambda/2}$ in the front integral of (4.2.3) and $v=\sqrt{\frac{C}{A+B}}u^{-\lambda/2}$ in the last integral, we find

$$k_\lambda=\frac{1}{\lambda}[\frac{2}{\sqrt{B(A+C)}}\int_0^{\sqrt{\frac{B}{A+C}}}\frac{1}{1+v^2}dv$$
$$+\frac{2}{\sqrt{C(A+B)}}\int_0^{\sqrt{\frac{C}{A+B}}}\frac{1}{1+v^2}dv]$$
$$=\frac{2}{\lambda}[\frac{1}{\sqrt{B(A+C)}}\arctan\sqrt{\frac{B}{A+C}}$$
$$+\frac{1}{\sqrt{C(A+B)}}\arctan\sqrt{\frac{C}{A+B}}];$$

(2) for $B=0,C>0,A>0$, setting $v=\sqrt{\frac{C}{A}}u^{-\lambda/2}$ in the last integral of (4.2.3), we find

$$k_\lambda=\frac{1}{A+C}\int_0^1u^{\frac{\lambda}{2}-1}du+\frac{2}{\lambda\sqrt{CA}}\int_0^{\sqrt{\frac{C}{A}}}\frac{dv}{1+v^2}$$

$$=\frac{2}{\lambda}(\frac{1}{A+C}+\frac{1}{\sqrt{AC}}\arctan\sqrt{\frac{C}{A}});$$

(3) for $C=0,B>0,A>0$, similar to (2), we find

$$k_{\lambda}=\frac{2}{\lambda}(\frac{1}{A+B}+\frac{1}{\sqrt{AB}}\arctan\sqrt{\frac{B}{A}});$$

(4) for $B=C=0,A>0$, by (4.2.3), we find $k_{\lambda}=\frac{4}{\lambda A}$. Hence it follows

$$k_{\lambda}:=\begin{cases}\frac{2}{\lambda}[\frac{\arctan\sqrt{\frac{B}{A+C}}}{\sqrt{B(A+C)}}+\frac{\arctan\sqrt{\frac{C}{A+B}}}{\sqrt{C(A+B)}}], B,C>0, \\ \qquad A>-\min\{B,C\}, \\ \frac{2}{\lambda}(\frac{1}{A+C}+\frac{1}{\sqrt{AC}}\arctan\sqrt{\frac{C}{A}}), B=0,C,A>0, \\ \frac{2}{\lambda}(\frac{1}{A+B}+\frac{1}{\sqrt{AB}}\arctan\sqrt{\frac{B}{A}}), C=0,B,A>0, \\ \frac{4}{\lambda A}, \qquad B=C=0,A>0.\end{cases} \quad (4.2.4)$$

By Corollary 4.2.1, if $\lambda>0,B,C\ge 0$, $A>-\min\{B,C\}$, then (1) for $p>1$, we have the following equivalent inequalities:

$$\int_0^{\infty}\int_0^{\infty}\frac{f(x)g(y)}{A\max\{x^{\lambda},y^{\lambda}\}+Bx^{\lambda}+Cy^{\lambda}}dxdy < k_{\lambda}\|f\|_{p,\phi_p}\|g\|_{q,\phi_q}, \quad (4.2.5)$$

$$\int_0^{\infty}y^{\frac{p\lambda}{2}-1}(\int_0^{\infty}\frac{f(x)}{A\max\{x^{\lambda},y^{\lambda}\}+Bx^{\lambda}+Cy^{\lambda}}dx)^p dy < k_{\lambda}^p\|f\|_{p,\phi_p}^p; \quad (4.2.6)$$

(2) for $0<p<1$, we have the equivalent reverses of (4.2.5) and (4.2.6).

Note 4.2.3 For $\lambda=1,A=0,B=C=1,p=q=2$ in (4.2.5), it deduces to Hilbert's integral inequality (1.3.8); for $\lambda=1$, $A=1$, $B=C=0,p=q=2$ in (4.2.5), it deduces to H-L-P integral inequality (1.3.9). Hence (4.2.5) is a relation of (1.3.8) and (1.3.9).

Particular case 4.2.4 Since

$$\max\{x,y\}-\tilde{A}|x-y| = (1-2\tilde{A})\max\{x,y\}+\tilde{A}(x+y),$$

setting $B=C=\tilde{A},A=1-2\tilde{A}$ in (4.2.4), we find

$$\tilde{k}_{\lambda}(\tilde{A}):=\begin{cases}\frac{4}{\lambda\sqrt{\tilde{A}(1-\tilde{A})}}\arctan\sqrt{\frac{\tilde{A}}{1-\tilde{A}}}, 0<\tilde{A}<1, \\ \frac{4}{\lambda}, \quad \tilde{A}=0.\end{cases}$$

By (4.2.5) and (4.2.6), if $\lambda>0,0\le\tilde{A}<1$, then (1) for $p>1$, we have the following equivalent inequalities:

$$\int_0^{\infty}\int_0^{\infty}\frac{f(x)g(y)}{\max\{x^{\lambda},y^{\lambda}\}-\tilde{A}|x^{\lambda}-y^{\lambda}|}dxdy < \tilde{k}_{\lambda}(\tilde{A})\|f\|_{p,\phi_p}\|g\|_{q,\phi_q}, \quad (4.2.7)$$

$$\int_0^{\infty}y^{\frac{p\lambda}{2}-1}(\int_0^{\infty}\frac{f(x)}{\max\{x^{\lambda},y^{\lambda}\}-\tilde{A}|x^{\lambda}-y^{\lambda}|}dx)^p dy < \tilde{k}_{\lambda}^p(\tilde{A})\|f\|_{p,\phi_p}^p; \quad (4.2.8)$$

(2) for $0<p<1$, we have the equivalent reverses of (4.2.7) and (4.2.8).

Particular case 4.2.5 Since

$$\max\{x,y\}+\tilde{A}\min\{x,y\} = (1-\tilde{A})\max\{x,y\}+\tilde{A}(x+y),$$

setting $B=C=\tilde{A},A=1-\tilde{A}$ in (4.2.4), we find

$$\tilde{K}_{\lambda}(\tilde{A}):=\begin{cases}\frac{4}{\lambda\sqrt{\tilde{A}}}\arctan\sqrt{\tilde{A}}, \ \tilde{A}>0, \\ \frac{4}{\lambda}, \quad \tilde{A}=0.\end{cases}$$

By (4.2.5) and (4.2.6), if $\lambda>0,\tilde{A}\ge 0$, then (1) for $p>1$, we have the following equivalent inequalities (Li JIA 2007)[1]:

$$\int_0^{\infty}\int_0^{\infty}\frac{f(x)g(y)}{\max\{x^{\lambda},y^{\lambda}\}+\tilde{A}\min\{x^{\lambda},y^{\lambda}\}}dxdy < \tilde{K}_{\lambda}(\tilde{A})\|f\|_{p,\phi_p}\|g\|_{q,\phi_q}, \quad (4.2.9)$$

$$\int_0^{\infty}y^{\frac{p\lambda}{2}-1}(\int_0^{\infty}\frac{f(x)}{\max\{x^{\lambda},y^{\lambda}\}+\tilde{A}\min\{x^{\lambda},y^{\lambda}\}}dx)^p dy < \tilde{K}_{\lambda}^p(\tilde{A})\|f\|_{p,\phi_p}^p, \quad (4.2.10)$$

(2) for $0<p<1$, we have the equivalent reverses of (4.2.9) and (4.2.10).

Particular case 4.2.6 Since

$$x+y+\tilde{A}|x-y| = 2\tilde{A}\max\{x,y\}+(1-\tilde{A})(x+y),$$

setting $B=C=1-\tilde{A},A=2\tilde{A}$ in (4.2.4), we find

$$k'_{\lambda}(\tilde{A}):=\begin{cases}\frac{4}{\lambda\sqrt{1-\tilde{A}^2}}\arctan\sqrt{\frac{1-\tilde{A}}{1+\tilde{A}}}, \ -1<\tilde{A}<1, \\ \frac{2}{\lambda}, \quad \tilde{A}=1.\end{cases}$$

By (4.2.5) and (4.2.6), if $\lambda>0,-1<\tilde{A}\le 1$, then (1) for $p>1$, we have the following equivalent inequalities:

$$\int_0^{\infty}\int_0^{\infty}\frac{f(x)g(y)}{x^{\lambda}+y^{\lambda}+\tilde{A}|x^{\lambda}-y^{\lambda}|}dxdy < k'_{\lambda}(\tilde{A})\|f\|_{p,\phi_p}\|g\|_{q,\phi_q}, \quad (4.2.11)$$

$$\int_0^\infty y^{\frac{p\lambda}{2}-1}\left(\int_0^\infty \frac{f(x)}{x^\lambda+y^\lambda+\tilde{A}|x^\lambda-y^\lambda|}dx\right)^p dy$$
$$< k_\lambda'^p(\tilde{A})\|f\|_{p,\phi_p}^p; \qquad (4.2.12)$$

(2) for $0<p<1$, we have the equivalent reverses of (4.2.11) and (4.2.12).

Particular case 4.2.7 Since

$$x+y+\tilde{A}\min\{x,y\}$$
$$=(1+\tilde{A})(x+y)-\tilde{A}\max\{x,y\},$$

setting $B=C=1+\tilde{A}, A=-\tilde{A}$ in (4.2.4), we have

$$K_\lambda'(\tilde{A}):=\begin{cases}\frac{4}{\lambda\sqrt{1+\tilde{A}}}\arctan\sqrt{1+\tilde{A}}, & \tilde{A}>-1,\\ \frac{4}{\lambda}, & \tilde{A}=-1.\end{cases}$$

By (4.2.5) and (4.2.6), if $\lambda>0, \tilde{A}\ge -1$, then (1) for $p>1$, we have the following equivalent inequalities:

$$\int_0^\infty\int_0^\infty \frac{f(x)g(y)}{x^\lambda+y^\lambda+\tilde{A}\min\{x^\lambda,y^\lambda\}}dxdy$$
$$< K_\lambda'(\tilde{A})\|f\|_{p,\phi_p}\|g\|_{q,\phi_q}, \qquad (4.2.13)$$

$$\int_0^\infty y^{\frac{p\lambda}{2}-1}\left(\int_0^\infty \frac{f(x)}{x^\lambda+y^\lambda+\tilde{A}\min\{x^\lambda,y^\lambda\}}dx\right)^p dy$$
$$< K_\lambda'^p(\tilde{A})\|f\|_{p,\phi_p}^p. \qquad (4.2.14)$$

(2) for $0<p<1$, we have the equivalent reverses of (4.2.13) and (4.2.14).

4.2.3. SOME EXAMPLES

Example 4.2.8 If $\lambda>0, A\ge 0$,

$$k_\lambda(x,y)=\frac{1}{\max\{x^\lambda,y^\lambda\}+A(xy)^{\lambda/2}},$$

the we find

$$k_\lambda(A):=\int_0^\infty k_\lambda(u,1)u^{\frac{\lambda}{2}-1}du$$
$$=\int_0^\infty \frac{1}{\max\{u^\lambda,1\}+Au^{\lambda/2}}u^{\frac{\lambda}{2}-1}du=2\int_0^1\frac{u^{\frac{\lambda}{2}-1}}{1+Au^{\lambda/2}}du$$
$$=\begin{cases}\frac{4}{\lambda A}\ln(1+A), A>0,\\ \frac{4}{\lambda}, \quad A=0.\end{cases}$$

By Corollary 4.2.1, if $\lambda>0, A\ge 0$, then (1) for $p>1$, we have the following equivalent inequalities:

$$\int_0^\infty\int_0^\infty \frac{f(x)g(y)}{\max\{x^\lambda,y^\lambda\}+A(xy)^{\lambda/2}}dxdy$$
$$< k_\lambda(A)\|f\|_{p,\phi_p}\|g\|_{q,\phi_q}, \qquad (4.2.15)$$

$$\int_0^\infty y^{\frac{p\lambda}{2}-1}\left[\int_0^\infty \frac{f(x)}{\max\{x^\lambda,y^\lambda\}+A(xy)^{\lambda/2}}dx\right]^p dy$$
$$< k_\lambda^p(A)\|f\|_{p,\phi_p}^p; \qquad (4.2.16)$$

(2) for $0<p<1$, we have the equivalent reverses of (4.2.15) and (4.2.16).

Example 4.2.9 If $\lambda>0, A>-2$,

$$k_\lambda(x,y)=\frac{1}{x^\lambda+y^\lambda+A(xy)^{\lambda/2}},$$

then we find

$$\tilde{k}_\lambda(A):=\int_0^\infty k_\lambda(u,1)u^{\frac{\lambda}{2}-1}du$$
$$=2\int_0^1\frac{1}{u^\lambda+1+Au^{\lambda/2}}u^{\frac{\lambda}{2}-1}du=\frac{4}{\lambda}\int_0^1\frac{1}{v^2+Av+1}dv$$
$$=\begin{cases}\frac{8}{\lambda\sqrt{4-A^2}}\left(\arctan\sqrt{\frac{2+A}{2-A}}-\arctan\frac{A}{\sqrt{4-A^2}}\right), & |A|<2,\\ \frac{2}{\lambda}, & A=2,\\ \frac{8}{\lambda\sqrt{A^2-4}}\ln\left(\frac{A+\sqrt{A^2-4}}{\sqrt{A+2}+\sqrt{A-2}}\right), & A>2.\end{cases}$$

By Corollary 4.2.1, if $\lambda>0, A>-2$, then (1) for $p>1$, we have the following equivalent inequalities:

$$\int_0^\infty\int_0^\infty \frac{f(x)g(y)}{x^\lambda+y^\lambda+A(xy)^{\lambda/2}}dxdy$$
$$< \tilde{k}_\lambda(A)\|f\|_{p,\phi_p}\|g\|_{q,\phi_q}, \qquad (4.2.17)$$

$$\int_0^\infty y^{\frac{p\lambda}{2}-1}\left[\int_0^\infty \frac{f(x)}{x^\lambda+y^\lambda+A(xy)^{\lambda/2}}dx\right]^p dy$$
$$< \tilde{k}_\lambda^p(A)\|f\|_{p,\phi_p}^p; \qquad (4.2.18)$$

(2) for $0<p<1$, we have the equivalent reverses of (4.1.17) and (4.1.18).

Example 4.2.10 If $\lambda>0, A\ge -1$,

$$k_\lambda(x,y)=\frac{1}{\max\{x^\lambda,y^\lambda\}+A(x^{-\lambda}+y^{-\lambda})^{-1}},$$

then we find

$$k_\lambda'(A):=\int_0^\infty k_\lambda(u,1)u^{\frac{\lambda}{2}-1}du$$
$$=\int_0^\infty \frac{1}{\max\{u^\lambda,1\}+A(u^{-\lambda}+1)^{-1}}u^{\frac{\lambda}{2}-1}du$$
$$=2\int_0^1\frac{(u^\lambda+1)}{(1+A)u^\lambda+1}u^{\frac{\lambda}{2}-1}du=\frac{4}{\lambda}\int_0^1\frac{v^2+1}{(1+A)v^2+1}dv$$
$$=\begin{cases}\frac{4}{\lambda(1+A)}\left(1+\frac{A}{\sqrt{1+A}}\arctan\sqrt{1+A}\right), & A>-1,\\ \frac{16}{3\lambda}, & A=-1.\end{cases}$$

By Corollary 4.2.1, if $\lambda>0, A\ge -1$, then (1) for $p>1$, we have the following equivalent inequalities:

$$\int_0^\infty\int_0^\infty \frac{f(x)g(y)}{\max\{x^\lambda,y^\lambda\}+A(x^{-\lambda}+y^{-\lambda})^{-1}}dxdy$$
$$< k_\lambda'(A)\|f\|_{p,\phi_p}\|g\|_{q,\phi_q}, \qquad (4.2.19)$$

$$\int_0^\infty y^{\frac{p\lambda}{2}-1}\left[\int_0^\infty \frac{f(x)}{\max\{x^\lambda,y^\lambda\}+A(x^{-\lambda}+y^{-\lambda})^{-1}}dx\right]^p dy$$
$$< k_\lambda'^p(A)\| f \|_{p,\phi_p}^p ; \quad (4.2.20)$$

(2) for $0 < p < 1$, we have the equivalent reverses of (4.2.19) and (4.2.20).

Example 4.2.11 If $\lambda > 0, A > -4$,

$$k_\lambda(x,y) = \frac{1}{x^\lambda+y^\lambda+A(x^{-\lambda}+y^{-\lambda})^{-1}},$$

then we find

$$K_\lambda(A) := \int_0^\infty k_\lambda(u,1)u^{\frac{\lambda}{2}-1}du$$
$$= \int_0^\infty \frac{1}{u^\lambda+1+A(u^{-\lambda}+1)^{-1}}u^{\frac{\lambda}{2}-1}du$$
$$= \frac{1}{\lambda}\int_{-\infty}^\infty \frac{v^2+1}{v^4+(2+A)v^2+1}dv.$$

For $A > -4$, since

$$v^4+(2+A)v^2+1$$
$$= v^4+2v^2+1+Av^2$$
$$= (v^2+1)^2-(\sqrt{-A}v)^2$$
$$= (v^2+1+\sqrt{-A}v)(v^2+1-\sqrt{-A}v),$$

we have the following four roots:

$$v_1 = \tfrac{1}{2}(\sqrt{-A}+\sqrt{4+A}i),$$
$$v_2 = \tfrac{1}{2}(\sqrt{-A}-\sqrt{4+A}i),$$
$$v_3 = \tfrac{1}{2}(-\sqrt{-A}+\sqrt{4+A}i),$$
$$v_4 = \tfrac{1}{2}(-\sqrt{-A}-\sqrt{4+A}i).$$

It is obvious that for $A \neq 0$, the above four roots are different imaginary numbers and only v_1 and v_3 are in the upper half plane. Since

$$f(z) = \frac{z^2+1}{z^4+(2+A)z^2+1}$$
$$= \frac{z^2+1}{(z-v_1)(z-v_2)(z-v_3)(z-v_4)},$$

we find

$$\operatorname*{Res}_{z=v_1} f(z) = \frac{v_1^2+1}{(v_1-v_2)(v_1-v_3)(v_1-v_4)}$$
$$= \frac{-A+\sqrt{-A(A+4)}i}{2\sqrt{-A(A+4)}(\sqrt{-A}+\sqrt{A+4}i)i} = \frac{1}{2\sqrt{A+4}i},$$
$$\operatorname*{Res}_{z=v_3} f(z) = \frac{v_3^2+1}{(v_3-v_1)(v_3-v_2)(v_3-v_4)}$$
$$= \frac{A+\sqrt{-A(A+4)}i}{2\sqrt{-A(A+4)}(-\sqrt{-A}+\sqrt{A+4}i)i} = \frac{1}{2\sqrt{A+4}i}.$$

Applying the theorem of obtaining real integral by the residue, we have

$$k_\lambda(A) = \frac{1}{\lambda}\int_{-\infty}^\infty f(v)dv$$
$$= \frac{1}{\lambda}2\pi i[\operatorname*{Res}_{z=v_1} f(z) + \operatorname*{Res}_{z=v_3} f(z)]$$
$$= \frac{2\pi i}{\lambda}\left(\frac{1}{2\sqrt{A+4}i}+\frac{1}{2\sqrt{A+4}i}\right) = \frac{2\pi}{\lambda\sqrt{A+4}}. \quad (4.2.21)$$

We conclude that (4.2.21) is valid for $A = 0$ by obtaining the integral straightway.

By Corollary 4.2.1, if $\lambda > 0, A > -4$, then (1) for $p > 1$, we have the following equivalent inequalities:

$$\int_0^\infty\int_0^\infty \frac{f(x)g(y)}{x^\lambda+y^\lambda+A(x^{-\lambda}+y^{-\lambda})^{-1}}dxdy$$
$$< \frac{2\pi}{\sqrt{A+4}}\| f \|_{p,\phi_p}\| g \|_{q,\phi_q}, \quad (4.2.22)$$
$$\int_0^\infty y^{\frac{p\lambda}{2}-1}\left[\int_0^\infty \frac{f(x)}{x^\lambda+y^\lambda+A(x^{-\lambda}+y^{-\lambda})^{-1}}dx\right]^p dy$$
$$< \left(\frac{2\pi}{\lambda\sqrt{A+4}}\right)^p \| f \|_{p,\phi_p}^p ; \quad (4.2.23)$$

(2) for $0 < p < 1$, we have the equivalent reverses of (4.2.22) and (4.2.23).

Example 4.2.12 If $\lambda > 0$,

$$k_\lambda(x,y) = \frac{(\max\{x,y\})^\lambda}{(x^\lambda+y^\lambda)^2},$$

then setting $v = u^{\lambda/2}$, we find

$$k_\lambda := \int_0^\infty k_\lambda(u,1)u^{\frac{\lambda}{2}-1}du = \int_0^\infty \frac{\max\{u^\lambda,1\}}{(u^\lambda+1)^2}u^{\frac{\lambda}{2}-1}du$$
$$= \frac{2}{\lambda}\int_0^\infty \frac{\max\{v^2,1\}}{(v^2+1)^2}dv = \frac{4}{\lambda}\int_0^1 \frac{dv}{(v^2+1)^2}$$
$$= \frac{4}{\lambda}\int_0^{\pi/4}\cos^2\theta d\theta$$
$$= \frac{4}{\lambda}\cdot\frac{1}{2}[\tfrac{1}{2}\sin 2\theta+\theta]_0^{\pi/4} = \frac{1}{\lambda}(1+\tfrac{\pi}{2}).$$

By Corollary 4.2.1, if $\lambda > 0$, then (1) for $p > 1$, we have the following equivalent inequalities:

$$\int_0^\infty\int_0^\infty \frac{(\max\{x,y\})^\lambda}{(x^\lambda+y^\lambda)^2}f(x)g(y)dxdy$$
$$< \frac{1}{\lambda}(1+\tfrac{\pi}{2})\| f \|_{p,\phi_p}\| g \|_{q,\phi_q}, \quad (4.2.24)$$
$$\int_0^\infty y^{\frac{p\lambda}{2}-1}\left[\int_0^\infty \frac{(\max\{x,y\})^\lambda}{(x^\lambda+y^\lambda)^2}f(x)dx\right]^p dy$$
$$< [\tfrac{1}{\lambda}(1+\tfrac{\pi}{2})]^p \| f \|_{p,\phi_p}^p ; \quad (4.2.25)$$

(2) for $0 < p < 1$, we have the equivalent reverses of (4.2.24) and (4.2.25).

Example 4.2.13 If $\lambda > 0$,

$$k_\lambda(x,y) = \frac{\min\{x^\lambda,y^\lambda\}}{(x^\lambda+y^\lambda)^2},$$

then setting $v = u^{\lambda/2}$, we find

$$\tilde{k}_\lambda := \int_0^\infty k_\lambda(u,1)u^{\frac{\lambda}{2}-1}du = \int_0^\infty \frac{\min\{u^\lambda,1\}}{(u^\lambda+1)^2}u^{\frac{\lambda}{2}-1}du$$
$$= \frac{2}{\lambda}\int_0^\infty \frac{\min\{v^2,1\}}{(v^2+1)^2}dv = \frac{4}{\lambda}\int_1^\infty \frac{dv}{(v^2+1)^2}$$

$$=\frac{4}{\lambda}\int_{\pi/4}^{\pi/2}\cos^2\theta d\theta$$

$$=\frac{4}{\lambda}\cdot\frac{1}{2}[\frac{1}{2}\sin 2\theta+\theta]_{\pi/4}^{\pi/2}=\frac{1}{\lambda}(\frac{\pi}{2}-1).$$

By Corollary 4.2.1, if $\lambda>0$, then (1) for $p>1$, we have the following equivalent inequalities:

$$\int_0^\infty\int_0^\infty\frac{(\min\{x,y\})^\lambda}{(x^\lambda+y^\lambda)^2}f(x)g(y)dxdy$$

$$<\frac{1}{\lambda}(\frac{\pi}{2}-1)\|f\|_{p,\phi_p}\|g\|_{q,\phi_q},\qquad(4.2.26)$$

$$\int_0^\infty y^{\frac{p\lambda}{2}-1}[\int_0^\infty\frac{(\max\{x,y\})^\lambda}{(x^\lambda+y^\lambda)^2}f(x)dx]^p dy$$

$$<[\frac{1}{\lambda}(\frac{\pi}{2}-1)]^p\|f\|_{p,\phi_p}^p;\qquad(4.2.27)$$

(2) for $0<p<1$, we have the equivalent reverses of (4.2.26) and (4.2.27).

Note 4.2.14 Since

$$\frac{1}{x+y}=\frac{\max\{x,y\}}{(x+y)^2}+\frac{\min\{x,y\}}{(x+y)^2},\quad x,y>0,$$

for $\lambda=1,p=q=2$, inequalities (4.2.26) and (4.2.27) reduce to

$$\int_0^\infty\int_0^\infty\frac{\max\{x,y\}}{(x+y)^2}f(x)g(y)dxdy$$

$$<(\frac{\pi}{2}+1)\|f\|_2\|g\|_2,\qquad(4.2.28)$$

$$\int_0^\infty\int_0^\infty\frac{\min\{x,y\}}{(x+y)^2}f(x)g(y)dxdy$$

$$<(\frac{\pi}{2}-1)\|f\|_2\|g\|_2,\qquad(4.2.29)$$

which are interesting decompositions of the following Hilbert's integral inequality

$$\int_0^\infty\int_0^\infty\frac{1}{x+y}f(x)g(y)dxdy$$

$$<\pi\|f\|_2\|g\|_2.\qquad(4.2.30)$$

4.3. HILBERT-TYPE INTEGRAL INEQUALITIES WITH THE HOMOGENEOUS KERNELS OF -2-DEGREE AND -3-DEGREE

4.3.1. HILBERT-TYPE INTEGRAL INEQUALITIES WITH THE HOMOGENEOUS KERNELS OF -2-DEGREE AND EXTENSIONS

Example 4.3.1 If $\lambda>0,C>0,B>-C$,

$$k_{2\lambda}(x,y)=\frac{1}{x^{2\lambda}+2Bx^\lambda y^\lambda+C^2y^{2\lambda}},$$

then setting $v=u^\lambda$, we find

$$k_{2\lambda}:=\int_0^\infty k_{2\lambda}(u,1)u^{\frac{2\lambda}{2}-1}du$$

$$=\int_0^\infty\frac{u^{\lambda-1}}{u^{2\lambda}+2Bu^\lambda+C^2}du$$

$$=\frac{1}{\lambda}\int_0^\infty\frac{1}{v^2+2Bv+C^2}dv$$

$$=\begin{cases}\frac{1}{\lambda\sqrt{C^2-B^2}}(\frac{\pi}{2}-\arctan\frac{B}{\sqrt{C^2-B^2}}), & |B|<C,\\ \frac{1}{\lambda B}, & B=C,\\ \frac{1}{\lambda\sqrt{B^2-C^2}}\ln\frac{B+\sqrt{B^2-C^2}}{C}, & B>C.\end{cases}$$

By Corollary 4.2.1, replacing λ by 2λ, and $\phi_r(x)=x^{r(1-\frac{\lambda}{2})-1}$ by

$$\tilde{\phi}_r(x)=x^{r(1-\lambda)-1}\ (r=p,q),$$

if $\lambda>0,C>0,B>-C$, then (1) for $p>1$, we have the following equivalent inequalities:

$$\int_0^\infty\int_0^\infty\frac{f(x)g(y)}{x^{2\lambda}+2Bx^\lambda y^\lambda+C^2y^{2\lambda}}dxdy$$

$$<k_{2\lambda}\|f\|_{p,\tilde{\phi}_p}\|g\|_{q,\tilde{\phi}_q},\qquad(4.3.1)$$

$$\int_0^\infty y^{p\lambda-1}(\int_0^\infty\frac{f(x)}{x^{2\lambda}+2Bx^\lambda y^\lambda+C^2y^{2\lambda}}dx)^p dy$$

$$<k_{2\lambda}^p\|f\|_{p,\tilde{\phi}_p}^p;$$

(4.3.2)

(2) for $0<p<1$, we have the equivalent reverses of (4.3.1) and (4.3.2). In particular,

(i) for $\lambda=1,B=0,C>0$, $k_2=\frac{\pi}{2C}$, (1) for $p>1$, we have the following equivalent inequalities:

$$\int_0^\infty\int_0^\infty\frac{1}{x^2+C^2y^2}f(x)g(y)dxdy$$

$$<\frac{\pi}{2C}\{\int_0^\infty\frac{f^p(x)}{x}dx\}^{\frac{1}{p}}\{\int_0^\infty\frac{g^q(x)}{x}dx\}^{\frac{1}{q}},\qquad(4.3.3)$$

$$\int_0^\infty y^{p-1}(\int_0^\infty\frac{f(x)}{x^2+C^2y^2}dx)^p dy$$

$$<(\frac{\pi}{2C})^p\int_0^\infty\frac{f^p(x)}{x}dx;\qquad(4.3.4)$$

(2) for $0<p<1$, we have the equivalent reverses of (4.3.3) and (4.3.4).

(ii) For $\lambda=1,B=\frac{\alpha+\beta}{2},C=\sqrt{\alpha\beta}(\alpha>\beta>0)$, $k_2=\frac{\ln(\alpha/\beta)}{\alpha-\beta}$, (1) for $p>1$, we have the following equivalent inequalities:

$$\int_0^\infty\int_0^\infty\frac{f(x)g(y)}{(x+\alpha y)(x+\beta y)}dxdy$$

$$<\frac{\ln(\alpha/\beta)}{\alpha-\beta}\{\int_0^\infty\frac{f^p(x)}{x}dx\}^{\frac{1}{p}}\{\int_0^\infty\frac{g^q(x)}{x}dx\}^{\frac{1}{q}},\qquad(4.3.5)$$

$$\int_0^\infty y^{p-1}[\int_0^\infty\frac{f(x)}{(x+\alpha y)(x+\beta y)}dx]^p dy$$

$$<[\frac{\ln(\alpha/\beta)}{\alpha-\beta}]^p\int_0^\infty\frac{f^p(x)}{x}dx;\qquad(4.3.6)$$

(2) for $0<p<1$, we have the equivalent reverses of (4.3.5) and (4.3.6).

4.3.2. HILBERT-TYPE INTEGRAL INEQUALITIES WITH THE HOMOGENEOUS KERNELS OF -3-DEGREE AND EXTENSIONS

Example 4.3.2 If $\lambda>0, A>0, C>0, B>-C$,

$$k_{3\lambda}(x,y)=\frac{1}{(x^\lambda+Ay^\lambda)(x^{2\lambda}+2Bx^\lambda y^\lambda+C^2y^{2\lambda})},$$

then setting $v=u^{\lambda/2}$, we find

$$k_{3\lambda}:=\int_0^\infty k_{3\lambda}(u,1)u^{\frac{3\lambda}{2}-1}du$$

$$=\int_0^\infty \frac{1}{(u^\lambda+A)(u^{2\lambda}+2Bu^\lambda+C^2)}u^{\frac{3\lambda}{2}-1}du$$

$$=\frac{1}{\lambda}\int_{-\infty}^\infty \frac{v^2}{(v^2+A)(v^4+2Bv^2+C^2)}dv. \tag{4.3.7}$$

In the following, we obtain the integral (4.3.7) by using the theory of residue:

(1) For $-C<B<C$,

$$v^4+2Bv^2+C^2=(v^2+\alpha)(v^2+\beta)$$

(α and β are imaginary numbers), since $\alpha+\beta=2B$ is a real number with $\operatorname{Im}\alpha+\operatorname{Im}\beta=0$, setting

$$\alpha=re^{i\theta}, \beta=re^{-i\theta}\ (r>0, 0<\theta<\pi),$$

and $\sqrt{\alpha}=\sqrt{r}e^{\frac{i\theta}{2}}, \sqrt{\beta}=\sqrt{r}e^{\frac{-i\theta}{2}}$, then we find

$$(v^2+\alpha)(v^2+\beta)$$

$$=(v+i\sqrt{\alpha})(v-i\sqrt{\alpha})(v+i\sqrt{\beta})(v-i\sqrt{\beta}),$$

where $v_1=i\sqrt{\alpha}=\sqrt{r}e^{i\frac{\theta+\pi}{2}}, v_2=i\sqrt{\beta}=\sqrt{r}e^{i\frac{-\theta+\pi}{2}}$ are in the upper half plane. Setting $v_0=i\sqrt{A}$, then applying the theorem of obtaining real integral by the residue, we find

$$k_{3\lambda}=\frac{1}{\lambda}\int_{-\infty}^\infty \frac{v^2}{(v^2+A)(v^4+2Bv^2+C^2)}dv$$

$$=\frac{2\pi}{\lambda}i\sum_{i=0}^{2}\operatorname{Res}_{z=v_i}f(z)$$

$$=\frac{1}{\lambda}2\pi i\Big[\frac{z^2}{2z(z^2+\alpha)(z^2+\beta)}\Big|_{z=i\sqrt{A}}$$

$$+\frac{z^2}{2z(z^2+A)(z^2+\beta)}\Big|_{z=i\sqrt{\alpha}}+\frac{z^2}{2z(z^2+A)(z^2+\alpha)}\Big|_{z=i\sqrt{\beta}}\Big]$$

$$=\frac{1}{\lambda}\pi\Big[\frac{-\sqrt{A}}{(-A+\alpha)(-A+\beta)}$$

$$+\frac{-\sqrt{\alpha}}{(-\alpha+A)(-\alpha+\beta)}+\frac{-\sqrt{\beta}}{(-\beta+A)(-\beta+\alpha)}\Big]$$

$$=\frac{\pi}{\lambda(\sqrt{\alpha}+\sqrt{A})(\sqrt{\beta}+\sqrt{A})(\sqrt{\beta}+\sqrt{\alpha})}. \tag{4.3.8}$$

(2) For $B\ge C$,

$$v^4+2Bv^2+C^2=(v^2+\alpha)(v^2+\beta)$$

(α,β are positive numbers), without lose of generality, suppose $0<A\le\alpha\le\beta$. For $n\in$ N, setting $\tilde{\alpha}=\alpha+\frac{1}{2n}, \tilde{\beta}=\beta+\frac{1}{n}$ and

$$\tilde{f}(z)=\frac{z^2}{(z^2+A)(z^2+\tilde{\alpha})(z^2+\tilde{\beta})}\ (0<A<\tilde{\alpha}<\tilde{\beta}),$$

by using the way of (1), we find

$$\tilde{k}_{3\lambda}=\frac{1}{\lambda}\int_{-\infty}^\infty \tilde{f}(v)dv$$

$$=\frac{\pi}{\lambda(\sqrt{\tilde{\alpha}}+\sqrt{A})(\sqrt{\tilde{\beta}}+\sqrt{A})(\sqrt{\tilde{\beta}}+\sqrt{\tilde{\alpha}})}.$$

For $n\to\infty$ in the above equality, by Levi Theorem, we still have (4.3.8).

Since $u^2+2Bu+C^2=(u+\alpha)(u+\beta)$, we find

$$\alpha+\beta=2B, \alpha\beta=C^2,\ \sqrt{\alpha\beta}=C,$$

$$(\sqrt{\beta}+\sqrt{\alpha})^2=\beta+\alpha+2\sqrt{\alpha\beta}=2(B+C),$$

$$\sqrt{\beta}+\sqrt{\alpha}=\sqrt{2(B+C)},$$

and by (4.3.8), it follows

$$k_{3\lambda}=\frac{\pi}{\lambda[\sqrt{\alpha\beta}+(\sqrt{\beta}+\sqrt{\alpha})\sqrt{A}+A](\sqrt{\beta}+\sqrt{\alpha})}$$

$$=\frac{\pi}{\lambda(C+\sqrt{2(B+C)A}+A)\sqrt{2(B+C)}}. \tag{4.3.9}$$

By Corollary 4.2.1, replacing λ by 3λ, and $\phi_r(x)=x^{r(1-\frac{\lambda}{2})-1}$ by

$$\tilde{\varphi}_r(x)=x^{r(1-\frac{3\lambda}{2})-1}\ (r=p,q),$$

if $\lambda>0, A>0, C>0, B>-C$, then (1) for $p>1$, we have the following equivalent inequalities:

$$\int_0^\infty\int_0^\infty \frac{f(x)g(y)}{(x^\lambda+Ay^\lambda)(x^{2\lambda}+2Bx^\lambda y^\lambda+C^2y^{2\lambda})}dxdy$$

$$<k_{3\lambda}\,||f||_{p,\tilde{\varphi}_p}||g||_{q,\tilde{\varphi}_q}, \tag{4.3.10}$$

$$\int_0^\infty y^{\frac{3p\lambda}{2}-1}\Big[\int_0^\infty \frac{f(x)}{(x^\lambda+Ay^\lambda)(x^{2\lambda}+2Bx^\lambda y^\lambda+C^2y^{2\lambda})}dx\Big]^p dy$$

$$<(k_{3\lambda})^p\,||f||^p_{p,\tilde{\varphi}_p}; \tag{4.3.11}$$

(2) for $0<p<1$, we have the equivalent reverses of (4.3.10) and (4.3.11). In particular,

(i) for $\lambda=1, A>0, C>0, B=0>-C$,

$$k_{3\lambda}=\frac{\pi}{(C+\sqrt{2CA}+A)\sqrt{2C}},$$

(1) if $p>1$, then we have the following equivalent inequalities:

$$\int_0^\infty\int_0^\infty \frac{f(x)g(y)}{(x+Ay)(x^2+C^2y^2)}dxdy<\frac{\pi}{(C+\sqrt{2CA}+A)\sqrt{2C}}$$

$$\times\Big\{\int_0^\infty \frac{f^p(x)}{x^{1+p/2}}dx\Big\}^{\frac{1}{p}}\Big\{\int_0^\infty \frac{g^q(x)}{x^{1+q/2}}dx\Big\}^{\frac{1}{q}}, \tag{4.3.12}$$

$$\int_0^\infty y^{\frac{3p}{2}-1}[\int_0^\infty \frac{f(x)}{(x+Ay)(x^2+C^2y^2)}dx]^p dy$$

$$<[\frac{\pi}{(C+\sqrt{2CA}+A)\sqrt{2C}}]^2 \int_0^\infty \frac{f^p(x)}{x^{1+p/2}}dx; \quad (4.3.13)$$

(2) if $0<p<1$, then we have the equivalent reverses of (4.3.12) and (4.3.13);

(ii)for $\lambda=1, B=\frac{1}{2}(\alpha+\beta), C=\sqrt{\alpha\beta}(\alpha,\beta>0)$,

$$k_3=\frac{\pi}{(\sqrt{\alpha}+\sqrt{A})(\sqrt{\beta}+\sqrt{A})(\sqrt{\beta}+\sqrt{\alpha})},$$

(1) if $p>1$, then we have the following equivalent inequalities:

$$\int_0^\infty\int_0^\infty \frac{f(x)g(y)}{(x+Ay)(x+\alpha y)(x+\beta y)}dxdy$$

$$<k_3\{\int_0^\infty \frac{f^p(x)}{x^{1+p/2}}dx\}^{\frac{1}{p}}\{\int_0^\infty \frac{g^q(x)}{x^{1+q/2}}dx\}^{\frac{1}{q}}, \quad (4.3.14)$$

$$\int_0^\infty y^{\frac{3p}{2}-1}[\int_0^\infty \frac{f(x)}{(x+Ay)(x+\alpha y)(x+\beta y)}dx]^p dy$$

$$<k_3^p\int_0^\infty \frac{f^p(x)}{x^{1+p/2}}dx; \quad (4.3.15)$$

(2) if $0<p<1$, then we have the equivalent reverses of (4.3.14) and (4.3.15).

4.4. SOME HILBERT-TYPE INTEGRAL INEQUALITIES WITH THE HOMOGENEOUS KERNELS OF -4-DEGREE

4.4.1. INTEGRAL INEQUALITIES WITH THE KERNEL $\frac{1}{(x+Ay)(x+By)(x+Cy)(x+Dy)}$

Example 4.4.1 If $\lambda>0, 0<A<B<C<D$,

$$k_{4\lambda}(x,y)=\frac{1}{(x^\lambda+Ay^\lambda)(x^\lambda+By^\lambda)(x^\lambda+Cy^\lambda)(x^\lambda+Dy^\lambda)},$$

then setting $v=u^\lambda$, we find

$$k_{4\lambda}:=\int_0^\infty k_{4\lambda}(u,1)u^{2\lambda-1}du$$

$$=\int_0^\infty \frac{u^{2\lambda-1}}{(u^\lambda+A)(u^\lambda+B)(u^\lambda+C)(u^\lambda+D)}du$$

$$=\frac{1}{\lambda}\int_0^\infty \frac{v}{(v+A)(v+B)(v+C)(v+D)}dv. \quad (4.4.1)$$

Setting

$$\frac{v}{(v+A)(v+B)(v+C)(v+D)}=\frac{\alpha}{v+A}+\frac{\beta}{v+B}+\frac{\chi}{v+C}+\frac{\delta}{v+D},$$

we have

$$v=\alpha(v+B)(v+C)(v+D)$$
$$+\beta(v+A)(v+C)(v+D)$$
$$+\chi(v+A)(v+B)(v+D)$$
$$+\delta(v+A)(v+B)(v+C)$$

For $v=-A,-B,-C,-D$, we find

$$\begin{cases}\alpha=\frac{-A}{(B-A)(C-A)(D-A)}, \beta=\frac{B}{(B-A)(C-B)(D-B)},\\ \chi=\frac{-C}{(C-A)(C-B)(D-C)}, \delta=\frac{D}{(D-A)(D-B)(D-C)}.\end{cases} \quad (4.4.2)$$

In view of $\alpha+\beta+\chi+\delta=0$, we obtain

$$k_{4\lambda}=\frac{1}{\lambda}\int_0^\infty(\frac{\alpha}{v+A}+\frac{\beta}{v+B}+\frac{\chi}{v+C}+\frac{\delta}{v+D})dv$$

$$=\frac{1}{\lambda}\ln(v+A)^\alpha(v+B)^\beta(v+C)^\chi(v+D)^\delta|_0^\infty$$

$$=\frac{1}{\lambda}\ln(A^\alpha B^\beta C^\chi D^\delta)^{-1}$$

$$=\frac{1}{\lambda}[\frac{A\ln A}{(B-A)(C-A)(D-A)}-\frac{B\ln B}{(B-A)(C-B)(D-B)}$$
$$+\frac{C\ln C}{(C-A)(C-B)(D-C)}-\frac{D\ln D}{(D-A)(D-B)(D-C)}].$$

(4.4.3)

By Corollary 4.2.1, replacing λ by 4λ, and $\phi_r(x)=x^{r(1-\frac{\lambda}{2})-1}$ by

$$\tilde{\psi}_r(x)=x^{r(1-2\lambda)-1} \quad (r=p,q),$$

if $\lambda>0, 0<A<B<C<D$, then (1) for $p>1$, we have the following equivalent inequalities:

$$\int_0^\infty\int_0^\infty \frac{f(x)g(y)}{(x^\lambda+Ay^\lambda)(x^\lambda+By^\lambda)(x^\lambda+Cy^\lambda)(x^\lambda+Dy^\lambda)}dxdy$$

$$<k_{4\lambda}\|f\|_{p,\tilde{\psi}_p}\|g\|_{q,\tilde{\psi}_q}, \quad (4.4.4)$$

$$\int_0^\infty y^{2p\lambda-1}[\int_0^\infty \frac{f(x)dx}{(x^\lambda+Ay^\lambda)(x^\lambda+By^\lambda)(x^\lambda+Cy^\lambda)(x^\lambda+Dy^\lambda)}]^p dy$$

$$<(k_{4\lambda})^p\|f\|_{p,\tilde{\psi}_p}^p. \quad (4.4.5)$$

(2) for $0<p<1$, we have the equivalent reverses of (4.4.4) and (4.4.5).

Particular case 4.4.2 For $D=C$, setting $D=C+t$ $(t>0)$ in (4.4.3), we find

$$\lim_{t\to 0^+}(\chi\ln C+\delta\ln D)$$

$$=\lim_{t\to 0^+}[\frac{-C\ln C}{(C-A)(C-B)t}+\frac{(t+C)\ln(t+C)}{(t+C-A)(t+C-B)t}]$$

$$=\lim_{t\to 0^+}\frac{-C\ln C}{(C-A)(C-B)}$$
$$\times[\frac{(t+C-A)(x+C-B)C\ln C-(C-A)(C-B)(t+C)\ln(t+C)}{t(t+C-A)(t+C-B)C\ln C}]$$

$$=\frac{-(C^2-AB)\ln C}{(C-A)^2(C-B)^2}+\frac{1}{(C-A)(C-B)},$$

$$\tilde{k}_{4\lambda}:=\lim_{t\to 0^+}k_{4\lambda}$$

$$=\frac{1}{\lambda}\lim_{t\to 0^+}[\ln(A^\alpha B^\beta)^{-1}-\chi\ln C-\delta\ln D]$$

$$=\frac{1}{\lambda}[\frac{A\ln A}{(B-A)(C-A)^2}-\frac{B\ln B}{(B-A)(C-B)^2}$$

$$+\frac{(C^2-AB)\ln C}{(C-A)^2(C-B)^2}-\frac{1}{(C-A)(C-B)}]. \quad (4.4.6)$$

Setting $D=C+\frac{1}{n}$ in (4.4.1), by Levi theorem, we find

$$\int_0^\infty \frac{u^{2\lambda-1}}{(u^\lambda+A)(u^\lambda+B)(u^\lambda+C)^2}du$$

$$=\lim_{n\to\infty}\int_0^\infty \frac{u^{2\lambda-1}}{(u^\lambda+A)(u^\lambda+B)(u^\lambda+C)(u^\lambda+C+\frac{1}{n})}du=\tilde{k}_{4\lambda}. \quad (4.4.7)$$

If $\lambda>0, 0<A<B<C$, similar to (4.4.4) and (4.5.5), then (1) for $p>1$, we have the following equivalent inequalities (Xie JMAA 2007) [2], (Yang JYU 2008) [3]:

$$\int_0^\infty\int_0^\infty \frac{f(x)g(y)}{(x^\lambda+Ay^\lambda)(x^\lambda+By^\lambda)(x^\lambda+Cy^\lambda)^2}dxdy$$

$$<\tilde{k}_{4\lambda}\|f\|_{p,\tilde{\psi}_p}\|g\|_{q,\tilde{\psi}_q}, \quad (4.4.8)$$

$$\int_0^\infty y^{2p\lambda-1}[\int_0^\infty \frac{f(x)}{(x^\lambda+Ay^\lambda)(x^\lambda+By^\lambda)(x^\lambda+Cy^\lambda)^2}dx]^p dy$$

$$<\tilde{k}_{4\lambda}^p\|f\|_{p,\tilde{\psi}_p}^p; \quad (4.4.9)$$

(2) for $0<p<1$, we have the equivalent reverses of (4.4.8) and (4.4.9).

Particular case 4.4.3 For $D=C=B$, setting $C=B+t\ (t>0)$ in (4.4.6), we find

$$k'_{4\lambda}:=\lim_{t\to0^+}\tilde{k}_{4\lambda}=\frac{1}{\lambda}\lim_{t\to0^+}[\frac{A\ln A}{(B-A)(B+t-A)^2}$$

$$-\frac{B\ln B}{(B-A)t^2}+\frac{[(B+t)^2-AB]\ln(B+t)}{(B+t-A)^2t^2}-\frac{1}{(B+t-A)t}]$$

$$=\frac{1}{\lambda}[\frac{A\ln(A/B)}{(B-A)^3}+\frac{B+A}{2(B-A)^2B}]. \quad (4.4.10)$$

Setting $C=B+\frac{1}{n}$ in (4.4.7), by Levi theorem, we find

$$\int_0^\infty \frac{u^{2\lambda-1}}{(u^\lambda+A)(u^\lambda+B)^3}du$$

$$=\lim_{n\to\infty}\int_0^\infty \frac{u^{2\lambda-1}du}{(u^\lambda+A)(u^\lambda+B)(u^\lambda+B+\frac{1}{n})^2}=k'_{4\lambda}. \quad (4.4.11)$$

If $\lambda>0, 0<A<B$, similar to (4.4.4) and (4.4.5), then (1) for $p>1$, we have the following equivalent inequalities:

$$\int_0^\infty\int_0^\infty \frac{f(x)g(y)}{(x^\lambda+Ay^\lambda)(x^\lambda+By^\lambda)^3}dxdy$$

$$<k'_{4\lambda}\|f\|_{p,\tilde{\psi}_p}\|g\|_{q,\tilde{\psi}_q}, \quad (4.4.12)$$

$$\int_0^\infty y^{2p\lambda-1}[\int_0^\infty \frac{f(x)}{(x^\lambda+Ay^\lambda)(x^\lambda+By^\lambda)^3}dx]^p dy$$

$$<k'^p_{4\lambda}\|f\|_{p,\tilde{\psi}_p}^p; \quad (4.4.13)$$

(2) for $0<p<1$, we have the equivalent reverses of (4.4.12) and (4.4.13).

Particular case 4.4.4 For $D=C, B=A$, setting $B=A+t\ (t>0)$ in (4.4.7), we find

$$K_{4\lambda}:=\lim_{t\to0^+}\tilde{k}_{4\lambda}=\frac{1}{\lambda}\lim_{t\to0^+}[\frac{A\ln A}{t(C-A)^2}$$

$$-\frac{(A+t)\ln(A+t)}{t(C-A-t)^2}+\frac{[C^2-A(A+t)]\ln C}{(C-A)^2(C-A-t)^2}-\frac{1}{(C-A)(C-A-t)}]$$

$$=\frac{1}{\lambda}[\frac{-2A\ln A+(C+A)\ln C}{(C-A)^3}-\frac{2+\ln A}{(C-A)^2}]. \quad (4.4.14)$$

Setting $B=A+\frac{1}{n}$, in (4.4.7), by Levi theorem, we find

$$\int_0^\infty \frac{u^{2\lambda-1}}{(u^\lambda+A)^2(u^\lambda+C)^2}du$$

$$=\lim_{n\to\infty}\int_0^\infty \frac{u^{2\lambda-1}du}{(u^\lambda+A)(u^\lambda+A+\frac{1}{n})(u^\lambda+C)^2}=K_{4\lambda}. \quad (4.4.15)$$

If $\lambda>0, 0<A<C$, similar to (4.4.4) and (4.4.5), then (1) for $p>1$, we have the following equivalent inequalities:

$$\int_0^\infty\int_0^\infty \frac{f(x)g(y)}{(x^\lambda+Ay^\lambda)^2(x^\lambda+Cy^\lambda)^2}dxdy$$

$$<K_{4\lambda}\|f\|_{p,\tilde{\psi}_p}\|g\|_{q,\tilde{\psi}_q}, \quad (4.4.16)$$

$$\int_0^\infty y^{2p\lambda-1}[\int_0^\infty \frac{f(x)}{(x^\lambda+Ay^\lambda)^2(x^\lambda+Cy^\lambda)^2}dx]^p dy$$

$$<K_{4\lambda}^p\|f\|_{p,\tilde{\psi}_p}^p. \quad (4.4.17)$$

(2) for $0<p<1$, we have the equivalent reverses of (4.4.16) and (4.4.17).

Particular case 4.4.5 For $D=C=B=A$, setting $B=A+t\ (t>0)$ in (4.4.10), we find

$$\tilde{K}_{4\lambda}:=\lim_{t\to0^+}k'_{4\lambda}$$

$$=\frac{1}{\lambda}\lim_{t\to0^+}[\frac{A\ln[A/(A+t)]}{t^3}+\frac{2A+t}{2t^2(A+t)}]$$

$$=\frac{1}{\lambda}\lim_{t\to0^+}[\frac{2A(A+t)\ln[A/(A+t)]+(2A+t)t}{2t^3(A+t)}]=\frac{1}{6\lambda A^2}. \quad (4.4.18)$$

Setting $B=A+\frac{1}{n}$ in (4.4.11), by Levi theorem, we have

$$\int_0^\infty \frac{u^{2\lambda-1}du}{(u^\lambda+A)^4}=\lim_{n\to\infty}\int_0^\infty \frac{u^{2\lambda-1}du}{(u^\lambda+A)(u^\lambda+A+\frac{1}{n})^3}=\tilde{K}_{4\lambda}. \quad (4.4.19)$$

If $\lambda>0, A>0$, similar to (4.4.4) and (4.4.5), then (1) for $p>1$, we have the following equivalent inequalities:

$$\int_0^\infty\int_0^\infty \frac{1}{(x^\lambda+Ay^\lambda)^4}f(x)g(y)dxdy$$

$$<\frac{1}{6\lambda A^2}\|f\|_{p,\tilde{\psi}_p}\|g\|_{q,\tilde{\psi}_q}, \quad (4.4.20)$$

$$\int_0^\infty y^{2p\lambda-1}[\int_0^\infty \frac{f(x)}{(x^\lambda+Ay^\lambda)^4}dx]^p dy$$

$$<(\frac{1}{6\lambda A^2})^p\|f\|_{p,\tilde{\psi}_p}^p; \quad (4.4.21)$$

(2) for $0 < p < 1$, we have the equivalent reverses of (4.4.20) and (4.4.21).

4.4.2. INTEGRAL INEQUALITIES WITH THE KERNEL $\frac{1}{(x+Ay)(x+By)(x^2+Cy^2)}$

Example 4.4.6 If $\lambda > 0, 0 < A < B, C > 0$,

$$k_{4\lambda}(x,y) = \frac{1}{(x^\lambda+Ay^\lambda)(x^\lambda+By^\lambda)(x^{2\lambda}+Cy^{2\lambda})},$$

then setting $v = u^\lambda$, we find

$$k_{4\lambda} := \int_0^\infty k_{4\lambda}(u,1)u^{2\lambda-1}du$$

$$= \int_0^\infty \frac{u^{2\lambda-1}}{(u^\lambda+A)(u^\lambda+B)(u^{2\lambda}+C)}du$$

$$= \frac{1}{\lambda}\int_0^\infty \frac{v}{(v+A)(v+B)(v^2+C)}dv. \quad (4.4.22)$$

Setting

$$\frac{v}{(v+A)(v+B)(v^2+C)} = \frac{\alpha}{v+A} + \frac{\beta}{v+B} + \frac{\chi v+\delta}{v^2+C},$$

we have

$$v = \alpha(v+B)(v^2+C) + \beta(v+A)(v^2+C) + (\chi v+\delta)(v+A)(v+B).$$

For $v = -A, -B$, we obtain

$$\begin{cases} \alpha = \frac{-A}{(B-A)(A^2+C)}, \beta = \frac{B}{(B-A)(B^2+C)}, \\ \chi = \frac{AB-C}{(A^2+C)(B^2+C)}, \delta = \frac{(B+A)C}{(A^2+C)(B^2+C)}. \end{cases} \quad (4.4.23)$$

In view of $\alpha + \beta + \chi = 0$, we find

$$k_{4\lambda} := \frac{1}{\lambda}\int_0^\infty [\frac{\alpha}{v+A} + \frac{\beta}{v+B} + \frac{\chi v+\delta}{v^2+C}]dv$$

$$= \frac{1}{\lambda}[\int_0^\infty (\frac{\alpha}{v+A} + \frac{\beta}{v+B} + \frac{\chi v}{v^2+C})dv + \int_0^\infty \frac{\delta}{v^2+C}dv]$$

$$= \frac{1}{\lambda}[\ln(v+A)^\alpha(v+B)^\beta(v^2+C)^{\chi/2}|_0^\infty + \frac{\delta}{\sqrt{C}}\arctan v|_0^\infty]$$

$$= \frac{1}{\lambda}[-\ln(A^\alpha B^\beta C^{\chi/2}) + \frac{\delta\pi}{2\sqrt{C}}]$$

$$= \frac{1}{\lambda}[-\alpha\ln A - \beta\ln B - \frac{\chi}{2}\ln C + \frac{\delta\pi}{2\sqrt{C}}]$$

$$= \frac{1}{\lambda}[\frac{A\ln A}{(B-A)(A^2+C)} - \frac{B\ln B}{(B-A)(B^2+C)} - \frac{(AB-C)\ln C}{2(A^2+C)(B^2+C)} + \frac{(A+B)\sqrt{C}\pi}{2(A^2+C)(B^2+C)}]. \quad (4.4.24)$$

By Corollary 4.2.1, replacing λ by 4λ, and $\phi_r(x) = x^{r(1-\frac{\lambda}{2})-1}$ by

$$\tilde{\psi}_r(x) = x^{r(1-2\lambda)-1} \ (r = p,q),$$

if $\lambda > 0, 0 < A < B, C > 0$, then (1) for $p > 1$, we have the following equivalent inequalities:

$$\int_0^\infty\int_0^\infty \frac{f(x)g(y)}{(x^\lambda+Ay^\lambda)(x^\lambda+By^\lambda)(x^{2\lambda}+Cy^{2\lambda})}dxdy < k_{4\lambda}\|f\|_{p,\tilde{\psi}_p}\|g\|_{q,\tilde{\psi}_q}, \quad (4.4.25)$$

$$\int_0^\infty y^{2p\lambda-1}[\int_0^\infty \frac{f(x)}{(x^\lambda+Ay^\lambda)(x^\lambda+By^\lambda)(x^{2\lambda}+Cy^{2\lambda})}dx]^p dy < (k_{4\lambda})^p\|f\|_{p,\tilde{\psi}_p}^p. \quad (4.4.26)$$

(2) for $0 < p < 1$, we have the equivalent reverses of (4.4.25) and (4.4.26).

Particular case 4.4.7 For $B = A$, setting $B = A + t \ (t > 0)$ in (4.4.24), we find

$$K'_{4\lambda} := \lim_{t\to0^+} k_{4\lambda} = \frac{1}{\lambda}\lim_{t\to0^+}\{\frac{A\ln A}{t(A^2+C)} - \frac{(A+t)\ln(A+t)}{t[(A+t)^2+C]} - \frac{[A(A+t)-C]\ln C}{2(A^2+C)[(A+t)^2+C]} + \frac{(2A+t)\sqrt{C}\pi}{2(A^2+C)[(A+t)^2+C]}\}$$

$$= \frac{1}{\lambda}\lim_{t\to0^+}\{\frac{A\ln A[(A+t)^2+C]-(A^2+C)(A+t)\ln(A+t)}{t(A^2+C)[(A+t)^2+C]} - \frac{[A(A+t)-C]\ln C}{2(A^2+C)[(A+t)^2+C]} + \frac{(2A+t)\sqrt{C}\pi}{2(A^2+C)[(A+t)^2+C]}\}$$

$$= \frac{1}{\lambda(A^2+C)^2}[(A^2-C)(\ln A - \frac{1}{2}\ln C) - A^2 - C + A\sqrt{C}\pi]. \quad (4.4.27)$$

Setting $B = A + \frac{1}{n}$ in (4.4.22), by Levi theorem, we have

$$\int_0^\infty \frac{u^{2\lambda-1}}{(u^\lambda+A)^2(u^{2\lambda}+C)}du = \lim_{n\to\infty}\int_0^\infty \frac{u^{2\lambda-1}}{(u^\lambda+A)(u^\lambda+A+\frac{1}{n})(u^{2\lambda}+C)}du = K'_{4\lambda}.$$

If $\lambda > 0, A, C > 0$, similar to (4.4.25) and (4.4.26), then (1) for $p > 1$, we have the following equivalent inequalities (Wang MTA 2008) [4], (Ge JZU 2009)[5]:

$$\int_0^\infty\int_0^\infty \frac{f(x)g(y)}{(x^\lambda+Ay^\lambda)^2(x^{2\lambda}+Cy^{2\lambda})}dxdy < K'_{4\lambda}\|f\|_{p,\tilde{\psi}_p}\|g\|_{q,\tilde{\psi}_q}, \quad (4.4.28)$$

$$\int_0^\infty y^{2p\lambda-1}[\int_0^\infty \frac{f(x)}{(x^\lambda+Ay^\lambda)^2(x^{2\lambda}+Cy^{2\lambda})}dx]^p dy < K'^p_{4\lambda}\|f\|_{p,\tilde{\psi}_p}^p; \quad (4.4.29)$$

(2) for $0 < p < 1$, we have the equivalent reverses of (4.4.28) and (4.4.29).

4.4.3. INTEGRAL INEQUALITIES WITH THE KERNEL $\frac{1}{(x^2+Ay^2)(x^2+By^2)}$

Example 4.4.8 If $\lambda > 0, 0 < A < B$,

$$k_{4\lambda}(x,y) = \frac{1}{(x^{2\lambda}+Ay^{2\lambda})(x^{2\lambda}+By^{2\lambda})},$$

then setting $v = u^\lambda$, we find

$$k_{4\lambda} = \int_0^\infty k_{4\lambda}(u,1)u^{2\lambda-1}du$$
$$= \int_0^\infty \frac{u^{2\lambda-1}}{(u^{2\lambda}+A)(u^{2\lambda}+B)}du$$
$$= \frac{1}{\lambda}\int_0^\infty \frac{v}{(v^2+A)(v^2+B)}dv$$
$$= \frac{1}{\lambda(B-A)}\int_0^\infty (\frac{v}{v^2+A} - \frac{v}{v^2+B})dv$$
$$= \frac{\ln(B/A)}{2\lambda(B-A)}. \qquad (4.4.30)$$

By Corollary 4.1.1, replacing λ by 4λ, and $\phi_r(x) = x^{r(1-\frac{\lambda}{2})-1}$ by

$$\tilde{\psi}_r(x) = x^{r(1-2\lambda)-1} \quad (r = p,q),$$

if $\lambda > 0, 0 < A < B$, then (1) for $p > 1$, we have the following equivalent inequalities:

$$\int_0^\infty \int_0^\infty \frac{f(x)g(y)}{(x^{2\lambda}+Ay^{2\lambda})(x^{2\lambda}+Bx^{2\lambda})}dxdy$$
$$< \frac{\ln(B/A)}{2\lambda(B-A)} \| f \|_{p,\tilde{\psi}_p} \| g \|_{q,\tilde{\psi}_q}, \qquad (4.4.31)$$
$$\int_0^\infty y^{2p\lambda-1}[\int_0^\infty \frac{f(x)}{(x^{2\lambda}+Ay^{2\lambda})(x^{2\lambda}+Bx^{2\lambda})}dx]^p dy$$
$$< [\frac{\ln(B/A)}{2\lambda(B-A)}]^p \| f \|_{p,\tilde{\psi}_p}^p. \qquad (4.4.32)$$

(2) for $0 < p < 1$, we have the equivalent reverses of (4.4.31) and (4.4.32).

Particular case 4.4.9 For $B = A$, setting $B = A+t \ (t > 0)$ in (4.4.30), we find

$$K'_{4\lambda} := \lim_{t \to 0^+} k_{4\lambda}$$
$$= \lim_{t\to 0^+} \frac{\ln[(A+t)/A]}{2\lambda t} = \frac{1}{2A\lambda}.$$

By Levi Theorem, we obtain

$$\int_0^\infty \frac{u^{2\lambda-1}}{(u^{2\lambda}+A)^2}du$$
$$= \lim_{n\to\infty}\int_0^\infty \frac{u^{2\lambda-1}du}{(u^{2\lambda}+A)(u^{2\lambda}+A+\frac{1}{n})} = K'_{4\lambda}.$$

If $\lambda > 0, A > 0$, similar to (4.4.31) and (4.4.32), (1) for $p > 1$, we have the following equivalent inequalities:

$$\int_0^\infty \int_0^\infty \frac{f(x)g(y)}{(x^{2\lambda}+Ay^{2\lambda})^2}dxdy$$
$$< \frac{1}{2\lambda A} \| f \|_{p,\tilde{\psi}_p} \| g \|_{q,\tilde{\psi}_q}, \qquad (4.4.33)$$
$$\int_0^\infty y^{2p\lambda-1}[\int_0^\infty \frac{f(x)dx}{(x^{2\lambda}+Ay^{2\lambda})^2}]^p dy$$
$$< \frac{1}{(2\lambda A)^p} \| f \|_{p,\tilde{\psi}_p}^p, \qquad (4.4.34)$$

(2) for $0 < p < 1$, we have the equivalent reverses of (4.4.33) and (4.4.34).

4.4.4. INTEGRAL INEQUALITIES WITH THE KERNEL $\frac{1}{(x^2+2Bxy+C^2y^2)^2}$

Example 4.4.10 If $\lambda > 0, C > 0, B > -C$,

$$k_{4\lambda}(x,y) = \frac{1}{(x^{2\lambda}+2Bx^\lambda y^\lambda + C^2y^{2\lambda})^2},$$

then setting $v = u^\lambda$, we find

$$k_{4\lambda} = \int_0^\infty k_{4\lambda}(u,1)u^{2\lambda-1}du$$
$$= \int_0^\infty \frac{u^{2\lambda-1}}{(u^{2\lambda}+2Bu^\lambda+C^2)^2}du$$
$$= \frac{1}{\lambda}\int_0^\infty \frac{v}{(v^2+2Bv+C^2)^2}dv = k_{4\lambda}$$
$$:= \begin{cases} \frac{1}{2\lambda(C^2-B^2)}[1 - \frac{B}{\sqrt{C^2-B^2}}(\frac{\pi}{2} - \arctan\frac{B}{\sqrt{C^2-B^2}})], & |B| < C, \\ \frac{1}{6\lambda B^2}, & B = C, \\ \frac{1}{2\lambda(B^2-C^2)}[1 - \frac{B}{\sqrt{B^2-C^2}}\ln(\frac{C}{B+\sqrt{B^2-C^2}})], & B > C. \end{cases} \qquad (4.4.35)$$

By Corollary 4.2.1, replacing λ by 4λ, and $\phi_r(x) = x^{r(1-\frac{\lambda}{2})-1}$ by

$$\tilde{\psi}_r(x) = x^{r(1-2\lambda)-1} \quad (r = p,q),$$

if $\lambda > 0, C > 0, B > -C$, then (1) for $p > 1$, we have the following equivalent inequalities:

$$\int_0^\infty \int_0^\infty \frac{f(x)g(y)}{(x^{2\lambda}+2Bx^\lambda y^\lambda+C^2y^{2\lambda})^2}dxdy$$
$$< k_{4\lambda} \| f \|_{p,\tilde{\psi}_p} \| g \|_{q,\tilde{\psi}_q}, \qquad (4.4.36)$$
$$\int_0^\infty y^{2p\lambda-1}[\int_0^\infty \frac{f(x)}{(x^{2\lambda}+2Bx^\lambda y^\lambda+C^2y^{2\lambda})^2}dx]^p dy$$
$$< k_{4\lambda}^p \| f \|_{p,\tilde{\psi}_p}^p; \qquad (4.4.37)$$

(2) for $0 < p < 1$, we have the equivalent reverses of (4.4.36) and (4.4.37).

4.5. TWO CLASSES OF NEW HILBERT-TYPE INTEGRAL INEQUA-LITIES

4.5.1. A CLASS OF HILBERT-TYPE INTEGRAL INEQUALITIES WITH THE HOMOGENEOUS KERNELS OF $-\lambda$-DEGREE

We introduce the formulas of the Beta function again as follows:

$$B(u,v) := \int_0^\infty \frac{t^{u-1}}{(1+t)^{u+v}}dt = \int_0^1 (1-t)^{u-1}t^{v-1}dt$$
$$= \int_1^\infty \frac{(t-1)^{u-1}}{t^{u+v}}dt, \quad u, v > 0. \qquad (4.5.1)$$

Lemma 4.5.1 If $r,s>1, 0<\lambda<1$, define the weight function $\omega(r,x)$ $(x\in(-\infty,\infty))$ as

$$\omega_\lambda(r,x):=\int_{-\infty}^{\infty}\frac{1}{|x+y|^\lambda}\cdot\frac{|x|^{\lambda/s}}{|y|^{1-\lambda/r}}dy. \tag{4.5.2}$$

Then for $x\in(-\infty,0)\cup(0,\infty)$, we have

$$\omega_\lambda(r,x)=k_\lambda(r)$$
$$:=B(\tfrac{\lambda}{r},\tfrac{\lambda}{s})+B(1-\lambda,\tfrac{\lambda}{r})+B(1-\lambda,\tfrac{\lambda}{s}). \tag{4.5.3}$$

Proof For $x\in(-\infty,0)$, since for $y<-x$, $x+y<0$; for $y\geq -x$, $x+y\geq 0$, we find

$$\omega(r,x)=\int_{-\infty}^{\infty}\frac{1}{|x+y|^\lambda}\frac{(-x)^{\lambda/s}}{|y|^{1-\lambda/r}}dy$$
$$=\int_{-\infty}^{0}\frac{1}{(-x-y)^\lambda}\frac{(-x)^{\lambda/s}}{(-y)^{1-\lambda/r}}dy$$
$$+\int_{0}^{-x}\frac{1}{(-x-y)^\lambda}\frac{(-x)^{\lambda/s}}{y^{1-\lambda/r}}dy+\int_{-x}^{\infty}\frac{1}{(x+y)^\lambda}\frac{(-x)^{\lambda/s}}{y^{1-\lambda/r}}dy. \tag{4.5.4}$$

Setting $t=\frac{y}{x}$, $t=-\frac{y}{x}$ and $t=-\frac{y}{x}$ respectively in the three integrals of (4.5.4), by (4.5.1), we find

$$\int_{-\infty}^{0}\frac{(-x)^{\lambda/s}dy}{(-x-y)^\lambda(-y)^{1-\lambda/r}}=\int_{\infty}^{0}\frac{(-x)^{\lambda/s}x}{(-x-tx)^\lambda(-tx)^{1-\lambda/r}}dt$$
$$=\int_{0}^{\infty}\frac{t^{\frac{\lambda}{r}-1}}{(1+t)^\lambda}dt=B(\tfrac{\lambda}{r},\tfrac{\lambda}{s}),$$
$$\int_{0}^{-x}\frac{(-x)^{\lambda/s}dy}{(-x-y)^\lambda y^{1-\lambda/r}}=\int_{0}^{1}\frac{(-x)^{\lambda/s}(-x)}{(-x+xt)^\lambda(-xt)^{1-\lambda/r}}dt$$
$$=\int_{0}^{1}(1-t)^{-\lambda}t^{\frac{\lambda}{r}-1}=B(1-\lambda,\tfrac{\lambda}{r}),$$
$$\int_{-x}^{\infty}\frac{(-x)^{\lambda/s}dy}{(x+y)^\lambda y^{1-\lambda/r}}=\int_{1}^{\infty}\frac{(-x)^{\lambda/s}(-x)}{(x-xt)^\lambda(-xt)^{1-\lambda/r}}dt$$
$$=\int_{1}^{\infty}\frac{(t-1)^{(1-\lambda)-1}}{t^{1-\lambda/r}}dt=B(1-\lambda,\tfrac{\lambda}{s}).$$

Then we have (4.5.3).

For $x\in(0,\infty)$, setting $t=-\frac{y}{x}, t=-\frac{y}{x}$ and $t=\frac{y}{x}$ in the three integrals of (4.5.4) respectively, by (4.5.1), we still obtain

$$\omega(r,x)=\int_{-\infty}^{-x}\frac{x^{\lambda/s}}{(-x-y)^\lambda(-y)^{1-\lambda/r}}dy$$
$$+\int_{-x}^{0}\frac{x^{\lambda/s}}{(x+y)^\lambda(-y)^{1-\lambda/r}}dy+\int_{0}^{\infty}\frac{x^{\lambda/s}}{(x+y)^\lambda y^{1-\lambda/r}}dy$$
$$=B(1-\lambda,\tfrac{\lambda}{s})+B(1-\lambda,\tfrac{\lambda}{r})+B(\tfrac{\lambda}{r},\tfrac{\lambda}{s}).$$

Hence (4.5.2) is valid for $x\in(0,\infty)$, and then (4.5.2) is valid for $x\in(-\infty,0)\cup(0,\infty)$. □

Note 4.5.2 It is obvious that

$$\omega_\lambda(s,0)=\omega_\lambda(r,0)=0,$$
$$\omega_\lambda(s,x)=k_\lambda(r)=\omega_\lambda(r,x)\ (x\neq 0).$$

Lemma 4.5.3 If $p>0(p\neq 1), r,s>1$, $0<\lambda<1$ and $0<\varepsilon<\frac{|q|\lambda}{2s}$, define the functions $\tilde{f}(x)$ and $\tilde{g}(x)$ as

$$\tilde{f}(x)=\begin{cases}x^{\frac{\lambda}{r}-1-\frac{2\varepsilon}{p}}, & x\in(1,\infty),\\ 0, & x\in[-1,1],\\ (-x)^{\frac{\lambda}{r}-1-\frac{2\varepsilon}{p}}, & x\in(-\infty,-1);\end{cases}$$

$$\tilde{g}(x)=\begin{cases}x^{\frac{\lambda}{s}-1-\frac{2\varepsilon}{q}}, & x\in(1,\infty),\\ 0, & x\in[-1,1],\\ (-x)^{\frac{\lambda}{s}-1-\frac{2\varepsilon}{q}}, & x\in(-\infty,-1).\end{cases}$$

Then we have

$$h(\varepsilon):=\varepsilon\{\int_{-\infty}^{\infty}|x|^{p(1-\frac{\lambda}{r})-1}\tilde{f}^p(x)dx\}^{\frac{1}{p}}$$
$$\times\{\int_{-\infty}^{\infty}|x|^{q(1-\frac{\lambda}{s})-1}\tilde{g}^q(x)dx\}^{\frac{1}{q}}=1,$$
$$I(\varepsilon):=\varepsilon\int_{-\infty}^{\infty}\int_{-\infty}^{\infty}\frac{1}{|x+y|^\lambda}\tilde{f}(x)\tilde{g}(y)dxdy$$
$$=k_\lambda(r)+o(1)\ (\varepsilon\to 0^+). \tag{4.5.5}$$

Proof We find

$$h(\varepsilon)=\varepsilon\{2\int_{1}^{\infty}x^{p(1-\frac{\lambda}{r})-1}x^{p(\frac{\lambda}{r}-1-\frac{2\varepsilon}{p})}dx\}^{\frac{1}{p}}$$
$$\times\{2\int_{1}^{\infty}x^{q(1-\frac{\lambda}{s})-1}x^{q(\frac{\lambda}{s}-1-\frac{2\varepsilon}{q})}dx\}^{\frac{1}{q}}=1.$$

Setting $y=-Y$, since

$$\tilde{f}(-x)=\tilde{f}(x),\ \tilde{g}(-y)=\tilde{g}(y),$$

we obtain

$$\tilde{f}(-x)\int_{-\infty}^{\infty}\frac{\tilde{g}(y)}{|-x+y|^\lambda}dy$$
$$=\tilde{f}(x)\int_{-\infty}^{\infty}\frac{\tilde{g}(-Y)}{|-x-Y|^\lambda}dY$$
$$=\tilde{f}(x)\int_{-\infty}^{\infty}\frac{\tilde{g}(y)}{|x+y|^\lambda}dy.$$

Hence $\tilde{f}(x)\int_{-\infty}^{\infty}\frac{\tilde{g}(y)}{|x+y|^\lambda}dy$ is an even function with respect to x, and

$$I(\varepsilon)=\varepsilon\int_{-\infty}^{\infty}\tilde{f}(x)[\int_{-\infty}^{\infty}\frac{\tilde{g}(y)}{|x+y|^\lambda}dy]dx$$
$$=2\varepsilon\int_{0}^{\infty}\tilde{f}(x)[\int_{-\infty}^{\infty}\frac{\tilde{g}(y)}{|x+y|^\lambda}dy]dx$$
$$=2\varepsilon\{\int_{1}^{\infty}x^{\frac{\lambda}{r}-1-\frac{2\varepsilon}{p}}[\int_{-\infty}^{-x}\frac{(-y)^{\frac{\lambda}{s}-1-\frac{2\varepsilon}{q}}}{(-x-y)^\lambda}dy]dx$$
$$+\int_{1}^{\infty}x^{\frac{\lambda}{r}-1-\frac{2\varepsilon}{p}}[\int_{-x}^{-1}\frac{(-y)^{\frac{\lambda}{s}-1-\frac{2\varepsilon}{q}}}{(x+y)^\lambda}dy]dx$$
$$+\int_{1}^{\infty}x^{\frac{\lambda}{r}-1-\frac{2\varepsilon}{p}}[\int_{1}^{\infty}\frac{1}{(x+y)^\lambda}y^{\frac{\lambda}{s}-1-\frac{2\varepsilon}{q}}dy]dx\}$$

$$=I_1+I_2+I_3. \qquad (4.5.6)$$

Setting $t=-\frac{y}{x}, t=-\frac{x}{y}$ and $t=\frac{x}{y}$ in I_1, I_2 and I_3, by (4.5.1), we obtain

$$I_1:=2\varepsilon\int_1^\infty x^{\frac{\lambda}{r}-1-\frac{2\varepsilon}{p}}\left[\int_{-\infty}^{-x}\frac{(-y)^{\frac{\lambda}{s}-1-\frac{2\varepsilon}{q}}}{(-x-y)^\lambda}dy\right]dx$$

$$=\int_1^\infty \frac{(t-1)^{(1-\lambda)-1}}{t^{1-\frac{\lambda}{s}+\frac{2\varepsilon}{q}}}dt=B(1-\lambda,\tfrac{\lambda}{r}+\tfrac{2\varepsilon}{q}),$$

$$I_2:=2\varepsilon\int_1^\infty x^{\frac{\lambda}{r}-1-\frac{2\varepsilon}{p}}\left[\int_{-x}^{-1}\frac{(-y)^{\frac{\lambda}{s}-1-\frac{2\varepsilon}{q}}}{(x+y)^\lambda}dy\right]dx$$

$$=2\varepsilon\int_1^\infty x^{-1-2\varepsilon}\left[\int_1^x\frac{t^{\frac{\lambda}{r}-1+\frac{2\varepsilon}{q}}}{(t-1)^\lambda}dt\right]dx$$

$$=2\varepsilon\int_1^\infty\left(\int_t^\infty x^{-1-2\varepsilon}dx\right)\frac{t^{\frac{\lambda}{r}-1+\frac{2\varepsilon}{q}}}{(t-1)^\lambda}dt$$

$$=\int_1^\infty \frac{(t-1)^{(1-\lambda)-1}}{t^{1-\frac{\lambda}{r}+\frac{2\varepsilon}{p}}}dt=B(1-\lambda,\tfrac{\lambda}{s}+\tfrac{2\varepsilon}{p}),$$

$$I_3:=2\varepsilon\int_1^\infty x^{\frac{\lambda}{r}-1-\frac{2\varepsilon}{p}}\left[\int_1^\infty\frac{y^{\frac{\lambda}{s}-1-\frac{2\varepsilon}{q}}}{(x+y)^\lambda}dy\right]dx$$

$$=2\varepsilon\int_1^\infty x^{-1-2\varepsilon}\left[\int_0^x\frac{1}{(1+t)^\lambda}t^{\frac{\lambda}{r}-1+\frac{2\varepsilon}{q}}dt\right]dx$$

$$=\int_0^1\frac{1}{(1+t)^\lambda}t^{\frac{\lambda}{r}-1+\frac{2\varepsilon}{q}}dt$$

$$+2\varepsilon\int_1^\infty x^{-1-2\varepsilon}\left[\int_1^x\frac{1}{(1+t)^\lambda}t^{\frac{\lambda}{r}-1+\frac{2\varepsilon}{q}}dt\right]dx$$

$$=\int_0^1\frac{1}{(1+t)^\lambda}t^{\frac{\lambda}{r}-1+\frac{2\varepsilon}{q}}dt$$

$$+2\varepsilon\int_1^\infty\left(\int_t^\infty x^{-1-2\varepsilon}dx\right)\frac{1}{(1+t)^\lambda}t^{\frac{\lambda}{r}-1+\frac{2\varepsilon}{q}}dt$$

$$=\int_0^1\frac{1}{(1+t)^\lambda}t^{\frac{\lambda}{r}-1+\frac{2\varepsilon}{q}}dt+\int_1^\infty\frac{1}{(1+t)^\lambda}t^{\frac{\lambda}{r}-1-\frac{2\varepsilon}{p}}dt.$$

Since for $p>0(\neq 1)$, we find

$$0\le\int_1^\infty\frac{1}{(1+t)^\lambda}t^{\frac{\lambda}{r}-1}dt-\int_1^\infty\frac{1}{(1+t)^\lambda}t^{\frac{\lambda}{r}-1-\frac{2\varepsilon}{p}}dt$$

$$\le\int_1^\infty\left(t^{\frac{-\lambda}{s}-1}-t^{\frac{-\lambda}{s}-1-\frac{2\varepsilon}{p}}\right)dt=\frac{s}{\lambda}-\frac{1}{\frac{\lambda}{s}+\frac{2\varepsilon}{p}};$$

for $q>1$, we obtain

$$0\le\int_0^1\frac{1}{(1+t)^\lambda}t^{\frac{\lambda}{r}-1}dt-\int_0^1\frac{1}{(1+t)^\lambda}t^{\frac{\lambda}{r}-1+\frac{2\varepsilon}{q}}dt$$

$$\le\int_0^1\left(t^{\frac{\lambda}{r}-1}-t^{\frac{\lambda}{r}-1+\frac{2\varepsilon}{q}}\right)dt=\frac{r}{\lambda}-\frac{1}{\frac{\lambda}{r}+\frac{2\varepsilon}{q}}$$

(for $q<0$, exchange the above two terms). Hence we find

$$I_3=\int_0^1\frac{t^{\frac{\lambda}{r}-1}dt}{(1+t)^\lambda}+o_1(1)+\int_1^\infty\frac{t^{\frac{\lambda}{r}-1}dt}{(1+t)^\lambda}+o_2(1)$$

$$=B(\tfrac{\lambda}{r},\tfrac{\lambda}{s})+o_3(1)\ (\varepsilon\to0^+). \qquad (4.5.7)$$

In view of the above results and (4.5.6), we have (4.5.5).□

Theorem 4.5.4 If $(p,q),(r,s)$ are two pairs of conjugate exponents with $p>1, r,s>1$, $0<\lambda<1$, $f,g\ge0$, such that

$$0<\int_{-\infty}^\infty|x|^{p(1-\frac{\lambda}{r})-1}f^p(x)dx<\infty$$

and

$$0<\int_{-\infty}^\infty|x|^{q(1-\frac{\lambda}{s})-1}g^q(x)dx<\infty,$$

then we have the following equivalent inequalities (Yang JJU 2008) [6]:

$$I:=\int_{-\infty}^\infty\int_{-\infty}^\infty\frac{f(x)g(y)}{|x+y|^\lambda}dxdy$$

$$<k_\lambda(r)\left\{\int_{-\infty}^\infty|x|^{p(1-\frac{\lambda}{r})-1}f^p(x)dx\right\}^{\frac{1}{p}}$$

$$\times\left\{\int_{-\infty}^\infty|x|^{q(1-\frac{\lambda}{s})-1}g^q(x)dx\right\}^{\frac{1}{q}}, \qquad (4.5.8)$$

$$J:=\int_{-\infty}^\infty|y|^{\frac{p\lambda}{s}-1}\left(\int_{-\infty}^\infty\frac{f(x)}{|x+y|^\lambda}dx\right)^p dy<k_\lambda^p(r)$$

$$\times\int_{-\infty}^\infty|x|^{p(1-\frac{\lambda}{r})-1}f^p(x)dx, \qquad (4.5.9)$$

where the constant factors

$$k_\lambda(r)=B(\tfrac{\lambda}{r},\tfrac{\lambda}{s})+B(1-\lambda,\tfrac{\lambda}{r})+B(1-\lambda,\tfrac{\lambda}{s})$$

and $k_\lambda^p(r)$ are the best possible. In particular, for $p=r=2$, we have the following equivalent inequalities:

$$\int_{-\infty}^\infty\int_{-\infty}^\infty\frac{f(x)g(y)}{|x+y|^\lambda}dxdy$$

$$<[B(\tfrac{\lambda}{2},\tfrac{\lambda}{2})+2B(1-\lambda,\tfrac{\lambda}{2})]$$

$$\times\left\{\int_{-\infty}^\infty|x|^{1-\lambda}f^2(x)dx\int_{-\infty}^\infty|x|^{1-\lambda}g^2(x)dx\right\}^{\frac{1}{2}}, \qquad (4.5.10)$$

$$\int_{-\infty}^\infty|y|^{\lambda-1}\left(\int_{-\infty}^\infty\frac{f(x)dx}{|x+y|^\lambda}\right)^2dy$$

$$<[B(\tfrac{\lambda}{2},\tfrac{\lambda}{2})+2B(1-\lambda,\tfrac{\lambda}{2})]^2$$

$$\times\int_{-\infty}^\infty|x|^{1-\lambda}f^2(x)dx. \qquad (4.5.11)$$

Proof By Hölder's inequality with weight, we have

$$I=\int_{-\infty}^\infty\int_{-\infty}^\infty\frac{1}{|x+y|^\lambda}\left[\frac{|x|^{(1-\frac{\lambda}{r})/q}}{|y|^{(1-\frac{\lambda}{s})/p}}f(x)\right]\left[\frac{|y|^{(1-\frac{\lambda}{s})/p}}{|x|^{(1-\frac{\lambda}{r})/q}}g(y)\right]dxdy$$

$$\le\left\{\int_{-\infty}^\infty\int_{-\infty}^\infty\frac{1}{|x+y|^\lambda}\frac{|x|^{(1-\frac{\lambda}{r})(p-1)}}{|y|^{1-\frac{\lambda}{s}}}f^p(x)dxdy\right\}^{\frac{1}{p}}$$

$$\times\left\{\int_{-\infty}^\infty\int_{-\infty}^\infty\frac{1}{|x+y|^\lambda}\frac{|y|^{(1-\frac{\lambda}{s})(q-1)}}{|x|^{1-\frac{\lambda}{r}}}g^q(y)dxdy\right\}^{\frac{1}{q}}. \qquad (4.5.12)$$

We conform that (4.5.12) keeps the strict sign-inequality, otherwise, there exist two constants a and b, which are not all zero satisfying

$$a\frac{|x|^{(1-\lambda/r)(p-1)}}{|y|^{1-\lambda/s}}f^p(x)=b\frac{|y|^{(1-\lambda/s)(q-1)}}{|x|^{1-\lambda/r}}g^q(y)$$

$$a.e. \text{ in } (-\infty,\infty)\times(-\infty,\infty),$$

and then there exists a constant c, such that

$$a|x|^{p(1-\frac{\lambda}{r})} f^p(x) = b|y|^{q(1-\frac{\lambda}{s})} g^q(y) = c$$

$$\text{a.e. in } (-\infty,\infty)\times(-\infty,\infty).$$

Without lose of generality, suppose $a \neq 0$. Then

$$|x|^{p(1-\frac{\lambda}{r})-1} f^p(x) = \frac{c}{a|x|} \text{ a.e. in } (-\infty,\infty),$$

which contradicts the fact that

$$0 < \int_{-\infty}^{\infty} |x|^{p(1-\frac{\lambda}{r})-1} f^p(x)dx < \infty.$$

Hence by (4.5.2), we may rewrite (4.5.12) as

$$I < \{\int_{-\infty}^{\infty} \omega_\lambda(s,x)|x|^{p(1-\frac{\lambda}{r})-1} f^p(x)dx\}^{\frac{1}{p}}$$

$$\times\{\int_{-\infty}^{\infty} \omega_\lambda(r,x)|x|^{q(1-\frac{\lambda}{s})-1} g^q(x)dx\}^{\frac{1}{q}},$$

and by (4.5.3), we have (4.5.8).

If there exists a constant $0 < k \leq k_\lambda(r)$, such that (4.5.8) is still valid as we replace $k_\lambda(r)$ by k, then by Lemma 4.5.3 and (4.5.5), for $0 < \varepsilon < \frac{|q|\lambda}{2s}$, we have

$$k_\lambda(r) + o(1) = I(\varepsilon) < kh(\varepsilon) = k$$

and $k_\lambda(r) \leq k$ $(\varepsilon \to 0^+)$. Hence $k = k_\lambda(r)$ is the best value of (4.5.8).

If $J = 0$, then (4.5.9) is naturally valid; if $J > 0$, for $x \in (-\infty,\infty)$, setting

$$[f(x)]_n = f(x), f(x) \leq n;$$

$$[f(x)]_n = n, f(x) > n,$$

and $E_n = [-n, -\frac{1}{n}] \cup [\frac{1}{n}, n]$, then there exists a $n_0 \in \mathbf{N}$, such that for any $n \geq n_0$,

$$\int_{E_n} |x|^{p(1-\frac{\lambda}{r})-1} [f(x)]_n^p dx > 0$$

and

$$J(n) := \int_{E_n} |y|^{\frac{p\lambda}{s}-1} (\int_{E_n} \frac{[f(x)]_n}{|x+y|^\lambda} dx)^p dy > 0.$$

Setting

$$g_n(y) := |y|^{\frac{p\lambda}{s}-1} (\int_{E_n} \frac{[f(x)]_n}{|x+y|^\lambda} dx)^{p-1} (n \geq n_0; y \in E_n),$$

$$I(n) = \int_{E_n}\int_{E_n} \frac{[f(x)]_n g_n(y)}{|x+y|^\lambda} dxdy,$$

then by (4.5.8), we find

$$0 < \int_{E_n} |y|^{q(1-\frac{\lambda}{s})-1} g_n^q(y)dy = J(n) = I(n)$$

$$< k_\lambda(r)\{\int_{E_n} |x|^{p(1-\frac{\lambda}{r})-1} [f(x)]_n^p dx\}^{\frac{1}{p}}$$

$$\times\{\int_{E_n} |y|^{q(1-\frac{\lambda}{s})-1} g_n^q(y)dy\}^{\frac{1}{q}} < \infty, \quad (4.5.13)$$

$$0 < \int_{E_n} |y|^{q(1-\frac{\lambda}{s})-1} g_n^q(y)dy = J(n)$$

$$< k_\lambda^p(r)\int_{-\infty}^{\infty} |x|^{p(1-\frac{\lambda}{r})-1} f^p(x)dx < \infty. \quad (4.5.14)$$

It follows

$$0 < \int_{-\infty}^{\infty} |x|^{q(1-\frac{\lambda}{s})-1} g_\infty^q(x)dx < \infty$$

and for $n \to \infty$, using (4.5.8), inequalities (4.5.13) and (4.5.14) still keep the strict sign-inequalities. Hence we have (4.5.9). On the other hand, suppose that (4.5.9) is valid. By Hölder's inequality, we find

$$I = \int_{-\infty}^{\infty} (|y|^{\frac{\lambda}{s}-\frac{1}{p}} \int_{-\infty}^{\infty} \frac{f(x)}{|x+y|^\lambda} dx)(|y|^{\frac{-\lambda}{s}+\frac{1}{p}} g(y))dy$$

$$\leq J^{\frac{1}{p}} \{\int_{-\infty}^{\infty} |y|^{q(1-\frac{\lambda}{s})-1} g^q(y)dy\}^{\frac{1}{q}}. \quad (4.5.15)$$

Then by (4.5.9), we have (4.5.8), which is equivalent to (4.5.9).

We conform that the constant factor in (4.5.9) is the best possible, otherwise we can get a contradiction by (4.5.15) that the constant factor in (4.5.8) is not the best possible. □

Theorem 4.5.5 If $(p,q),(r,s)$ are two pairs of conjugate exponents with $0 < p < 1, r, s > 1$, $0 < \lambda < 1, f, g \geq 0$, such that

$$0 < \int_{-\infty}^{\infty} |x|^{p(1-\frac{\lambda}{r})-1} f^p(x)dx < \infty$$

and $0 < \int_{-\infty}^{\infty} |x|^{q(1-\frac{\lambda}{s})-1} g^q(x)dx < \infty$, then we have the following equivalent inequalities:

$$I > k_\lambda(r)\ \{\int_{-\infty}^{\infty} |x|^{p(1-\frac{\lambda}{r})-1} f^p(x)dx\}^{\frac{1}{p}}$$

$$\times\{\int_{-\infty}^{\infty} |x|^{q(1-\frac{\lambda}{s})-1} g^q(x)dx\}^{\frac{1}{q}}, \quad (4.5.16)$$

$$J = \int_{-\infty}^{\infty} |y|^{\frac{p\lambda}{s}-1} (\int_{-\infty}^{\infty} \frac{f(x)}{|x+y|^\lambda} dx)^p dy > k_\lambda^p(r)$$

$$\times\int_{-\infty}^{\infty} |x|^{p(1-\frac{\lambda}{r})-1} f^p(x)dx, \quad (4.5.17)$$

$$L := \int_{-\infty}^{\infty} |x|^{\frac{q\lambda}{r}-1} (\int_{-\infty}^{\infty} \frac{g(y)}{|x+y|^\lambda} dy)^q dx < k_\lambda^q(r)$$

$$\times\int_{-\infty}^{\infty} |y|^{q(1-\frac{\lambda}{s})-1} g^q(y)dy, \quad (4.5.18)$$

where the constant factors

$$k_\lambda(r) = B(\tfrac{\lambda}{r}, \tfrac{\lambda}{s}) + B(1-\lambda, \tfrac{\lambda}{r}) + B(1-\lambda, \tfrac{\lambda}{s})$$

and $k_\lambda^\rho(r) (\rho = p, q)$ are the best possible (I is indicated by (4.5.8)).

Proof By using the reverse Hölder's with weight and the same way of Theorem 4.5.4, we obtain the following reverse inequality:

$$I > \{\int_{-\infty}^{\infty} \omega_\lambda(s,x)|x|^{p(1-\frac{\lambda}{r})-1} f^p(x)dx\}^{\frac{1}{p}}$$
$$\times\{\int_{-\infty}^{\infty} \omega_\lambda(r,x)|x|^{q(1-\frac{\lambda}{s})-1} g^q(x)dx\}^{\frac{1}{q}} \quad .$$

(4.5.19)

Then by (4.5.3), we have (4.5.16).

If there exists a constant $k \geq k_\lambda(r)$, such that (4.5.16) is valid as we replace $k_\lambda(r)$ by k, then in particular, by Lemma 4.5.3 and (4.5.5), for $0 < \varepsilon < \frac{|q|\lambda}{2s}$, we have

$$k_\lambda(r) + o(1) = I(\varepsilon) > k \cdot h(\varepsilon) = k$$

and $k_\lambda(r) \geq k$ $(\varepsilon \to 0^+)$. Hence $k = k_\lambda(r)$ is the best value of (4.5.16).

Since

$$\int_{-\infty}^{\infty} |x|^{p(1-\frac{\lambda}{r})-1} f^p(x)dx > 0,$$

it follows $J > 0$. If $J = \infty$, then (4.5.17) is naturally valid; if $0 < J < \infty$, setting

$$g(y) = |y|^{\frac{p\lambda}{s}-1}(\int_{-\infty}^{\infty} \frac{f(x)}{|x+y|^\lambda}dx)^{p-1}, y \in (-\infty,\infty),$$

then by (4.5.16), we have

$$\infty > \int_{-\infty}^{\infty} |y|^{q(1-\frac{\lambda}{s})-1} g^q(y)dy = J = I$$
$$> k_\lambda(r)\{\int_{-\infty}^{\infty} |x|^{p(1-\frac{\lambda}{r})-1} f^p(x)dx\}^{\frac{1}{p}}$$
$$\times\{\int_{-\infty}^{\infty} |y|^{q(1-\frac{\lambda}{s})-1} g^q(y)dy\}^{\frac{1}{q}} > 0,$$
$$J^{\frac{1}{p}} = \{\int_{-\infty}^{\infty} |y|^{q(1-\frac{\lambda}{s})-1} g^q(y)dy\}^{\frac{1}{p}}$$
$$> k_\lambda(r)\{\int_{-\infty}^{\infty} |x|^{p(1-\frac{\lambda}{r})-1} f^p(x)dx\}^{\frac{1}{p}}.$$

And (4.5.17) is valid. On the other hand, suppose that (4.5.17) is valid. By the reverse Hölder's inequality, we find

$$I = \int_{-\infty}^{\infty} [|y|^{\frac{-1}{p}+\frac{\lambda}{s}} \int_{-\infty}^{\infty} \frac{f(x)}{|x+y|^\lambda}dx][|y|^{\frac{1}{p}-\frac{\lambda}{s}} g(y)]dy$$
$$\geq J^{\frac{1}{p}}\{\int_{-\infty}^{\infty} |y|^{q(1-\frac{\lambda}{s})-1} g^q(y)dy\}^{\frac{1}{q}}. \quad (4.5.20)$$

Then by (4.5.17), we have (4.5.16), which is equivalent to (4.5.17).

If $L = 0$, then (4.5.18) is naturally valid; if $L > 0$, we set $E_n = [-n, -\frac{1}{n}] \cup [\frac{1}{n}, n]$ and for $y \in (-\infty,\infty)$,

$$[g(y)]_n = \begin{cases} \frac{1}{n}, & g(y) < \frac{1}{n}, \\ g(y), & \frac{1}{n} \leq g(y) \leq n, \\ n, & g(y) > n. \end{cases}$$

There exists a $n_0 \in \mathbf{N}$, such that for any $n \geq n_0$,

$$L(n) := \int_{E_n} |x|^{\frac{q\lambda}{r}-1}(\int_{E_n} \frac{[g(y)]_n}{|x+y|^\lambda}dy)^q dx > 0,$$
$$\int_{E_n} |y|^{q(1-\frac{\lambda}{s})-1} [g(y)]_n^q dy > 0.$$

Setting

$$f_n(x) := |x|^{\frac{q\lambda}{r}-1}(\int_{E_n} \frac{[g(y)]_n}{|x+y|^\lambda}dy)^{q-1}$$

and

$$I(n) := \int_{E_n}\int_{E_n} \frac{f_n(x)[g(y)]_n}{|x+y|^\lambda}dxdy \ (n \geq n_0),$$

then by (4.5.16), for $q < 0$, we have

$$\infty > \int_{E_n} |x|^{p(1-\frac{\lambda}{r})-1} f_n^p(x)dx = L(n) = I(n)$$
$$> k_\lambda(r)\{\int_{E_n} |x|^{p(1-\frac{\lambda}{r})-1} f_n^p(x)dx\}^{\frac{1}{p}}$$
$$\times\{\int_{E_n} |x|^{q(1-\frac{\lambda}{s})-1} [g(x)]_n^q dx\}^{\frac{1}{q}} > 0,$$
$$0 < \int_{E_n} |x|^{p(1-\frac{\lambda}{r})-1} f_n^p(x)dx = L(n)$$
$$< k_\lambda^q(r)\int_{-\infty}^{\infty} |y|^{q(1-\frac{\lambda}{s})-1} g^q(y)dy.$$

It follows

$$0 < \int_{-\infty}^{\infty} |x|^{p(1-\frac{\lambda}{r})-1} f_\infty^p(x)dx < \infty$$

and for $n \to \infty$, by (4.5.16), the above two inequalities still keep the strict sign-inequalities. Hence we have (4.5.18). By the reverse Hölder's inequality, we have

$$I = \int_{-\infty}^{\infty} [|x|^{\frac{1}{q}-\frac{\lambda}{r}} f(x)][|x|^{\frac{-1}{q}+\frac{\lambda}{r}} \int_{-\infty}^{\infty} \frac{g(y)}{|x+y|^\lambda}dy]dx$$
$$\geq \{\int_{-\infty}^{\infty} |x|^{p(1-\frac{\lambda}{r})-1} f^p(x)dx\}^{\frac{1}{p}} L^{\frac{1}{q}}. \quad (4.5.21)$$

Hence by (4.8.18), we have (4.5.16), which is equivalent to (4.5.18), and then inequalities (4.5.16), (4.5.17) and (4.5.18) are equivalent.

We conform that the constant factors in (4.5.17) and (4.5.18) are all the best possible, otherwise we can get a contradiction by (4.5.20) or (4.5.21) that the constant factor in (4.5.16) is not the best possible. □

4.5.2. A CLASS OF HILBERT-TYPE INTEGRAL INEQUALITIES WITH THE NON-HOMOGENEOUS KERNELS

Lemma 4.5.6 If $0 < \lambda < 1$, define the weight function $\omega(x)$ $(x \in (-\infty,\infty))$ as

$$\omega_\lambda(x) := \int_{-\infty}^{\infty} \frac{1}{|1+xy|^\lambda} \cdot \frac{|x|^{\lambda/2}}{|y|^{1-\lambda/2}} dy. \qquad (4.5.22)$$

Then for $x \in (-\infty, 0) \cup (0, \infty)$, we have

$$\omega_\lambda(x) = k_\lambda := B(\tfrac{\lambda}{2}, \tfrac{\lambda}{2}) + 2B(1-\lambda, \tfrac{\lambda}{2}). \qquad (4.5.23)$$

Proof (1) For $x \in (-\infty, 0)$, since for $y < -\frac{1}{x}$, $1+xy > 0$; for $y \ge -\frac{1}{x}, 1+xy \le 0$, we find

$$\omega_\lambda(x) = \int_{-\infty}^{\infty} \frac{1}{|1+xy|^\lambda} \frac{(-x)^{\lambda/2}}{|y|^{1-\lambda/2}} dy$$
$$= \int_{-\infty}^{0} \frac{1}{(1+xy)^\lambda} \frac{(-x)^{\lambda/2}}{(-y)^{1-\lambda/2}} dy$$
$$+ \int_{0}^{-1/x} \frac{1}{(1+xy)^\lambda} \frac{(-x)^{\lambda/2}}{y^{1-\lambda/2}} dy$$
$$+ \int_{-1/x}^{\infty} \frac{1}{(-1-xy)^\lambda} \frac{(-x)^{\lambda/2}}{y^{1-\lambda/2}} dy. \qquad (4.5.24)$$

Setting $t = xy$, $t = -xy$ and $t = -xy$ in the three integrals of (4.5.24) respectively, by (4.5.1), we obtain

$$\int_{-\infty}^{0} \frac{1}{(1+xy)^\lambda} \frac{(-x)^{\lambda/2}}{(-y)^{1-\lambda/2}} dy$$
$$= \int_{0}^{\infty} \frac{1}{(1+t)^\lambda} t^{\frac{\lambda}{2}-1} dt = B(\tfrac{\lambda}{2}, \tfrac{\lambda}{2}),$$
$$\int_{0}^{-1/x} \frac{1}{(1+xy)^\lambda} \frac{(-x)^{\lambda/2}}{y^{1-\lambda/2}} dy$$
$$= \int_{0}^{1} (1-t)^{(1-\lambda)-1} t^{\frac{\lambda}{2}-1} dt = B(1-\lambda, \tfrac{\lambda}{2}),$$
$$\int_{-1/x}^{\infty} \frac{1}{(-1-xy)^\lambda} \frac{(-x)^{\lambda/2}}{y^{1-\lambda/2}} dy$$
$$= \int_{1}^{\infty} \frac{(t-1)^{(1-\lambda)-1}}{t^{1-\lambda/2}} dt = B(1-\lambda, \tfrac{\lambda}{2}).$$

Hence we have (4.5.23).

(2) For $x \in (0, \infty)$, setting $t = -xy, t = -xy$ and $t = xy$ in the three integrals of (4.5.24) respectively, we still obtain

$$\omega_\lambda(x) = \int_{-\infty}^{-1/x} \frac{x^{\lambda/2}}{(-1-xy)^\lambda (-y)^{1-\lambda/2}} dy$$
$$+ \int_{-1/x}^{0} \frac{x^{\lambda/2}}{(1+xy)^\lambda (-y)^{1-\lambda/2}} dy + \int_{0}^{\infty} \frac{x^{\lambda/2}}{(1+xy)^\lambda y^{1-\lambda/2}} dy$$
$$= \int_{1}^{\infty} \frac{(t-1)^{(1-\lambda)-1}}{t^{1-\lambda/2}} dt + \int_{0}^{1} (1-t)^{(1-\lambda)-1} t^{\frac{\lambda}{2}-1} dt$$
$$+ \int_{0}^{\infty} \frac{t^{\frac{\lambda}{2}-1}}{(1+t)^\lambda} dt = k_\lambda.$$

Hence (4.5.23) is valid for $x \in (0, \infty)$ and then for $x \in (-\infty, 0) \cup (0, \infty)$, (4.5.23) is still valid. □

Lemma 4.5.7 If $p > 0 (\neq 1)$, $0 < \lambda < 1$, $0 < \varepsilon < \frac{|q|\lambda}{4}$, define the real functions $\tilde{f}(x)$ and $\tilde{g}(x)$ as

$$\tilde{f}(x) = \begin{cases} x^{\frac{\lambda}{2}-1-\frac{2\varepsilon}{p}}, & x \in (1, \infty), \\ 0, & x \in [-1, 1], \\ (-x)^{\frac{\lambda}{2}-1-\frac{2\varepsilon}{p}}, & x \in (-\infty, -1); \end{cases}$$

$$\tilde{g}(x) = \begin{cases} x^{\frac{\lambda}{2}-1+\frac{2\varepsilon}{q}}, & x \in (0, 1), \\ 0, & x \in (-\infty, -1] \cup [1, \infty), \\ (-x)^{\frac{\lambda}{2}-1+\frac{2\varepsilon}{q}}, & x \in (-1, 0). \end{cases}$$

Then we have

$$\tilde{h}(\varepsilon) := \varepsilon \{ \int_{-\infty}^{\infty} |x|^{p(1-\frac{\lambda}{2})-1} \tilde{f}^p(x) dx \}^{\frac{1}{p}}$$
$$\times \{ \int_{-\infty}^{\infty} |x|^{q(1-\frac{\lambda}{2})-1} \tilde{g}^q(x) dx \}^{\frac{1}{q}} = 1,$$
$$\tilde{I}(\varepsilon) := \varepsilon \int_{-\infty}^{\infty} \int_{-\infty}^{\infty} \frac{1}{|1+xy|^\lambda} \tilde{f}(x) \tilde{g}(y) dx dy$$
$$= k_\lambda + \tilde{o}(1) \ (\varepsilon \to 0^+). \qquad (4.5.25)$$

Proof We find

$$\tilde{h}(\varepsilon) = \varepsilon \{ 2 \int_{1}^{\infty} x^{p(1-\frac{\lambda}{2})-1} x^{p(\frac{\lambda}{2}-1-\frac{2\varepsilon}{p})} dx \}^{\frac{1}{p}}$$
$$\times \{ 2 \int_{0}^{1} x^{q(1-\frac{\lambda}{2})-1} x^{q(\frac{\lambda}{2}-1+\frac{2\varepsilon}{q})} dx \}^{\frac{1}{q}} = 1.$$

For $x > 0$, setting $y = -Y$, since $\tilde{f}(-x) = \tilde{f}(x)$ and $\tilde{g}(-y) = \tilde{g}(y)$, then

$$\tilde{f}(-x) \int_{-\infty}^{\infty} \frac{\tilde{g}(y)}{|1+(-x)y|^\lambda} dy$$
$$= \tilde{f}(x) \int_{-\infty}^{\infty} \frac{\tilde{g}(-Y)}{|1+xY|^\lambda} dY$$
$$= \tilde{f}(x) \int_{-\infty}^{\infty} \frac{\tilde{g}(y)}{|1+xy|^\lambda} dy,$$

and $\tilde{f}(x) \int_{-\infty}^{\infty} \frac{\tilde{g}(y)}{|1+xy|^\lambda} dy$ is an even function with respect to x. We find

$$\tilde{I}(\varepsilon) = 2\varepsilon \int_{0}^{\infty} \tilde{f}(x) [\int_{-\infty}^{\infty} \frac{\tilde{g}(y)}{|1+xy|^\lambda} dy] dx$$
$$= 2\varepsilon \{ \int_{1}^{\infty} x^{\frac{\lambda}{2}-1-\frac{2\varepsilon}{p}} [\int_{-1}^{-1/x} \frac{(-y)^{\frac{\lambda}{2}-1+\frac{2\varepsilon}{q}}}{(-1-xy)^\lambda} dy] dx$$
$$+ \int_{1}^{\infty} x^{\frac{\lambda}{2}-1-\frac{2\varepsilon}{p}} [\int_{-1/x}^{0} \frac{(-y)^{\frac{\lambda}{2}-1+\frac{2\varepsilon}{q}}}{(1+xy)^\lambda} dy] dx$$
$$+ \int_{1}^{\infty} x^{\frac{\lambda}{2}-1-\frac{2\varepsilon}{p}} [\int_{0}^{1} \frac{1}{(1+xy)^\lambda} y^{\frac{\lambda}{2}-1+\frac{2\varepsilon}{q}} dy] dx \}$$
$$= I_1 + I_2 + I_3. \qquad (4.5.26)$$

Setting $t = -xy, t = -xy$ and $t = xy$ in I_1, I_2 and I_3 of (4.5.26) respectively, by (4.5.1), we find

$$I_1 := 2\varepsilon \int_{1}^{\infty} x^{\frac{\lambda}{2}-1-\frac{2\varepsilon}{p}} [\int_{-1}^{-1/x} \frac{1}{(-1-xy)^\lambda} (-y)^{\frac{\lambda}{2}-1+\frac{2\varepsilon}{q}} dy] dx$$

$$= 2\varepsilon\int_1^\infty x^{-1-2\varepsilon}\Big[\int_1^x \frac{1}{(t-1)^\lambda} t^{\frac{\lambda}{2}-1+\frac{2\varepsilon}{q}}dt\Big]dx$$

$$= 2\varepsilon\int_1^\infty \Big(\int_t^\infty x^{-1-2\varepsilon}dx\Big)\frac{1}{(t-1)^\lambda} t^{\frac{\lambda}{2}-1+\frac{2\varepsilon}{q}}dt$$

$$= \int_1^\infty \frac{1}{(t-1)^\lambda} t^{\frac{\lambda}{2}-1-\frac{2\varepsilon}{p}}dt$$

$$= \int_1^\infty \frac{1}{t^{(1-\lambda)+(\frac{\lambda}{2}+\frac{2\varepsilon}{p})}}(t-1)^{(1-\lambda)-1}dt$$

$$= B\Big(1-\lambda,\frac{\lambda}{2}+\frac{2\varepsilon}{p}\Big),$$

$$I_2 := 2\varepsilon\int_1^\infty x^{\frac{\lambda}{2}-1-\frac{2\varepsilon}{p}}\Big[\int_{-1/x}^0 \frac{(-y)^{\frac{\lambda}{2}-1+\frac{2\varepsilon}{q}}}{(1+xy)^\lambda}dy\Big]dx$$

$$= 2\varepsilon\int_1^\infty x^{-1-2\varepsilon}\Big[\int_0^1 \frac{t^{\frac{\lambda}{2}-1+\frac{2\varepsilon}{q}}}{(1-t)^\lambda}dt\Big]dx$$

$$= \int_0^1 \frac{t^{\frac{\lambda}{2}-1+\frac{2\varepsilon}{q}}}{(1-t)^\lambda}dt = B\Big(1-\lambda,\frac{\lambda}{2}+\frac{2\varepsilon}{q}\Big);$$

$$I_3 := 2\varepsilon\int_1^\infty x^{\frac{\lambda}{2}-1-\frac{2\varepsilon}{p}}\Big[\int_0^1 \frac{y^{\frac{\lambda}{2}-1+\frac{2\varepsilon}{q}}}{(1+xy)^\lambda}dy\Big]dx$$

$$= 2\varepsilon\int_1^\infty x^{-1-2\varepsilon}\Big[\int_0^x \frac{t^{\frac{\lambda}{2}-1+\frac{2\varepsilon}{q}}}{(1+t)^\lambda}dt\Big]dx$$

$$= 2\varepsilon\int_1^\infty x^{-1-2\varepsilon}\Big[\int_0^1 \frac{t^{\frac{\lambda}{2}-1+\frac{2\varepsilon}{q}}}{(1+t)^\lambda}dt\Big]dx$$

$$+ 2\varepsilon\int_1^\infty x^{-1-2\varepsilon}\Big[\int_1^x \frac{t^{\frac{\lambda}{2}-1+\frac{2\varepsilon}{q}}}{(1+t)^\lambda}dt\Big]dx$$

$$= \int_0^1 \frac{t^{\frac{\lambda}{2}-1+\frac{2\varepsilon}{q}}}{(1+t)^\lambda}dt$$

$$+ 2\varepsilon\int_1^\infty \Big(\int_t^\infty x^{-1-2\varepsilon}dx\Big)\frac{t^{\frac{\lambda}{2}-1+\frac{2\varepsilon}{q}}}{(1+t)^\lambda}dt$$

$$= \int_0^1 \frac{t^{\frac{\lambda}{2}-1+\frac{2\varepsilon}{q}}}{(1+t)^\lambda}dt + \int_1^\infty \frac{t^{\frac{\lambda}{2}-1-\frac{2\varepsilon}{p}}}{(1+t)^\lambda}dt.$$

Similar to the way of obtaining (4.5.7), we find

$$I_3 = \int_0^1 \frac{t^{\frac{\lambda}{2}-1+\frac{2\varepsilon}{q}}}{(1+t)^\lambda}dt + \int_1^\infty \frac{t^{\frac{\lambda}{2}-1-\frac{2\varepsilon}{p}}}{(1+t)^\lambda}dt$$

$$= B\Big(\frac{\lambda}{2},\frac{\lambda}{2}\Big) + o(1)\ (\varepsilon \to 0^+).$$

By the above results and (4.5.26), we have (4.5.25).

Theorem 4.5.8 If (p,q) is one pair of conjugate exponents with $p>1, 0<\lambda<1$, $f,g\ge 0$, such that

$$0<\int_{-\infty}^\infty |x|^{p(1-\frac{\lambda}{2})-1} f^p(x)dx<\infty$$

and $0<\int_{-\infty}^\infty |x|^{q(1-\frac{\lambda}{2})-1} g^q(x)dx<\infty$, then we have the following equivalent inequalities (Yang JXU 2009) [7]:

$$\tilde I := \int_{-\infty}^\infty\int_{-\infty}^\infty \frac{f(x)g(y)}{|1+xy|^\lambda}dxdy$$

$$< k_\lambda \Big\{\int_{-\infty}^\infty |x|^{p(1-\frac{\lambda}{2})-1} f^p(x)dx\Big\}^{\frac{1}{p}}$$

$$\times\Big\{\int_{-\infty}^\infty |x|^{q(1-\frac{\lambda}{2})-1} g^q(x)dx\Big\}^{\frac{1}{q}}, \qquad (4.5.27)$$

$$\tilde J := \int_{-\infty}^\infty |y|^{\frac{p\lambda}{2}-1}\Big(\int_{-\infty}^\infty \frac{f(x)}{|1+xy|^\lambda}dx\Big)^p dy < k_\lambda^p$$

$$\times\int_{-\infty}^\infty |x|^{p(1-\frac{\lambda}{2})-1} f^p(x)dx, \qquad (4.5.28)$$

where the constant factors

$$k_\lambda = B\Big(\frac{\lambda}{2},\frac{\lambda}{2}\Big)+2B\Big(1-\lambda,\frac{\lambda}{2}\Big)$$

and k_λ^p are the best possible.

Proof By Hölder's inequality, we find

$$\tilde I = \int_{-\infty}^\infty\int_{-\infty}^\infty \frac{1}{|1+xy|^\lambda}\Big[\frac{|x|^{(1-\lambda/2)/q}}{|y|^{(1-\lambda/2)/p}} f(x)\Big]$$

$$\times\Big[\frac{|y|^{(1-\lambda/2)/p}}{|x|^{(1-\lambda/2)/q}} g(y)\Big]dxdy$$

$$\le \Big\{\int_{-\infty}^\infty\int_{-\infty}^\infty \frac{1}{|1+xy|^\lambda}\frac{|x|^{(1-\frac{\lambda}{2})(p-1)}}{|y|^{1-\lambda/2}} f^p(x)dxdy\Big\}^{\frac{1}{p}}$$

$$\times\Big\{\int_{-\infty}^\infty\int_{-\infty}^\infty \frac{1}{|1+xy|^\lambda}\frac{|y|^{(1-\frac{\lambda}{2})(q-1)}}{|x|^{1-\lambda/2}} g^q(y)dxdy\Big\}^{\frac{1}{q}}. \qquad (4.4.29)$$

We conform that (4.5.29) keeps the strict sign-inequality, otherwise, there exist constants a and b, which are not all zero and satisfying

$$a\frac{|x|^{(1-\lambda/2)(p-1)}}{|y|^{1-\lambda/2}} f^p(x) = b\frac{|y|^{(1-\lambda/2)(q-1)}}{|x|^{1-\lambda/2}} g^q(y)$$

a.e. in $(-\infty,\infty)\times(-\infty,\infty)$,

and there exists a constant c, such that

$$a|x|^{p(1-\frac{\lambda}{2})} f^p(x) = b|y|^{q(1-\frac{\lambda}{2})} g^q(y) = c$$

a.e. in $(-\infty,\infty)\times(-\infty,\infty)$.

Without lose of generality, suppose $a \ne 0$. Then

$$|x|^{p(1-\frac{\lambda}{2})-1} f^p(x) = \frac{c}{a|x|} \text{ a.e. in } (-\infty,\infty),$$

which contradicts the fact that

$$0<\int_{-\infty}^\infty |x|^{p(1-\frac{\lambda}{2})-1} f^p(x)dx<\infty.$$

Hence by (4.5.22), we may rewrite (4.5.29) as

$$\tilde I < \Big\{\int_{-\infty}^\infty \omega_\lambda(x)|x|^{p(1-\frac{\lambda}{2})-1} f^p(x)dx\Big\}^{\frac{1}{p}}$$

$$\times\Big\{\int_{-\infty}^\infty \omega_\lambda(x)|x|^{q(1-\frac{\lambda}{2})-1} g^q(x)dx\Big\}^{\frac{1}{q}}. \qquad (4.5.30)$$

By (4.5.23), we have (4.5.27).

If there exists a positive constant $k(\le k_\lambda)$, such that (4.5.26) is still valid as we replace k_λ by k, then in particular, by Lemma 4.5.7 and (4.5.25), for $0<\varepsilon<\frac{q\lambda}{4}$, we have

$$k_\lambda + \tilde o(1) = \tilde I(\varepsilon) < k\cdot \tilde h(\varepsilon) = k,$$

and then $k_\lambda \le k$ $(\varepsilon \to 0^+)$. Hence $k = k_\lambda$ is the best value of (4.5.27).

If $\tilde{J}=0$, then (4.5.28) is naturally valid; if $\tilde{J}>0$, for $x\in(-\infty,\infty)$, we set

$$[f(x)]_n = f(x), f(x)\le n;$$
$$[f(x)]_n = n, f(x) > n$$

and $E_n=[-n,-\frac{1}{n}]\cup[\frac{1}{n},n]$. Then there exists a $n_0\in\mathbf{N}$, such that for any $n\ge n_0$,

$$\int_{E_n}|x|^{p(1-\frac{\lambda}{2})-1}[f(x)]_n^p dx>0$$

and

$$\tilde{J}(n):=\int_{E_n}|y|^{\frac{p\lambda}{2}-1}\left(\int_{E_n}\frac{[f(x)]_n}{|1+xy|^{\lambda}}dx\right)^p dy>0.$$

Setting

$$g_n(y):=|y|^{\frac{p\lambda}{2}-1}\left(\int_{E_n}\frac{[f(x)]_n}{|1+xy|^{\lambda}}dx\right)^{p-1},$$
$$n\ge n_0, y\in E_n$$

and

$$\tilde{I}(n)=\int_{E_n}\int_{E_n}\frac{[f(x)]_n g_n(y)}{|1+xy|^{\lambda}}dxdy,$$

then by (4.5.27), we obtain

$$0<\int_{E_n}|y|^{q(1-\frac{\lambda}{2})-1}g_n^q(y)dy=\tilde{J}(n)=\tilde{I}(n)$$
$$<k_\lambda\{\int_{E_n}|x|^{p(1-\frac{\lambda}{2})-1}[f(x)]_n^p dx\}^{\frac{1}{p}}$$
$$\times\{\int_{E_n}|y|^{q(1-\frac{\lambda}{2})-1}g_n^q(y)dy\}^{\frac{1}{q}}<\infty, \quad (4.5.31)$$

$$0<\int_{E_n}|y|^{q(1-\frac{\lambda}{2})-1}g_n^q(y)dy$$
$$<k_\lambda^p\int_{-\infty}^{\infty}|x|^{p(1-\frac{\lambda}{2})-1}f^p(x)dx<\infty. \quad (4.5.32)$$

It follows $0<\int_{-\infty}^{\infty}|y|^{q(1-\frac{\lambda}{2})-1}g_\infty^q(y)dy<\infty$ and for $n\to\infty$, using (4.5.27), both (4.5.31) and (4.5.32) still keep the strict sign-inequalities, and we have (4.5.28). On the other hand, suppose (4.5.28) is valid. By Hölder's inequality, we find

$$\tilde{I}=\int_{-\infty}^{\infty}(|y|^{\frac{\lambda}{2}-\frac{1}{p}}\int_{-\infty}^{\infty}\frac{f(x)}{|1+xy|^{\lambda}}dx)(|y|^{\frac{-\lambda}{2}+\frac{1}{p}}g(y))dy$$
$$\le\tilde{J}^{\frac{1}{p}}\{\int_{-\infty}^{\infty}|y|^{q(1-\frac{\lambda}{2})-1}g^q(y)dy\}^{\frac{1}{q}}. \quad (4.5.33)$$

Then by (4.5.28), we have (4.5.27), which is equivalent to (4.5.28).

We conform that the constant factor in (4.5.28) is the best possible, otherwise we can get a contradiction by (4.5.33) that the constant factor in (4.5.27) is not the best possible.

Theorem 4.5.9 If (p,q) is one pair of conjugate exponents with $0<p<1,\ 0<\lambda<1,\ f,g\ge0$, such that

$$0<\int_{-\infty}^{\infty}|x|^{p(1-\frac{\lambda}{2})-1}f^p(x)dx<\infty$$

and

$$0<\int_{-\infty}^{\infty}|x|^{q(1-\frac{\lambda}{2})-1}g^q(x)dx<\infty,$$

then we have the following equivalent inequalities:

$$\tilde{I}>k_\lambda\{\int_{-\infty}^{\infty}|x|^{p(1-\frac{\lambda}{2})-1}f^p(x)dx\}^{\frac{1}{p}}$$
$$\times\{\int_{-\infty}^{\infty}|x|^{q(1-\frac{\lambda}{2})-1}g^q(x)dx\}^{\frac{1}{q}}, \quad (4.5.34)$$

$$\tilde{J}=\int_{-\infty}^{\infty}|y|^{\frac{p\lambda}{2}-1}\left(\int_{-\infty}^{\infty}\frac{f(x)}{|1+xy|^{\lambda}}dx\right)^p dy>k_\lambda^p$$
$$\times\int_{-\infty}^{\infty}|x|^{p(1-\frac{\lambda}{r})-1}f^p(x)dx, \quad (4.5.35)$$

$$\tilde{L}:=\int_{-\infty}^{\infty}|x|^{\frac{q\lambda}{2}-1}\left(\int_{-\infty}^{\infty}\frac{g(y)}{|1+xy|^{\lambda}}dy\right)^q dx<k_\lambda^q$$
$$\times\int_{-\infty}^{\infty}|y|^{q(1-\frac{\lambda}{2})-1}g^q(y)dy, \quad (4.5.36)$$

where the constant factor

$$k_\lambda=B(\tfrac{\lambda}{2},\tfrac{\lambda}{2})+2B(1-\lambda,\tfrac{\lambda}{2})$$

and $k_\lambda^\rho\ (\rho=p,q)$ are the best possible ($\tilde{I}$ is indicated by (4.5.27)).

Proof Similar to the way of Theorem 4.5.6, we obtain the following inequality

$$\tilde{I}>\{\int_{-\infty}^{\infty}\omega_\lambda(x)|x|^{p(1-\frac{\lambda}{2})-1}f^p(x)dx\}^{\frac{1}{p}}$$
$$\times\{\int_{-\infty}^{\infty}\omega_\lambda(x)|x|^{q(1-\frac{\lambda}{2})-1}g^q(x)dx\}^{\frac{1}{q}}. \quad (4.5.37)$$

Then by (4.5.23), we have (4.5.34).

If there exists a constant $k\ge k_\lambda$, such that (4.5.34) is still valid as we replace k_λ by k, then in particular, by Lemma 4.5.7 and (4.5.25), for $0<\varepsilon<\frac{|q|\lambda}{4}$, we find

$$k_\lambda+\tilde{o}(1)=\tilde{I}(\varepsilon)>k\tilde{h}(\varepsilon)=k$$

and $k_\lambda\ge k(\varepsilon\to0^+)$. Hence $k=k_\lambda$ is the best value of (4.5.34).

Since

$$\int_{-\infty}^{\infty}|x|^{p(1-\frac{\lambda}{r})-1}f^p(x)dx>0,$$

it follows $\tilde{J}>0$. If $\tilde{J}=\infty$, then (4.5.35) is naturally valid; if $0<\tilde{J}<\infty$, setting

$$g(y)=|y|^{\frac{p\lambda}{s}-1}\left(\int_{-\infty}^{\infty}\frac{f(x)}{|1+xy|^{\lambda}}dx\right)^{p-1}, y\in(-\infty,\infty),$$

then by (4.5.34), we find

$$\infty>\int_{-\infty}^{\infty}|y|^{q(1-\frac{\lambda}{2})-1}g^q(y)dy=\tilde{J}=\tilde{I}$$

$$> k_\lambda\{\int_{-\infty}^{\infty}|x|^{p(1-\frac{\lambda}{2})-1} f^p(x)dx\}^{\frac{1}{p}}$$

$$\times\{\int_{-\infty}^{\infty}|x|^{q(1-\frac{\lambda}{2})-1} g^q(x)dx\}^{\frac{1}{q}} > 0,$$

$$\tilde{J}^{\frac{1}{p}} = \{\int_{-\infty}^{\infty}|y|^{q(1-\frac{\lambda}{2})-1} g^q(y)dy\}^{\frac{1}{p}}$$

$$> k_\lambda\{\int_{-\infty}^{\infty}|x|^{p(1-\frac{\lambda}{2})-1} f^p(x)dx\}^{\frac{1}{p}}.$$

Hence (4.5.35) is valid. On the other hand, suppose (4.5.35) is valid. By the reverse H$\ddot{o}$ lder's inequality, we obtain

$$\tilde{I} = \int_{-\infty}^{\infty}[|y|^{\frac{-1}{p}+\frac{\lambda}{2}}\int_{-\infty}^{\infty}\frac{f(x)}{|1+xy|^\lambda}dx][|y|^{\frac{1}{p}-\frac{\lambda}{2}} g(y)]dy$$

$$\geq \tilde{J}^{\frac{1}{p}}\{\int_{-\infty}^{\infty}|y|^{q(1-\frac{\lambda}{2})-1} g^q(y)dy\}^{\frac{1}{q}}. \quad (7.5.38)$$

Then by (4.5.35), we have (4.5.34), which is equivalent to (4.5.35).

If $\tilde{L} = 0$, then (4.5.36) is naturally valid; if $\tilde{L} > 0$, we set $E_n = [-n, -\frac{1}{n}] \cup [\frac{1}{n}, n]$, and for $y \in (-\infty, \infty)$,

$$[g(y)]_n = \begin{cases} \frac{1}{n}, & g(y) < \frac{1}{n}, \\ g(y), & \frac{1}{n} \leq g(y) \leq n, \\ n, & g(y) > n. \end{cases}$$

There exists a $n_0 \in \mathbf{N}$, such that for any $n \geq n_0$,

$$\tilde{L}(n) := \int_{E_n}|x|^{\frac{q\lambda}{2}-1}(\int_{E_n}\frac{[g(y)]_n}{|1+xy|^\lambda}dy)^q dx > 0,$$

$$\int_{E_n}|y|^{q(1-\frac{\lambda}{2})-1}[g(y)]_n^q dy > 0.$$

Setting

$$f_n(x) := |x|^{\frac{q\lambda}{2}-1}(\int_{E_n}\frac{[g(y)]_n}{|1+xy|^\lambda}dy)^{q-1}$$

and

$$\tilde{I}(n) := \int_{E_n}\int_{E_n}\frac{f_n(x)[g(y)]_n}{|1+xy|^\lambda}dxdy \ (n \geq n_0),$$

then by (4.5.34), for $q < 0$, we find

$$\infty > \int_{E_n}|x|^{p(1-\frac{\lambda}{2})-1} f_n^p(x)dx = \tilde{L}(n) = \tilde{I}(n)$$

$$> k_\lambda\{\int_{E_n}|x|^{p(1-\frac{\lambda}{2})-1} f_n^p(x)dx\}^{\frac{1}{p}}$$

$$\times\{\int_{E_n}|y|^{q(1-\frac{\lambda}{2})-1}[g(y)]_n^q dy\}^{\frac{1}{q}} > 0,$$

$$0 < \int_{E_n}|x|^{p(1-\frac{\lambda}{2})-1} f_n^p(x)dx = \tilde{L}(n)$$

$$< k_\lambda^q\int_{-\infty}^{\infty}|y|^{q(1-\frac{\lambda}{2})-1} g^q(y)dy.$$

It follows

$$0 < \int_{-\infty}^{\infty}|x|^{p(1-\frac{\lambda}{2})-1} f_\infty^p(x)dx < \infty$$

and for $n \to \infty$, the above two inequalities still keep the strict sign-inequalities, and we have (4.5.36). On the other hand, suppose (4.5.36) is valid.

By the reverse H$\ddot{o}$ lder's inequality, we find

$$\tilde{I} = \int_{-\infty}^{\infty}[|x|^{\frac{1}{q}-\frac{\lambda}{2}} f(x)][|x|^{\frac{-1}{q}+\frac{\lambda}{2}}\int_{-\infty}^{\infty}\frac{g(y)}{|1+xy|^\lambda}dy]dx$$

$$\geq \{\int_{-\infty}^{\infty}|x|^{p(1-\frac{\lambda}{2})-1} f^p(x)dx\}^{\frac{1}{p}}\tilde{L}^{\frac{-1}{q}} \quad .$$

(4.5.39)

Then by (4.5.36), we have (4.5.34), which is equivalent to (4.5.36). Hence inequalities (4.5.34), (4.5.36) and (4.5.36) are equivalent.

We conform that the constant factors in (4.5.35) and (4.5.36) are the best possible, otherwise, we can get a contradiction by (4.5.38) or (4.5.39) that the constant factor in (4.5.34) is not the best possible.

4.6. REFERENCES

1. Li YJ, Wang ZP, He B. Hilbert's type integral linear operator and some extensions of Hilbert's inequality, Journal of Inequalities and Applications, Volume 2007; Article ID 82138, 10 pages, doi:10.1155/2007/82138.
2. Xie ZT, Zeng Z, A Hilbert-type integral inequality whose kernel is a homogeneous form of degree -3, J. Math. Anal. Appl., 2007; 339: 324-331.
3. Yang BC, A Hilbert-type integral inequality with the kernel of -3-order homogeneous, Journal of Yunnan University, 2008; 30 (4): 325-330.
4. Wang WH, A Hardy-Hilbert's integral inequality with the homogeneous kernel of -4-order, Mathematical Theory and Applications, 2008; 28(1): 73-77.
5. Ge XK, A Hilbert-type integral inequality with the homogeneous kernel of -4-order, Journal of Zhejiang University, 2009; 36(1): 13-16.
6. Yang BC, A new Hilbert-type integral inequality with some parameters, Journal of Jilin University, 2008; 46(6): 1085-1090.
7. Yang BC, A Hilbert-type integral inequality with a non-homogeneous kernel, Journal of Xiamen University, 2009; 48 (2): 165-169.

5 Multiple Hilbert-Type Integral Inequalities

Bicheng Yang

Department of Mathematics, Guangdong Education Institute, Guangzhou, Guangdong 510303, P. R. China; E-mail: bcyang@pub.guangzhou.gd.cn

Abstract: In this chapter, we establish some lemmas and obtain two equivalent multiple Hilbert-type integral inequalities with the homogeneous kernels of real number-degree and the reverses, which are the best extensions of the corresponding results in Chapter 2. As applications, two equivalent multiple integral inequalities with the non-homogeneous kernels, two classes of multiple Hardy-type integral inequalities and some particular examples are also considered.

5.1 TWO EQUIVALENT MULTIPLE HILBERT-TYPE INTEGRAL INEQUALITIES AND APPLICATIONS

5.1.1 SOME LEMMAS

Lemma 5.1.1 If $n \in \mathbf{N} \setminus \{1\}$, $p_i \neq 0,1,\ r_i \neq 0$ $(i=1,\cdots,n), \sum_{i=1}^{n} \frac{1}{p_i}=1, \sum_{i=1}^{n} \frac{1}{r_i}=1$ ($\frac{1}{r_i}=0$, for $r_i = \pm\infty$) and $\lambda \in \mathbf{R}$, then we have the following equality:

$$A := \prod_{i=1}^{n} [x_i^{(\frac{\lambda}{r_i}-1)(1-p_i)} \prod_{j=1(j\neq i)}^{n} x_j^{\frac{\lambda}{r_j}-1}]^{\frac{1}{p_i}} = 1. \quad (5.1.1)$$

Proof We find

$$A = \prod_{i=1}^{n} [x_i^{(\frac{\lambda}{r_i}-1)(1-p_i)+1-\frac{\lambda}{r_i}} \prod_{j=1}^{n} x_j^{\frac{\lambda}{r_j}-1}]^{\frac{1}{p_i}}$$

$$= \prod_{i=1}^{n} [x_i^{(\frac{\lambda}{r_i}-1)(-p_i)}]^{\frac{1}{p_i}} (\prod_{j=1}^{n} x_j^{\frac{\lambda}{r_j}-1})^{\frac{1}{p_i}}$$

$$= \prod_{i=1}^{n} x_i^{1-\frac{\lambda}{r_i}} (\prod_{j=1}^{n} x_j^{\frac{\lambda}{r_j}-1})^{\sum_{i=1}^{n}\frac{1}{p_i}}$$

$$= \prod_{i=1}^{n} x_i^{1-\frac{\lambda}{r_i}} \prod_{j=1}^{n} x_j^{\frac{\lambda}{r_j}-1} = 1.$$

Hence (5.1.1) is valid.□

Definition 5.1.2 If $\lambda \in \mathbf{R}, n \in \mathbf{N}$,

$$R_+^n = \{(x_1,\cdots,x_n) \mid x_i > 0\ (i=1,\cdots,n)\},$$

$k_\lambda(x_1,\cdots,x_n)$ is a measurable function in R_+^n, satisfying for any $u>0$ and $(x_1,\cdots,x_n) \in R_+^n$,

$$k_\lambda(ux_1,\cdots,ux_n) = u^{-\lambda} k_\lambda(x_1,\cdots,x_n),$$

then we call $k_\lambda(x_1,\cdots,x_n)$ the homogeneous function of $-\lambda$-degree in R_+^n.

Lemma 5.1.3 As the assumption of Lemma 5.1.1, (1) if $k_\lambda(x_1,\cdots,x_n)(\geq 0)$ is a homogeneous function of $-\lambda$-degree, such that

$$k_\lambda(r_1,\cdots,r_{n-1}) := \int_0^\infty \cdots \int_0^\infty k_\lambda(u_1,\cdots,u_{n-1},1) \times \prod_{j=1}^{n-1} u_j^{\frac{\lambda}{r_j}-1} du_1 \cdots du_{n-1} \quad (5.1.2)$$

is a finite number, then for $i=1,\cdots,n-1$, the following multiple integral

$$H(i) := \int_0^\infty \cdots \int_0^\infty k_\lambda(u_1,\cdots,u_{i-1},1,u_{i+1},\cdots,u_n) \times \prod_{j=1(j\neq i)}^{n} u_j^{\frac{\lambda}{r_j}-1} du_1 \cdots du_{i-1} du_{i+1} \cdots du_n$$

are all equal to $k_\lambda(r_1,\cdots,r_{n-1})$; (2) For $i=1,\cdots,n$, define the weight functions $\omega_i(x_i)$ as:

$$\omega_i(x_i) := x_i^{\frac{\lambda}{r_i}} \int_0^\infty \cdots \int_0^\infty k_\lambda(x_1,\cdots,x_n) \times \prod_{j=1(j\neq i)}^{n} x_j^{\frac{\lambda}{r_j}-1} dx_1 \cdots dx_{i-1} dx_{i+1} \cdots dx_n, \quad (5.1.3)$$

Then each $\omega_i(x_i)$ is a constant independent of x_i and i, that is

$$\omega_i(x_i) = k_\lambda(r_1,\cdots,r_{n-1})\ (i=1,\cdots,n).$$

In particular, for $r_n = 2$ $(\sum_{i=1}^{n-1} \frac{1}{r_i} = \frac{1}{2})$, we still have

$$\tilde{\omega}_i(x_i) := x_i^{\frac{\lambda}{r_i}} \int_0^\infty \cdots \int_0^\infty k_\lambda(x_1 x_n,\cdots,x_{n-1}x_n,1) \times \prod_{j=1(j\neq i)}^{n} x_j^{\frac{\lambda}{r_j}-1} dx_1 \cdots dx_{i-1} dx_{i+1} \cdots dx_n$$

$$= k_\lambda(r_1,\cdots,r_{n-1}) \quad (i = 1,2,\cdots,n) .(5.1.4)$$

Proof (1) Setting $u_j = u_n v_j (j \neq i,n)$ in $H(i)$, we find

$$H(i) = \int_0^\infty \cdots \int_0^\infty k_\lambda(v_1,\cdots,v_{i-1},\tfrac{1}{u_n},v_{i+1},\cdots,v_{n-1},1)$$

$$\times \prod_{j=1(j\neq i)}^{n-1} v_j^{\frac{\lambda}{r_j}-1} u_n^{-1-\frac{\lambda}{r_i}} dv_1\cdots dv_{i-1}dv_{i+1}\cdots dv_{n-1}du_n .$$

Setting $v_i = \frac{1}{u_n}$ in the above integral, by (5.1.2), we obtain $H(i) = k_\lambda(r_1,\cdots,r_{n-1})$. (2) Setting

$$u_j = x_j / x_i \quad (j = 1,\cdots,i-1,i+1,\cdots n)$$

in (5.1.3), we find $\omega_i(x_i) = H(i) = k_\lambda(r_1,\cdots,r_{n-1})$. In particular, for $r_n = 2$, setting $x_n' = x_n^{-1}$ in $\tilde{\omega}_i(x_i)$, due to $\lambda - \frac{\lambda}{r_n} = \frac{\lambda}{r_n}$, it follows

$$\tilde{\omega}_i(x_i) = x_i^{\frac{\lambda}{r_i}} \int_0^\infty \cdots \int_0^\infty k_\lambda(\tfrac{x_1}{x_n'},\cdots,\tfrac{x_{n-1}}{x_n'},1) x_n'^{\frac{-\lambda}{r_n}+1-2}$$

$$\times \prod_{\substack{j=1\\(j\neq i)}}^{n-1} x_j^{\frac{\lambda}{r_j}-1} dx_1\cdots dx_{i-1}dx_{i+1}\cdots dx_{n-1}dx_n'$$

$$= x_i^{\frac{\lambda}{r_i}} \int_0^\infty \cdots \int_0^\infty k_\lambda(x_1,\cdots,x_{n-1},x_n') x_n'^{(\lambda-\frac{\lambda}{r_n})-1}$$

$$\times \prod_{\substack{j=1\\(j\neq i)}}^{n-1} x_j^{\frac{\lambda}{r_j}-1} dx_1\cdots dx_{i-1}dx_{i+1}\cdots dx_{n-1}dx_n'$$

$$= \omega_i(x_i) = k_\lambda(r_1,\cdots,r_{n-1}) \ (i = 1,\cdots,n-1);$$

for $i = n$, we find

$$\tilde{\omega}_n(x_n) = \omega_n(x_n^{-1}) = k_\lambda(r_1,\cdots,r_{n-1}).$$

Hence we have (5.1.4).□

Lemma 5.1.4 As the assumption of Lemma 5.1.3, if for $(r_1,\cdots,r_n)(\sum_{i=1}^n \frac{1}{r_i} = 1)$ and $i = 1,\cdots,n$, there exists $\delta > 0$, such that for $|\eta_j| < \delta$ $(j = 1,\cdots,n-1)$,

$$K_\lambda(\eta_1,\cdots,\eta_{n-1}) := \int_0^\infty \cdots \int_0^\infty k_\lambda(u_1,\cdots,u_{n-1},1)$$

$$\times \prod_{j=1}^{n-1} u_j^{\frac{\lambda}{r_j}-\frac{\eta_j}{p_j}-1} du_1\cdots du_{n-1}$$

is a positive constant, setting $\varepsilon_j > 0$ $(j = 1,\cdots,n-1)$ small enough, then we have

$$K_\lambda(\varepsilon_1,\cdots,\varepsilon_{n-1})$$

$$= \int_0^\infty \cdots \int_0^\infty k_\lambda(u_1,\cdots,u_{n-1},1)$$

$$\times \prod_{j=1}^{n-1} u_j^{\frac{\lambda}{r_j}-\frac{\varepsilon_j}{p_j}-1} du_1\cdots du_{n-1}$$

$$= k_\lambda(r_1,\cdots,r_{n-1}) + o(1) \ (\varepsilon_j \to 0^+). \quad (5.1.5)$$

In particular, (5.1.5) is valid for $\varepsilon_j = \varepsilon \to 0^+$.

Proof We prove (5.1.5) by mathematical induction. For $n = 2$, By Lemma 2.2.5, we have

$$\lim_{\varepsilon_1\to 0^+} K_\lambda(\varepsilon_1)$$

$$= \lim_{\varepsilon_1\to 0^+} \int_0^\infty k_\lambda(u_1,1)\, u_1^{\frac{\lambda}{r_1}-\frac{\varepsilon_1}{p_1}-1} du_1$$

$$= \int_0^\infty k_\lambda(u_1,1)\, u_1^{\frac{\lambda}{r_1}-1} du_1 = k_\lambda(r_1),$$

and (5.1.5) is valid. Assuming that for $n(\geq 2)$, (5.1.5) is valid, then for $n+1$, following the way of $n = 2$, we have

$$\lim_{\varepsilon_n\to 0^+} K_\lambda(\varepsilon_1,\cdots,\varepsilon_n)$$

$$= \lim_{\varepsilon_n\to 0^+} \int_0^\infty [\int_0^\infty \cdots \int_0^\infty k_\lambda(u_1,\cdots,u_n,1)$$

$$\times \prod_{j=1}^{n-1} u_j^{\frac{\lambda}{r_j}-\frac{\varepsilon_j}{p_j}-1} du_1\cdots du_{n-1}] u_n^{\frac{\lambda}{r_n}-\frac{\varepsilon_n}{p_n}-1} du_n$$

$$= \int_0^\infty [\int_0^\infty \cdots \int_0^\infty k_\lambda(u_1,\cdots,u_n,1)$$

$$\times \prod_{j=1}^{n-1} u_j^{\frac{\lambda}{r_j}-\frac{\varepsilon_j}{p_j}-1} du_1\cdots du_{n-1}] u_n^{\frac{\lambda}{r_n}-1} du_n$$

$$= \int_0^\infty \cdots \int_0^\infty [\int_0^\infty k_\lambda(u_1,\cdots,u_n,1) u_n^{\frac{\lambda}{r_n}-1} du_n]$$

$$\times \prod_{j=1}^{n-1} u_j^{\frac{\lambda}{r_j}-\frac{\varepsilon_j}{p_j}-1} du_1\cdots du_{n-1} .$$

Hence by the assumption, for $n+1$, we have (5.1.5). By mathematical induction, we prove that (5.1.5) is valid for $n \in \mathbf{N}\setminus\{1\}$.□

Lemma 5.1.5 As the assumption of Lemma 5.1.4, if $\lambda \in \mathbf{R}$, indicating $k_\lambda := k_\lambda(r_1,\cdots,r_{n-1})$, then for $\varepsilon > 0$ small enough, we have

$$I_\varepsilon := \varepsilon \int_1^\infty \cdots \int_1^\infty k_\lambda(x_1,\cdots,x_n) \prod_{j=1}^n x_j^{\frac{\lambda}{r_j}-\frac{\varepsilon}{p_j}-1} dx_1\cdots dx_n$$

$$= k_\lambda + o(1), \varepsilon \to 0^+. \quad (5.1.6)$$

In particular, for $r_n = 2$, we have

$$\tilde{I}_\varepsilon := \varepsilon \int_1^\infty \cdots \int_1^\infty [\int_0^1 x_n^{\frac{\lambda}{2}+\frac{\varepsilon}{p_n}-1} k_\lambda(x_1x_n,\cdots,x_{n-1}x_n,1)dx_n]$$

$$\times\prod_{j=1}^{n-1} x_j^{\frac{\lambda}{r_j}-\frac{\varepsilon}{p_j}-1} dx_1 \cdots dx_{n-1}$$

$$= k_\lambda + o(1)\,(\varepsilon \to 0^+). \qquad (5.1.7)$$

Proof Setting $u_i = \frac{x_i}{x_n}$ $(i = 1, \cdots, n-1)$, by (5.1.5), we find

$$I_\varepsilon = \varepsilon \int_1^\infty x_n^{-1-\varepsilon} [\int_{x_n^{-1}}^\infty \cdots \int_{x_n^{-1}}^\infty k_\lambda(u_1, \cdots, u_{n-1}, 1)$$

$$\times \prod_{j=1}^{n-1} u_j^{\frac{\lambda}{r_j}-\frac{\varepsilon}{p_j}-1} du_1 \cdots du_{n-1}] dx_n$$

$$\le \int_0^\infty \cdots \int_0^\infty k_\lambda(u_1, \cdots, u_{n-1}, 1)$$

$$\times \prod_{j=1}^{n-1} u_j^{\frac{\lambda}{r_j}-\frac{\varepsilon}{p_j}-1} du_1 \cdots du_{n-1}$$

$$= k_\lambda + o_1(1)(\varepsilon \to 0^+). \qquad (5.1.8)$$

Still setting $u_i = \frac{x_i}{x_n}$ $(i = 1, \cdots, n-1)$ in I_ε, we find

$$I_\varepsilon = \varepsilon \int_1^\infty x_n^{-1-\varepsilon} [\int_{x_n^{-1}}^\infty \cdots \int_{x_n^{-1}}^\infty k_\lambda(u_1, \cdots, u_{n-1}, 1)$$

$$\times \prod_{j=1}^{n-1} u_j^{\frac{\lambda}{r_j}-\frac{\varepsilon}{p_j}-1} du_1 \cdots du_{n-1}] dx_n$$

$$\ge \int_0^\infty \cdots \int_0^\infty k_\lambda(u_1, \cdots, u_{n-1}, 1)$$

$$\times \prod_{j=1}^{n-1} u_j^{\frac{\lambda}{r_j}-\frac{\varepsilon}{p_j}-1} du_1 \cdots du_{n-1}$$

$$-\varepsilon \sum_{j=1}^{n-1} \int_1^\infty x_n^{-1} A_j(x_n) dx_n, \qquad (5.1.9)$$

where, for $j = 1, \cdots, n-1$, define $A_j(x_n)$ as

$$A_j(x_n) := \int \cdots \int_{D_j} k_\lambda(u_1, \cdots, u_{n-1}, 1)$$

$$\times \prod_{i=1}^{n-1} u_i^{\frac{\lambda}{r_i}-\frac{\varepsilon}{p_i}-1} du_1 \cdots du_{n-1},$$

$$D_j := \{(u_1, u_2, \cdots, u_{n-1}) \mid 0 < u_j \le x_n^{-1},$$

$$0 < u_k < \infty \ (k \ne j)\}.$$

Without lose of generality, we estimate integral $\int_1^\infty x_n^{-1} A_{n-1}(x_n) dx_n$ (for $j = n-1$).

By Fubini Theorem, we find

$$\int_1^\infty x_n^{-1} A_{n-1}(x_n) dx_n$$

$$= \int_1^\infty x_n^{-1} \int \cdots \int_{D_{n-1}} k_\lambda(u_1, \cdots, u_{n-1}, 1)$$

$$\times \prod_{i=1}^{n-1} u_i^{\frac{\lambda}{r_i}-\frac{\varepsilon}{p_i}-1} du_1 \cdots du_{n-1} dx_n$$

$$= \int_1^\infty x_n^{-1} \int_0^\infty \cdots \int_0^\infty \int_0^{x_n^{-1}} k_\lambda(u_1, \cdots, u_{n-1}, 1)$$

$$\times \prod_{i=1}^{n-1} u_i^{\frac{\lambda}{r_i}-\frac{\varepsilon}{p_i}-1} du_1 \cdots du_{n-2} du_{n-1} dx_n$$

$$= \int_0^1 \int_0^\infty \cdots \int_0^\infty k_\lambda(u_1, \cdots, u_{n-1}, 1)$$

$$\times \prod_{i=1}^{n-1} u_i^{\frac{\lambda}{r_i}-\frac{\varepsilon}{p_i}-1} du_1 \cdots du_{n-2} (\int_1^{u_{n-1}^{-1}} x_n^{-1} dx_n) du_{n-1}$$

$$= \int_0^1 \int_0^\infty \cdots \int_0^\infty k_\lambda(u_1, \cdots, u_{n-1}, 1)$$

$$\times \prod_{i=1}^{n-1} u_i^{\frac{\lambda}{r_i}-\frac{\varepsilon}{p_i}-1} (-\ln u_{n-1}) du_1 \cdots du_{n-2} du_{n-1}. \qquad (5.1.10)$$

Setting $\alpha > 0$ small enough, such that $|\alpha p_{n-1} + \varepsilon| < \delta$, since

$$\lim_{u_{n-1} \to 0^+} u_{n-1}^\alpha (-\ln u_{n-1}) = 0,$$

there exists $M_{n-1} > 0$, such that

$$0 < -u_{n-1}^\alpha \ln u_{n-1} \le M_{n-1},\ u_{n-1} \in (0, 1].$$

In view of (5.1.10) and (5.1.5), we find

$$0 \le \int_0^1 \int_0^\infty \cdots \int_0^\infty k_\lambda(u_1, \cdots, u_{n-1}, 1)$$

$$\times \prod_{i=1}^{n-1} u_i^{\frac{\lambda}{r_i}-\frac{\varepsilon}{p_i}-1} (-\ln u_{n-1}) du_1 \cdots du_{n-2} du_{n-1}$$

$$= \int_0^1 (-u_{n-1}^\alpha \ln u_{n-1}) [\int_0^\infty \cdots \int_0^\infty k_\lambda(u_1, \cdots, u_{n-1}, 1)$$

$$\times \prod_{i=1}^{n-2} u_i^{\frac{\lambda}{r_i}-\frac{\varepsilon}{p_i}-1} du_1 \cdots du_{n-2}] u_{n-1}^{\frac{\lambda}{r_{n-1}}-(\frac{\varepsilon}{p_{n-1}}+\alpha)-1} du_{n-1}$$

$$\le M_{n-1} \int_0^1 [\int_0^\infty \cdots \int_0^\infty k_\lambda(u_1, \cdots, u_{n-1}, 1)$$

$$\times \prod_{i=1}^{n-2} u_i^{\frac{\lambda}{r_i}-\frac{\varepsilon}{p_i}-1} du_1 \cdots du_{n-2}] u_{n-1}^{\frac{\lambda}{r_{n-1}}-\frac{\alpha p_{n-1}+\varepsilon}{p_{n-1}}-1} du_{n-1}$$

$$\le M_{n-1} \int_0^\infty \int_0^\infty \cdots \int_0^\infty k_\lambda(u_1, \cdots, u_{n-1}, 1)$$

$$\times \prod_{i=1}^{n-2} u_i^{\frac{\lambda}{r_i}-\frac{\varepsilon}{p_i}-1} u_{n-1}^{\frac{\lambda}{r_{n-1}}-\frac{\alpha p_{n-1}+\varepsilon}{p_{n-1}}-1} du_1 \cdots du_{n-1} < \infty.$$

Then by (5.1.5) and (5.1.9), we find

$$I_\varepsilon \ge k_\lambda + o_1(1) - \varepsilon \sum_{j=1}^{n-1} O_j = k_\lambda + o_2(1). \qquad (5.1.11)$$

In view of (5.1.8) and (5.1.11), we have (5.1.6).

In particular, for $r_n = 2$, setting $x_n' = x_n^{-1}$, we find

$$\tilde{I}_\varepsilon = \varepsilon \int_1^\infty \cdots \int_1^\infty [\int_0^1 x_n^{\frac{\lambda}{2}-\lambda+\frac{\varepsilon}{p_n}-1} k_\lambda(x_1,\cdots,x_{n-1},x_n^{-1}) dx_n] \times \prod_{j=1}^{n-1} x_j^{\frac{\lambda}{r_j}-\frac{\varepsilon}{p_j}-1} dx_1 \cdots dx_{n-1}$$

$$= \varepsilon \int_1^\infty \cdots \int_1^\infty \int_1^\infty x_n'^{\frac{\lambda}{r_n}-\frac{\varepsilon}{p_n}-1} k_\lambda(x_1,\cdots,x_{n-1},x_n') \times \prod_{j=1}^{n-1} x_j^{\frac{\lambda}{r_j}-\frac{\varepsilon}{p_j}-1} dx_1 \cdots dx_{n-1} dx_n' = I_\varepsilon .$$

Then by (5.1.6), we have (5.1.7).□

5.1.2 A MULTIPLE HILBERT-TYPE INTEGRAL INEQUALITY AND THE REVERSE

Theorem 5.1.6 Suppose that $n \in \mathbf{N} \setminus \{1\}$, $p_i \neq 0,1, r_i \neq 0 \ (i=1,\cdots,n)$, $\sum_{i=1}^n \frac{1}{p_i} = 1$ $\sum_{i=1}^n \frac{1}{r_i} = 1$, $\lambda \in \mathbf{R}$, $k_\lambda(x_1,\cdots,x_n)(\geq 0)$ is a homogeneous function of $-\lambda$-degree in R_+^n, and for $i=1,\cdots,n$, there exists $\delta > 0$, such that for $|\eta_j| < \delta (j=1,\cdots,n-1)$,

$$K_\lambda(\eta_1,\cdots,\eta_{n-1}) = \int_0^\infty \cdots \int_0^\infty k_\lambda(u_1,\cdots,u_{n-1},1) \times \prod_{j=1}^{n-1} u_j^{\frac{\lambda}{r_j}-\frac{\eta_j}{p_j}-1} du_1 \cdots du_{n-1}$$

is a positive constant,

$$k_\lambda = k_\lambda(r_1,\cdots,r_{n-1}) = K_\lambda(0,\cdots,0).$$

If $f_i \geq 0$,

$$0 < \int_0^\infty x^{p_i(1-\frac{\lambda}{r_i})-1} f_i^{p_i}(x) dx < \infty$$

$(i=1,\cdots,n)$, then (1) for $p_i > 1 (i=1,\cdots,n)$, we have the following inequality:

$$H := \int_0^\infty \cdots \int_0^\infty k_\lambda(x_1,\cdots,x_n) \times \prod_{i=1}^n f_i(x_i) dx_1 \cdots dx_n < k_\lambda \prod_{i=1}^n \{\int_0^\infty x^{p_i(1-\frac{\lambda}{r_i})-1} f_i^{p_i}(x) dx\}^{\frac{1}{p_i}}, \quad (5.1.12)$$

where the constant factor k_λ is the best possible; (2) for $0 < p_1 < 1, p_i < 0 \ (i=2,\cdots,n)$, we have the reverse of (5.1.12) with the best constant factor k_λ. In particular, for $r_n = 2$ $(\sum_{i=1}^{n-1} \frac{1}{r_i} = \frac{1}{2})$, if $p_i > 1 \ (i=1,2,\cdots,n)$, we have

$$\tilde{H} := \int_0^\infty \cdots \int_0^\infty k_\lambda(x_1 x_n,\cdots,x_{n-1}x_n,1) \times \prod_{i=1}^n f_i(x_i) dx_1 \cdots dx_n < k_\lambda \prod_{i=1}^n \{\int_0^\infty x^{p_i(1-\frac{\lambda}{r_i})-1} f_i^{p_i}(x) dx\}^{\frac{1}{p_i}}, \quad (5.1.13)$$

where the constant factor k_λ is still the best possible; if $0 < p_1 < 1, p_i < 0 \quad (i=2,\cdots,n)$, we have the reverse of (5.1.13) still with the best constant factor.

Proof By (5.1.1) and Hölder's inequality, we find

$$H = \int_0^\infty \cdots \int_0^\infty k_\lambda(x_1,\cdots,x_n) \times \prod_{i=1}^n [x_i^{(\frac{\lambda}{r_i}-1)(1-p_i)} \prod_{j=1(j\neq i)}^n x_j^{\frac{\lambda}{r_j}-1}]^{\frac{1}{p_i}} \times f_i(x_i) dx_1 \cdots dx_n$$

$$\leq \prod_{i=1}^n \{\int_0^\infty [\int_0^\infty \cdots \int_0^\infty k_\lambda(x_1,\cdots,x_n) \times x_i^{\frac{\lambda}{r_i}} \prod_{j=1(j\neq i)}^n x_j^{\frac{\lambda}{r_j}-1} dx_1 \cdots dx_{i-1} dx_{i+1} \cdots dx_n] \times x_i^{p_i(1-\frac{\lambda}{r_i})-1} f_i^{p_i}(x_i) dx_i\}^{\frac{1}{p_i}}. \quad (5.1.14)$$

We conform that (5.1.14) keeps the strict sign-inequality, otherwise, since $k_\lambda > 0$, $k_\lambda(x_1,\cdots,x_n)$ is not zero almost everywhere, then there exist $C_i, C_k \ (i \neq k)$, such that they are not all zero and

$$C_i x_i^{\frac{\lambda}{r_i}} \prod_{j=1(j\neq i)}^n x_j^{\frac{\lambda}{r_j}-1} x_i^{p_i(1-\frac{\lambda}{r_i})-1} f_i^{p_i}(x_i) = C_k x_k^{\frac{\lambda}{r_k}} \prod_{j=1(j\neq k)}^n x_j^{\frac{\lambda}{r_j}-1} x_k^{p_k(1-\frac{\lambda}{r_k})-1} f_k^{p_k}(x_k)$$

a.e. in R_+^n,

e.t.

$$C_i x_i^{p_i(1-\frac{\lambda}{r_i})} f_i^{p_i}(x_i) = C_k x_k^{p_k(1-\frac{\lambda}{r_k})} f_k^{p_k}(x_k) = C$$

a.e. in R_+^n.

Assuming $C_i \neq 0$, then we find

$$x_i^{p_i(1-\frac{\lambda}{r_i})-1} f_i^{p_i}(x_i) = \frac{C}{C_i x_i} \text{ a.e. in } (0,\infty),$$

which contradicts the fact that

$$0 < \int_0^\infty x_i^{p_i(1-\frac{\lambda}{r_i})-1} f_i^{p_i}(x_i) dx_i < \infty.$$

Then by (5.1.13), we may rewrite (5.1.14) as

$$H < \prod_{i=1}^n \{\int_0^\infty \omega_i(x_i) x_i^{p_i(1-\frac{\lambda}{r_i})-1} f^{p_i}(x_i) dx_i\}^{\frac{1}{p_i}}. (5.1.15)$$

By Lemma 5.1.3, we have (5.1.12).

If there exists a positive constant $k \le k_\lambda$, such that (5.1.12) is still valid as we replace k_λ by k, the in particular, for $\varepsilon > 0$ small enough, setting

$$\tilde{f}_i(x) = \begin{cases} 0, & x \in (0,1], \\ x^{\frac{\lambda}{r_i} - \frac{\varepsilon}{p_i} - 1}, & x \in (1,\infty), \end{cases} \quad i = 1,\cdots,n, \tag{5.1.16}$$

by (5.1.6), it follows

$$k_\lambda + o(1) = I_\varepsilon$$

$$< \varepsilon k \prod_{i=1}^{n} \{\int_0^\infty x^{p_i(1-\frac{\lambda}{r_i})-1} \tilde{f}_i^{p_i}(x)dx\}^{\frac{1}{p_i}} = k.$$

and $k_\lambda \le k$, $\varepsilon \to 0^+$. Hence $k = k_\lambda$ is the best value of (5.1.12).

By the reverse Hölder's inequality and the same way of obtaining (5.1.15), we can find the reverse of (5.1.12).

If there exists a positive constant $K \ge k_\lambda$, such that the reverse of (5.1.12) is still valid as we replace k_λ by K, then in particular, for $\varepsilon > 0$ small enough, still using (5.1.13), in view of (5.1.6), we have

$$k_\lambda + o_1(1) = I_\varepsilon$$

$$> \varepsilon K \prod_{i=1}^{n} \{\int_0^\infty x^{p_i(1-\frac{\lambda}{r_i})-1} \tilde{f}_i^{p_i}(x)dx\}^{\frac{1}{p_i}} = K,$$

and $k_\lambda \ge K$, $\varepsilon \to 0^+$. Hence $K = k_\lambda$ is the best value of the reverse of (5.1.12).

In particular, for $r_n = 2$, if $p_i > 1$ $(i = 1,\cdots,n)$, similar to (5.1.15), w have

$$\tilde{H} < \prod_{i=1}^{n-1} \{\int_0^\infty \tilde{\omega}_i(x_i) x_i^{p_i(1-\frac{\lambda}{r_i})-1} f^{p_i}(x_i)dx_i\}^{\frac{1}{p_i}}$$

$$\times \{\int_0^\infty \tilde{\omega}_n(x_n^{-1}) x_n^{p_n(1-\frac{\lambda}{2})-1} f^{p_n}(x_i)dx_i\}^{\frac{1}{p_n}}.$$

Then by (5.1.4), we have (5.1.13). If there exists a positive constant $k \le k_\lambda$, such that (5.1.13) is still valid as we replace k_λ by k, then in particular, for $\varepsilon > 0$ small enough, setting

$$\tilde{f}_n(x) = \begin{cases} x^{\frac{\lambda}{2} + \frac{\varepsilon}{p_n} - 1}, & x \in (0,1], \\ 0, & x \in (1,\infty), \end{cases}$$

$$\tilde{f}_i(x) = \begin{cases} 0, & x \in (0,1], \\ x^{\frac{\lambda}{r_i} - \frac{\varepsilon}{p_i} - 1}, & x \in (1,\infty), \end{cases} \quad i = 1,\cdots,n-1,$$

by (5.1.7), it follows

$$k_\lambda + o(1) = \tilde{I}_\varepsilon$$

$$< \varepsilon k \prod_{i=1}^{n} \{\int_0^\infty x^{p_i(1-\frac{\lambda}{r_i})-1} \tilde{f}_i^{p_i}(x)dx\}^{\frac{1}{p_i}} = k$$

and $k_\lambda \le k, \varepsilon \to 0^+$. Hence $k = k_\lambda$ is the best value of (5.1.13). Similarly, if $0 < p_1 < 1, p_i < 0$ $(i = 2,\cdots,n)$, we obtain the reverse of (5.1.13) with the best constant factor.□

5.1.3 AN EQUIVALENT FORM AND THE REVERSE

Theorem 5.1.7 Suppose that $n \in \mathbf{N} \setminus \{1\}$, $p_i \ne 0,1,\ r_i \ne 0$ $(i = 1,\cdots,n)$, $\sum_{i=1}^{n} \frac{1}{p_i} = 1$,

$$\frac{1}{q_n} = 1 - \frac{1}{p_n} = \sum_{i=1}^{n-1} \frac{1}{p_i},\ \sum_{i=1}^{n} \frac{1}{r_i} = 1,$$

$\lambda \in \mathbf{R}$, $k_\lambda(x_1,\cdots,x_n)(\ge 0)$ is a homogeneous function of $-\lambda$-degree in R_+^n, for $i = 1,\cdots,n$, there exists $\delta > 0$, such that for $|\eta_j| < \delta(j = 1,\cdots,n-1)$,

$$K_\lambda(\eta_1,\cdots,\eta_{n-1}):$$

$$= \int_0^\infty \cdots \int_0^\infty k_\lambda(u_1,\cdots,u_{n-1},1) \prod_{j=1}^{n-1} u_j^{\frac{\lambda}{r_j} - \frac{\eta_j}{p_j} - 1} du_1 \cdots du_{n-1}$$

is a positive constant,

$$k_\lambda = k_\lambda(r_1,\cdots,r_{n-1}) = K_\lambda(0,\cdots,0).$$

If $f_i \ge 0$,

$$0 < \int_0^\infty x^{p_i(1-\frac{\lambda}{r_i})-1} f_i^{p_i}(x)dx < \infty$$

$(i = 1,\cdots,n-1)$, then (1) for $p_i > 1$ $(i = 1,\cdots,n)$, we have the following inequality equivalent to (5.1.12):

$$J := \{\int_0^\infty x_n^{\frac{q_n\lambda}{r_n} - 1} [\int_0^\infty \cdots \int_0^\infty k_\lambda(x_1,\cdots,x_n)$$

$$\times \prod_{i=1}^{n-1} f_i(x_i)dx_1 \cdots dx_{n-1}]^{q_n} dx_n\}^{\frac{1}{q_n}}$$

$$< k_\lambda \prod_{i=1}^{n-1} \{\int_0^\infty x^{p_i(1-\frac{\lambda}{r_i})-1} f_i^{p_i}(x)dx\}^{\frac{1}{p_i}}, \tag{5.1.17}$$

where the constant factor k_λ is the best possible; (2) for $0 < p_1 < 1, p_i < 0$ $(i = 2,\cdots,n)$, we have the reverse of (5.1.17) equivalent to the reverse of (5.1.12) with the best constant factor k_λ. In particular, for $r_n = 2$, if $p_i > 1$ $(i = 1, \cdots, n)$, then we have the following inequality

$$\tilde{J} := \{\int_0^\infty x_n^{\frac{q_n\lambda}{2}-1}[\int_0^\infty \cdots \int_0^\infty k_\lambda(x_1 x_n, \cdots, x_{n-1}x_n, 1)$$
$$\times \prod_{i=1}^{n-1} f_i(x_i)dx_1 \cdots dx_{n-1}]^{q_n} dx_n\}^{\frac{1}{q_n}}$$
$$< k_\lambda \prod_{i=1}^{n-1}\{\int_0^\infty x^{p_i(1-\frac{\lambda}{r_i})-1} f_i^{p_i}(x)dx\}^{\frac{1}{p_i}}, \quad (5.1.18)$$

which equivalent to (5.1.13) and with the best constant factor; if $0 < p_1 < 1, p_i < 0$ $(i = 2, \cdots, n)$, then we have the reverse of (5.1.18) with the best constant factor equivalent to the reverse of (5.1.13).

Proof (1) For $n = 2$, due to (2.2.20), we have (5.1.17); For $n \ge 3$, since $q_n > 1, \frac{1}{p_n} + \frac{1}{q_n} = 1$, by Hölder's inequality, (5.1.1) and (5.1.3), we obtain

$$\int_0^\infty \cdots \int_0^\infty k_\lambda(x_1, \cdots, x_n) \prod_{i=1}^{n-1} f_i(x_i) dx_1 \cdots dx_{n-1}$$
$$= \int_0^\infty \cdots \int_0^\infty k_\lambda(x_1, \cdots, x_n)$$
$$\times\{\prod_{i=1}^{n-1}[x_i^{(\frac{\lambda}{r_i}-1)(1-p_i)} \prod_{j=1(j\neq i)}^{n} x_j^{\frac{\lambda}{r_j}-1}]^{\frac{1}{p_i}} f_i(x_i)\}$$
$$\times[x_n^{(\frac{\lambda}{r_n}-1)(1-p_n)} \prod_{j=1}^{n-1} x_j^{\frac{\lambda}{r_j}-1}]^{\frac{1}{p_n}} dx_1 \cdots dx_{n-1}$$
$$\le \{\int_0^\infty \cdots \int_0^\infty k_\lambda(x_1, \cdots, x_n)$$
$$\times \prod_{i=1}^{n-1}[x_i^{(\frac{\lambda}{r_i}-1)(1-p_i)} \prod_{j=1(j\neq i)}^{n} x_j^{\frac{\lambda}{r_j}-1}]^{\frac{q_n}{p_i}}$$
$$\times f_i^{q_n}(x_i)dx_1 \cdots dx_{n-1}\}^{\frac{1}{q_n}}$$
$$\times\{\int_0^\infty [\int_0^\infty \cdots \int_0^\infty k_\lambda(x_1, \cdots, x_n)$$
$$\times x_n^{(\frac{\lambda}{r_n}-1)(1-p_n)} \prod_{j=1}^{n-1} x_j^{\frac{\lambda}{r_j}-1} dx_1 \cdots dx_{n-1}\}^{\frac{1}{p_n}}$$
$$= \{k_\lambda x_n^{p_n(1-\frac{\lambda}{r_n})-1}\}^{\frac{1}{p_n}} \{\int_0^\infty \cdots \int_0^\infty k_\lambda(x_1, \cdots, x_n)$$
$$\times \prod_{i=1}^{n-1}[x_i^{(\frac{\lambda}{r_i}-1)(1-p_i)} \prod_{j=1(j\neq i)}^{n} x_j^{\frac{\lambda}{r_j}-1}]^{\frac{q_n}{p_i}}$$
$$\times f_i^{q_n}(x_i)dx_1 \cdots dx_{n-1}\}^{\frac{1}{q_n}}. \quad (5.1.19)$$

Then by (5.1.12), we find

$$\tilde{J} := \{\int_0^\infty x_n^{\frac{q_n\lambda}{r_n}-1}[\int_0^\infty \cdots \int_0^\infty k_\lambda(x_1, \cdots, x_n)$$
$$\times \prod_{i=1}^{n-1} f_i(x_i)dx_1 \cdots dx_{n-1}]^{q_n} dx_n\}^{\frac{1}{q_n}}$$
$$\le k_\lambda^{\frac{1}{p_n}}\{\int_0^\infty \int_0^\infty \cdots \int_0^\infty k_\lambda(x_1, \cdots, x_n)$$
$$\times \prod_{i=1}^{n-1}[x_i^{(\frac{\lambda}{r_i}-1)(1-p_i)} \prod_{j=1(j\neq i)}^{n} x_j^{\frac{\lambda}{r_j}-1}]^{\frac{q_n}{p_i}}$$
$$\times f_i^{q_n}(x_i)dx_1 \cdots dx_{n-1}dx_n\}^{\frac{1}{q_n}}$$
$$= k_\lambda^{\frac{1}{p_n}}\{\int_0^\infty \cdots \int_0^\infty [\int_0^\infty k_\lambda(x_1, \cdots, x_n) x_n^{\frac{\lambda}{r_n}-1} dx_n]$$
$$\times \prod_{i=1}^{n-1}[x_i^{(\frac{\lambda}{r_i}-1)(1-p_i)} \prod_{j=1(j\neq i)}^{n-1} x_j^{\frac{\lambda}{r_j}-1}]^{\frac{q_n}{p_i}}$$
$$\times f_i^{q_n}(x_i)dx_1 \cdots dx_{n-1}\}^{\frac{1}{q_n}}$$
$$< k_\lambda^{\frac{1}{p_n}}\{\tilde{k}_\lambda \prod_{i=1}^{n-1}[\int_0^\infty x_i^{p_i(1-\frac{\lambda}{r_i})-1} f_i^{p_i}(x_i)dx_i]^{\frac{q_n}{p_i}}\}^{\frac{1}{q_n}},$$

where, since

$$\tilde{k}_\lambda := \int_0^\infty \cdots \int_0^\infty [\int_0^\infty k_\lambda(x_1, \cdots, x_n) x_n^{\frac{\lambda}{r_n}-1} dx_n]$$
$$\times \prod_{j=1(j\neq i)}^{n-1} x_j^{\frac{\lambda}{r_j}-1} dx_1 \cdots dx_{n-1}$$
$$= \int_0^\infty \cdots \int_0^\infty k_\lambda(x_1, \cdots, x_n)$$
$$\times \prod_{j=1(j\neq i)}^{n} x_j^{\frac{\lambda}{r_j}-1} dx_1 \cdots dx_{n-1}dx_n = k_\lambda,$$

we obtain (5.1.17).

By Hölder's inequality, we have

$$H = \int_0^\infty x_n^{\frac{1}{p_n}+\frac{\lambda}{r_n}-1}[\int_0^\infty \cdots \int_0^\infty k_\lambda(x_1, \cdots, x_n)$$
$$\times \prod_{i=1}^{n-1} f_i(x_i)dx_1 \cdots dx_{n-1}][x^{1-\frac{\lambda}{r_n}-\frac{1}{p_n}} f_n(x_n)]dx_n$$
$$\le \{\int_0^\infty x_n^{\frac{q_n\lambda}{r_n}-1}[\int_0^\infty \cdots \int_0^\infty k_\lambda(x_1, \cdots, x_n)$$
$$\times \prod_{i=1}^{n-1} f_i(x_i)dx_1 \cdots dx_{n-1}]^{q_n} dx_n\}^{\frac{1}{q_n}}$$
$$\times\{\int_0^\infty x^{p_n(1-\frac{\lambda}{r_n})-1} f_n^{p_n}(x_n)dx_n\}^{\frac{1}{p_n}}. \quad (5.1.20)$$

Then by (5.1.17), we have (5.1.12), which is equivalent to (5.1.17). We conform that the constant factor k_λ in (5.1.17) is the best possible, otherwise, by (5.1.20), we can get a contradiction that the constant factor in (5.1.12) is not the best possible.

(2) For $n = 2$, due to (2.2.22), we have the reverse of (5.1.17); for $n \ge 3$, by the reverse Hölder's

inequality, we can obtain the reverses of (5.1.20), (5.1.17) and (5.1.19). Hence the reverses of (5.1.17) and the reverses of (5.1.13) are equivalent. We conform that the constant factor k_λ in the reverse of (5.1.17) is the best possible, otherwise, by the reverse of (5.1.20), we can get a contradiction that the constant factor in the reverse of (5.1.13) is not the best possible. By the same way, we can prove the cases in $r_n = 2$.□

Remark 5.1.8 For $n = 2, r_2 = r_1 = 2$, setting $p_1 = p,\ p_2 = q$, by Theorem 5.1.6 and Theorem 5.1.7, it follows (1) for $p > 1$, we have the following equivalent inequalities

$$\int_0^\infty \int_0^\infty k_\lambda(xy,1) f(x) g(y) dx dy$$
$$< k_\lambda \{ \int_0^\infty x^{p(1-\frac{\lambda}{2})-1} f^p(x) dx \}^{\frac{1}{p}}$$
$$\times \{ \int_0^\infty x^{q(1-\frac{\lambda}{2})-1} g^q(x) dx \}^{\frac{1}{q}}, \qquad (5.1.21)$$
$$\{ \int_0^\infty y^{\frac{p\lambda}{2}-1} [\int_0^\infty k_\lambda(xy,1) f(x) dx]^p dy \}^{\frac{1}{p}}$$
$$< k_\lambda \{ \int_0^\infty x^{p(1-\frac{\lambda}{2})-1} f^p(x) dx \}^{\frac{1}{p}}, \qquad (5.1.22)$$

where the constant factor $k_\lambda = \int_0^\infty k_\lambda(u,1) u^{\frac{\lambda}{2}-1} du$ is the best possible; (2) for $0 < p < 1$, we have the equivalent reverses of (5.1.21) and (5.1.22) with the same best constant factor.

5.1.4 OPERATOR EXPREESIONS OF TWO EQUIVALENT INEGUALITIES

As the conditions of Theorem 5.1.6, for $q_n = \frac{p_n}{p_n - 1}$,

$$\phi_i(x) := x^{p_i(1-\frac{\lambda}{r_i})-1} (x \in (0,\infty); i = 1,\cdots,n),$$

and $\phi_n^{\frac{1}{1-p_n}}(x) = x^{\frac{q_n\lambda}{r_n}-1}$, define the following real function spaces:

$$L_{\phi_i}^{p_i}(0,\infty):$$
$$= \{ f; \| f \|_{p_i,\phi_i} := \{ \int_0^\infty \phi_i(x) | f(x) |^{p_i} dx \}^{\frac{1}{p_i}} < \infty \}$$
$$(i = 1,\cdots,n),$$
$$\prod_{i=1}^{n-1} L_{\phi_i}^{p_i}(0,\infty):$$
$$= L_{\phi_1}^{p_1}(0,\infty) \times L_{\phi_2}^{p_2}(0,\infty) \times \cdots \times L_{\phi_{n-1}}^{p_{n-1}}(0,\infty),$$

and a multiple Hilbert-type integral operator

$$T_n : \prod_{i=1}^{n-1} L_{\phi_i}^{p_i}(0,\infty) \to L_{\phi_n^{1/(1-p_n)}}^{q_n}(0,\infty)$$

as: for any $f = (f_1,\cdots,f_{n-1}) \in \prod_{i=1}^{n-1} L_{\phi_i}^{p_i}(0,\infty)$, there exists

$$(T_n f)(x_n) = \int_0^\infty k_\lambda(x_1,\cdots,x_n)$$
$$\times \prod_{i=1}^{n-1} f_i(x_i) dx_1 \cdots dx_{n-1},\ x_n \in (0,\infty). \qquad (5.1.23)$$

Hence by (5.1.17), it follows $T_n f \in L_{\phi_n^{1/(1-p_n)}}^{q_n}(0,\infty)$.

For $f_n \in L_{\phi_n}^{p_n}(0,\infty)$, define the formal inner product of $T_n f$ and f_n as

$$(T_n f, f_n) := \int_0^\infty \cdots \int_0^\infty k_\lambda(x_1,\cdots,x_n)$$
$$\times \prod_{i=1}^{n} f_i(x_i) dx_1 \cdots dx_n. \qquad (5.1.24)$$

We may rewrite (5.1.12) and (5.1.17) to the following equivalent forms:

$$(T_n f, f_n) < k_\lambda \prod_{i=1}^{n} \| f_i \|_{p_i,\phi_i}, \qquad (5.1.25)$$
$$\| T_n f \|_{q_n,\phi^{1/(1-p_n)}} < k_\lambda \prod_{i=1}^{n-1} \| f_i \|_{p_i,\phi_i}. \qquad (5.1.26)$$

In view of (5.1.20), it follows that T_n is a bounded operator with the norm $\| T_n \| \le k_\lambda$, where

$$\| T_n \| := \sup_{f \ne 0} \frac{\| T_n f \|_{q_n,\phi^{1/(1-p_n)}}}{\prod_{i=1}^{n-1} \| f_i \|_{p_i,\phi_i}}. \qquad (5.1.27)$$

Since the constant factor k_λ in (5.1.26) is the best possible, we conclude that $\| T_n \| = k_\lambda$.

5.1.5 TWO MULTIPLE EQUIVALENT INTEGRAL INEQUALITIES WITH VARIABLES AND THE REVERSE

Theorem 5.1.9 Suppose that $n \in \mathbf{N} \setminus \{1\}$, $p_i \ne 0,1$, $r_i \ne 0\ (i = 1,2,\cdots,n)$, $\sum_{i=1}^{n} \frac{1}{p_i} = 1$, $q_n = \frac{p_n}{p_n - 1}$, $\sum_{i=1}^{n} \frac{1}{r_i} = 1$, $\lambda \in \mathbf{R}$, $k_\lambda(x_1,\cdots,x_n) (\ge 0)$ is a homogeneous function of $-\lambda$-degree in R_+^n, and for $i = 1,2,\cdots,n$, there exists $\delta > 0$, such that for $| \eta_j | < \delta (j = 1,\cdots,n-1)$,

$$K_\lambda(\eta_1,\cdots,\eta_{n-1}) := \int_0^\infty \cdots \int_0^\infty k_\lambda(u_1,\cdots,u_{n-1})$$

$$\times\prod_{j=1}^{n-1}u_j^{\frac{\lambda}{r_j}-\frac{\eta_j}{p_j}-1}du_1\cdots du_{n-1}$$

is a positive constant,

$$k_\lambda = k_\lambda(r_1,\cdots,r_{n-1}) = K_\lambda(0,\cdots,0).$$

If $v_i(x)$ are strict increasing deliverable functions in (a,b), $v_i(a^+)=0$, $v_i(b^-)=\infty$, $f_i \geq 0$, satisfying

$$0<\int_a^b \frac{(v_i(x))^{p_i(1-\frac{\lambda}{r_i})-1}}{(v_i'(x))^{p_i-1}}f_i^{p_i}(x)dx<\infty \ (i=1,\cdots,n),$$

setting

$$\tilde{k}_\lambda(x_1,\cdots,x_n):=k_\lambda(v_1(x_1),\cdots,v_n(x_n)),$$

then (1) For $p_i>1$ $(i=1,\cdots,n)$, we have the following equivalent inequalities:

$$\int_a^b\cdots\int_a^b \tilde{k}_\lambda(x_1,\cdots,x_n)\prod_{i=1}^n f_i(x_i)dx_1\cdots dx_n$$

$$<k_\lambda\prod_{i=1}^n\{\int_a^b\frac{[v_i(x)]^{p_i(1-\frac{\lambda}{r_i})-1}}{[v_i'(x)]^{p_i-1}}f_i^{p_i}(x)dx\}^{\frac{1}{p_i}}, \quad (5.1.28)$$

$$\{\int_a^b\frac{v_n'(x_n)}{[v_n(x_n)]^{1-\frac{q_n\lambda}{r_n}}}[\int_a^b\cdots\int_a^b\tilde{k}_\lambda(x_1,\cdots,x_n)$$

$$\times\prod_{i=1}^{n-1}f_i(x_i)dx_1\cdots dx_{n-1}]^{q_n}dx_n\}^{\frac{1}{q_n}}$$

$$<k_\lambda\prod_{i=1}^{n-1}\{\int_a^b\frac{[v_i(x)]^{p_i(1-\frac{\lambda}{r_i})-1}}{[v_i'(x)]^{p_i-1}}f_i^{p_i}(x)dx\}^{\frac{1}{p_i}}, \quad (5.1.29)$$

where the constant factor k_λ is the best possible; (2) for $0<p_1<1, p_i<0$ $(i=2,\cdots,n)$, we have the equivalent reverses of (5.1.28) and (5.1.29) with the same best constant factor k_λ. In particular, for $r_n=2$, setting

$$K_\lambda(x_1,\cdots,x_n):$$
$$=k_\lambda(v_n(x_n)v_1(x_1),\cdots,v_n(x_n)v_{n-1}(x_{n-1}),1),$$

if $p_i>1$ $(i=1,\cdots,n)$, then we have the following equivalent inequalities:

$$\int_a^b\cdots\int_a^b K_\lambda(x_1,\cdots,x_n)\prod_{i=1}^n f_i(x_i)dx_1\cdots dx_n$$

$$<k_\lambda\prod_{i=1}^n\{\int_a^b\frac{[v_i(x)]^{p_i(1-\frac{\lambda}{r_i})-1}}{[v_i'(x)]^{p_i-1}}f_i^{p_i}(x)dx\}^{\frac{1}{p_i}}, \quad (5.1.30)$$

$$\{\int_a^b\frac{v_n'(x_n)}{[v_n(x_n)]^{1-\frac{q_n\lambda}{2}}}[\int_a^b\cdots\int_a^b K_\lambda(x_1,\cdots,x_n)$$

$$\times\prod_{i=1}^{n-1}f_i(x_i)dx_1\cdots dx_{n-1}]^{q_n}dx_n\}^{\frac{1}{q_n}}$$

$$<k_\lambda\prod_{i=1}^{n-1}\{\int_a^b\frac{[v_i(x)]^{p_i(1-\frac{\lambda}{r_i})-1}}{[v_i'(x)]^{p_i-1}}f_i^{p_i}(x)dx\}^{\frac{1}{p_i}}, \quad (5.1.31)$$

where the constant factor k_λ is the best possible; (2) if $0<p_1<1, p_i<0$ $(i=2,\cdots,n)$, we have the equivalent reverses of (5.1.30) and (5.1.31) with the same best constant factor k_λ.

Proof (1) For $p_i>1 (i=1,\cdots,n)$, replacing x_i by $v_i(x_i)$ in (5.1.12) and (5.1.17), by calculations, then replacing $v_i'(x_i)f_i(v_i(x_i))$ by $f_i(x_i)$, we have (5.1.28) and (5.1.29), which are equivalent respectively to (5.1.12) and (5.1.17). We can show that (5.1.28) and (5.1.29) are equivalent and the constant factor k_λ is the best possible in (5.1.28) and (5.1.29) by their equivalent relationship. (2) By the same way, we can show the cases in $0<p_1<1,\ p_i<0$ $(i=2,\cdots,n)$.□

For example, (i) Setting $v_i(x)=x^{\alpha_i} (\alpha_i>0)$, $a=0, b=\infty$, (1) for $p_i>1$ $(i=1,\cdots,n)$, we have the following equivalent inequalities:

$$\int_0^\infty\cdots\int_0^\infty k_\lambda(x_1^{\alpha_1},\cdots,x_n^{\alpha_n})\prod_{i=1}^n f_i(x_i)dx_1\cdots dx_n$$

$$<k_\lambda\prod_{i=1}^n\frac{1}{\alpha_i^{1-1/p_i}}\{\int_0^\infty x^{p_i(1-\frac{\alpha_i\lambda}{r_i})-1}f_i^{p_i}(x)dx\}^{\frac{1}{p_i}}, \quad (5.1.32)$$

$$\{\int_0^\infty x_n^{\frac{q_n\lambda\alpha_n}{r_n}-1}[\int_0^\infty\cdots\int_0^\infty k_\lambda(x_1^{\alpha_1},\cdots,x_n^{\alpha_n})$$

$$\times\prod_{i=1}^{n-1}f_i(x_i)dx_1\cdots dx_{n-1}]^{q_n}dx_n\}^{\frac{1}{q_n}}$$

$$<k_\lambda\prod_{i=1}^{n-1}\frac{1}{\alpha_i^{1-1/p_i}}\{\int_0^\infty x^{p_i(1-\frac{\alpha_i\lambda}{r_i})-1}f_i^{p_i}(x)dx\}^{\frac{1}{p_i}}; \quad (5.1.33)$$

(2) for $0<p_1<1, p_i<0$ $(i=2,\cdots,n)$, we have the equivalent reverses of (5.1.32) and (5.1.33). In particular, for $r_n=2$, if $p_i>1$ $(i=1,\cdots,n)$, then we have the following equivalent inequalities:

$$\int_0^\infty\cdots\int_0^\infty k_\lambda(x_n^{\alpha_n}x_1^{\alpha_1},\cdots,x_n^{\alpha_n}x_{n-1}^{\alpha_{n-1}},1)$$

$$\times\prod_{i=1}^n f_i(x_i)dx_1\cdots dx_n$$

$$<k_\lambda\prod_{i=1}^n\frac{1}{\alpha_i^{1-1/p_i}}\{\int_0^\infty x^{p_i(1-\frac{\alpha_i\lambda}{r_i})-1}f_i^{p_i}(x)dx\}^{\frac{1}{p_i}}, \quad (5.1.34)$$

$$\{\int_0^\infty x_n^{\frac{q_n\lambda\alpha_n}{2}-1}[\int_0^\infty\cdots\int_0^\infty k_\lambda(x_n^{\alpha_n}x_1^{\alpha_1},\cdots,x_n^{\alpha_n}x_{n-1}^{\alpha_{n-1}},1)$$

$$\times\prod_{i=1}^{n-1} f_i(x_i)dx_1\cdots dx_{n-1}]^{q_n} dx_n\}^{\frac{1}{q_n}}$$

$$< k_\lambda \prod_{i=1}^{n-1} \frac{1}{\alpha_i^{1-1/p_i}}\{\int_0^\infty x^{p_i(1-\frac{\alpha_i\lambda}{r_i})-1} f_i^{p_i}(x)dx\}^{\frac{1}{p_i}}; \quad (5.1.35)$$

(2) if $0 < p_1 < 1, p_i < 0$ $(i = 2,\cdots,n)$, then we have the equivalent reverses of (5.1.34) and (5.1.35).

(ii) Setting $v_i(x) = \ln x, a = 1, b = \infty$, (1) for $p_i > 1$ $(i = 1,\cdots,n)$, we have the following equivalent inequalities:

$$\int_1^\infty\cdots\int_1^\infty k_\lambda(\ln x_1,\cdots,\ln x_n)\times\prod_{i=1}^{n} f_i(x_i)dx_1\cdots dx_n$$

$$< k_\lambda \prod_{i=1}^{n}\{\int_1^\infty x^{p_i-1}(\ln x)^{p_i(1-\frac{\lambda}{r_i})-1} f_i^{p_i}(x)dx\}^{\frac{1}{p_i}}, \quad (5.1.36)$$

$$\{\int_1^\infty \frac{1}{x_n}(\ln x_n)^{\frac{q_n\lambda}{r_n}}[\int_1^\infty\cdots\int_1^\infty k_\lambda(\ln x_1,\cdots,\ln x_n)\times\prod_{i=1}^{n-1} f_i(x_i)dx_1\cdots dx_{n-1}]^{q_n} dx_n\}^{\frac{1}{q_n}}$$

$$< k_\lambda \prod_{i=1}^{n-1}\{\int_1^\infty x^{p_i-1}(\ln x)^{p_i(1-\frac{\lambda}{r_i})-1} f_i^{p_i}(x)dx\}^{\frac{1}{p_i}}; \quad (5.1.37)$$

(2) for $0 < p_1 < 1, p_i < 0$ $(i = 2,\cdots,n)$, we have the equivalent reverses of (5.1.36) and (5.1.37). In particular, for $r_n = 2$, (1) if $p_i > 1$ $(i = 1,\cdots,n)$, we have the following equivalent inequalities:

$$\int_1^\infty\cdots\int_1^\infty k_\lambda(\ln x_n \ln x_1,\cdots,\ln x_n \ln x_{n-1},1)\times\prod_{i=1}^{n} f_i(x_i)dx_1\cdots dx_n$$

$$< k_\lambda \prod_{i=1}^{n}\{\int_1^\infty x^{p_i-1}(\ln x)^{p_i(1-\frac{\lambda}{r_i})-1} f_i^{p_i}(x)dx\}^{\frac{1}{p_i}}, \quad (5.1.38)$$

$$\{\int_1^\infty \frac{1}{x_n}(\ln x_n)^{\frac{q_n\lambda}{2}}\times[\int_1^\infty\cdots\int_1^\infty k_\lambda(\ln x_n \ln x_1,\cdots,\ln x_n \ln x_{n-1},1)\times\prod_{i=1}^{n-1} f_i(x_i)dx_1\cdots dx_{n-1}]^{q_n} dx_n\}^{\frac{1}{q_n}}$$

$$< k_\lambda \prod_{i=1}^{n-1}\{\int_1^\infty x^{p_i-1}(\ln x)^{p_i(1-\frac{\lambda}{r_i})-1} f_i^{p_i}(x)dx\}^{\frac{1}{p_i}}; \quad (5.1.39)$$

(2) for $0 < p_1 < 1, p_i < 0$ $(i = 2,\cdots,n)$, we have the equivalent reverses of (5.1.38) and (5.1.39).

5.2 SOME PARTICULAR RESULTS

5.2.1 TWO MULTIPLE INTEGRAL INEQUALITIES WITH THE HOMOGENEOUS KERNELS OF NEGATIVE NUMBER-DEGREE

Lemma 5.2.1 If $n \in \mathbf{N}\setminus\{1\}, P_i, a_i, \alpha_i > 0$ $(i = 1,\cdots,n-1), \psi(u) \geq 0$ is measurable,

$$D = \{(x_1,\cdots,x_{n-1}) \mid x_i > 0 (i = 1,\cdots,n-1), \sum_{i=1}^{n-1}(\tfrac{x_i}{a_i})^{\alpha_i} \leq 1\},$$

then (Wang SP 1979) [1]

$$\int\cdots\int_D \psi(\sum_{i=1}^{n-1}(\tfrac{x_i}{a_i})^{\alpha_i})\prod_{i=1}^{n-1} x_i^{P_i-1}dx_1\cdots dx_{n-1}$$

$$= \frac{\prod_{i=1}^{n-1} a_i^{P_i}\Gamma(\frac{P_i}{\alpha_i})}{\prod_{i=1}^{n-1}\alpha_i\Gamma(\sum_{i=1}^{n-1}\frac{P_i}{\alpha_i})}\int_0^1 \psi(u)u^{\sum_{i=1}^{n-1}\frac{P_i}{\alpha_i}-1}du. \quad (5.2.1)$$

In particular, for $a_i = r, \alpha_i = 1$ $(i = 1,\cdots,n-1)$, setting

$$\tilde{D} = \{(x_1,\cdots,x_{n-1}) \mid x_i > 0 (i = 1,\cdots,n-1), \sum_{i=1}^{n-1} x_i \leq r\},$$

then by (5.2.1), we have

$$\int\cdots\int_{\tilde{D}} \psi(\sum_{i=1}^{n-1}(\tfrac{x_i}{r}))\prod_{i=1}^{n-1} x_i^{P_i-1}dx_1\cdots dx_{n-1}$$

$$= \frac{r^{\sum_{i=1}^{n-1}P_i}\prod_{i=1}^{n-1}\Gamma(P_i)}{\Gamma(\sum_{i=1}^{n-1}P_i)}\int_0^1 \psi(u)u^{\sum_{i=1}^{n-1}P_i-1}du. \quad (5.2.2)$$

If $\psi(u) = \frac{1}{(1+ru)^\lambda}$, $r_i > 1, \sum_{i=1}^{n}\frac{1}{r_i} = 1, \lambda > 0$, $P_i = \frac{\lambda}{r_i}$ $(i = 1,\cdots,n-1)$, then $\sum_{i=1}^{n-1} P_i = \lambda - \frac{\lambda}{r_n}$. Since

$$r^{\sum_{i=1}^{n-1}P_i}\int_0^1 \psi(u)u^{\sum_{i=1}^{n-1}P_i-1}du$$

$$= r^{\lambda(1-\frac{1}{r_n})}\int_0^1 \frac{1}{(1+ru)^\lambda}u^{\lambda(1-\frac{1}{r_n})-1}du$$

$$= \int_0^r \frac{1}{(1+v)^\lambda}v^{\lambda(1-\frac{1}{r_n})-1}dv,$$

By (5.2.2), for $r \to \infty$, we find

$$\int\cdots\int_{\mathbb{R}_+^{n-1}} \frac{1}{(1+\sum_{j=1}^{n-1}x_j)^\lambda}\prod_{i=1}^{n-1} x_i^{\frac{\lambda}{r_i}-1}dx_1\cdots dx_{n-1}$$

$$= \frac{\prod_{i=1}^{n-1}\Gamma(\frac{\lambda}{r_i})}{\Gamma(\sum_{i=1}^{n-1}\frac{\lambda}{r_i})}\int_0^\infty \frac{v^{\lambda(1-\frac{1}{r_n})-1}}{(1+v)^\lambda}dv = \frac{\prod_{i=1}^{n}\Gamma(\frac{\lambda}{r_i})}{\Gamma(\lambda)} .$$

(5.2.3)

Example 5.2.2 If $\lambda > 0, r_i > 1$,

$$k_\lambda(x_1,\cdots,x_n) = \frac{1}{(x_1+\cdots+x_n)^\lambda},$$

then by (5.2.3), we have

$$k_\lambda = \frac{\prod_{i=1}^{n}\Gamma(\frac{\lambda}{r_i})}{\Gamma(\lambda)}.$$

By Theorem 5.1.6 and Theorem 5.1.7, (1) for $p_i > 1$ $(i = 1,2,\cdots,n)$, we have the following equivalent inequalities (Yang MIA 2005) [2]:

$$\int_0^\infty\cdots\int_0^\infty \frac{1}{(\sum_{i=1}^n x_i)^\lambda}\prod_{i=1}^n f_i(x_i)dx_1\cdots dx_n$$
$$< \frac{1}{\Gamma(\lambda)}\prod_{i=1}^n \Gamma(\tfrac{\lambda}{r_i})\{\int_0^\infty x^{p_i(1-\frac{\lambda}{r_i})-1} f_i^{p_i}(x)dx\}^{\frac{1}{p_i}}, \quad (5.2.4)$$

$$\{\int_0^\infty x_n^{\frac{q_n\lambda}{r_n}-1}[\int_0^\infty\cdots\int_0^\infty \frac{1}{(\sum_{i=1}^n x_i)^\lambda}\prod_{i=1}^{n-1} f_i(x_i)$$
$$\times dx_1\cdots dx_{n-1}]^{q_n} dx_n\}^{\frac{1}{q_n}}$$
$$< \frac{\Gamma(\frac{\lambda}{r_n})}{\Gamma(\lambda)}\prod_{i=1}^{n-1} \Gamma(\tfrac{\lambda}{r_i})\{\int_0^\infty x^{p_i(1-\frac{\lambda}{r_i})-1} f_i^{p_i}(x)dx\}^{\frac{1}{p_i}}, \quad (5.2.5)$$

where the constant factor $\frac{1}{\Gamma(\lambda)}\prod_{i=1}^n\Gamma(\frac{\lambda}{r_i})$ is the best possible; (2) for $0 < p_1 < 1, p_i < 0$, $i = 2,\cdots,n$, we have the equivalent reverses of (5.2.4) and (5.2.5) with the same best constant factor.

In particular, for $r_n = 2$, (1) if $p_i > 1$, $i = 1,\cdots,n$, then we have the following equivalent inequalities:

$$\int_0^\infty\cdots\int_0^\infty \frac{1}{(x_n\sum_{i=1}^{n-1} x_i+1)^\lambda}\prod_{i=1}^n f_i(x_i)dx_1\cdots dx_n$$
$$< \frac{1}{\Gamma(\lambda)}\prod_{i=1}^n \Gamma(\tfrac{\lambda}{r_i})\{\int_0^\infty x^{p_i(1-\frac{\lambda}{r_i})-1} f_i^{p_i}(x)dx\}^{\frac{1}{p_i}}, \quad (5.2.6)$$

$$\{\int_0^\infty x_n^{\frac{q_n\lambda}{2}-1}[\int_0^\infty\cdots\int_0^\infty \frac{1}{(x_n\sum_{i=1}^{n-1} x_i+1)^\lambda}\prod_{i=1}^{n-1} f_i(x_i)$$
$$\times dx_1\cdots dx_{n-1}]^{q_n} dx_n\}^{\frac{1}{q_n}}$$
$$< \frac{\Gamma(\frac{\lambda}{2})}{\Gamma(\lambda)}\prod_{i=1}^{n-1} \Gamma(\tfrac{\lambda}{r_i})\{\int_0^\infty x^{p_i(1-\frac{\lambda}{r_i})-1} f_i^{p_i}(x)dx\}^{\frac{1}{p_i}}; \quad (5.2.7)$$

(2) if $0 < p_1 < 1, p_i < 0$ $(i = 2,\cdots,n)$, then we have the equivalent reverses of (5.2.6) and (5.2.7).

(i) By (5.1.32)- (5.1.35), setting $\alpha_i = \alpha > 0$, (1) for $p_i > 1$ $(i = 1,\cdots,n)$, we have the following two pairs of equivalent inequalities:

$$\int_0^\infty\cdots\int_0^\infty \frac{1}{(\sum_{j=1}^n x_j^\alpha)^\lambda}\prod_{i=1}^n f_i(x_i)dx_1\cdots dx_n$$
$$< \frac{1}{\alpha^{n-1}\Gamma(\lambda)}\prod_{i=1}^n \Gamma(\tfrac{\lambda}{r_i})\{\int_0^\infty x^{p_i(1-\frac{\alpha\lambda}{r_i})-1} f_i^{p_i}(x)dx\}^{\frac{1}{p_i}},$$

(5.2.8)

$$\{\int_0^\infty x_n^{\frac{q_n\lambda\alpha}{r_n}-1}[\int_0^\infty\cdots\int_0^\infty \frac{1}{(\sum_{i=1}^n x_i^\alpha)^\lambda}\prod_{i=1}^{n-1} f_i(x_i)$$
$$\times dx_1\cdots dx_{n-1}]^{q_n} dx_n\}^{\frac{1}{q_n}}$$
$$< \frac{\Gamma(\frac{\lambda}{r_n})}{\alpha^{n-1}\Gamma(\lambda)}\prod_{i=1}^{n-1} \Gamma(\tfrac{\lambda}{r_i})\{\int_0^\infty x^{p_i(1-\frac{\lambda\alpha}{r_i})-1} f_i^{p_i}(x)dx\}^{\frac{1}{p_i}};$$

(5.2.9)

$$\int_0^\infty\cdots\int_0^\infty \frac{1}{(x_n^\alpha\sum_{j=1}^{n-1} x_j^\alpha+1)^\lambda}\prod_{i=1}^n f_i(x_i)dx_1\cdots dx_n$$
$$< \frac{\Gamma(\frac{\lambda}{2})}{\alpha^{n-1}\Gamma(\lambda)}\prod_{i=1}^{n-1} \Gamma(\tfrac{\lambda}{r_i})\{\int_0^\infty x^{p_i(1-\frac{\alpha\lambda}{r_i})-1} f_i^{p_i}(x)dx\}^{\frac{1}{p_i}}$$
$$\times\{\int_0^\infty x^{p_n(1-\frac{\alpha\lambda}{2})-1} f_n^{p_n}(x)dx\}^{\frac{1}{p_n}}, \quad (5.2.10)$$

$$\{\int_0^\infty x_n^{\frac{q_n\lambda\alpha}{2}-1}[\int_0^\infty\cdots\int_0^\infty \frac{1}{(x_n^\alpha\sum_{i=1}^{n-1} x_i^\alpha+1)^\lambda}\prod_{i=1}^{n-1} f_i(x_i)$$
$$\times dx_1\cdots dx_{n-1}]^{q_n} dx_n\}^{\frac{1}{q_n}}$$
$$< \frac{\Gamma(\frac{\lambda}{2})}{\alpha^{n-1}\Gamma(\lambda)}\prod_{i=1}^{n-1} \Gamma(\tfrac{\lambda}{r_i})\{\int_0^\infty x^{p_i(1-\frac{\lambda\alpha}{r_i})-1} f_i^{p_i}(x)dx\}^{\frac{1}{p_i}},$$

(5.2.11)

where the constant factors are the best possible ; (2) for $0 < p_1 < 1, p_i < 0$ $(i = 2,\cdots,n)$, we have the reverses of (5.2.8) - (5.2.11) with the best constant factors (in (5.2.10), (5.2.11) and the reverses, notice $\sum_{i=1}^{n-1}\frac{1}{r_i} = \frac{1}{2}$).

(ii) By (5.1.36) - (5.1.39), (1) for $p_i > 1$ $(i = 1,\cdots,n)$, we have the following two pairs of equivalent inequalities:

$$\int_1^\infty\cdots\int_1^\infty \frac{\prod_{i=1}^n f_i(x_i)}{\ln^\lambda(x_1\cdots x_n)}dx_1\cdots dx_n$$
$$< \frac{1}{\Gamma(\lambda)}\prod_{i=1}^n \Gamma(\tfrac{\lambda}{r_i})\{\int_1^\infty x^{p_i-1}(\ln x)^{p_i(1-\frac{\lambda}{r_i})-1} f_i^{p_i}(x)dx\}^{\frac{1}{p_i}},$$

(5.2.12)

$$\{\int_1^\infty \tfrac{1}{x_n}(\ln x_n)^{\frac{q_n\lambda}{r_n}}[\int_1^\infty\cdots\int_1^\infty \tfrac{1}{\ln^\lambda(x_1\cdots x_n)}\prod_{i=1}^{n-1} f_i(x_i)$$

$$\times dx_1\cdots dx_{n-1}]^{q_n} dx_n\}^{\frac{1}{q_n}}$$

$$<\tfrac{\Gamma(\frac{\lambda}{r_n})}{\Gamma(\lambda)}\prod_{i=1}^{n-1}\Gamma(\tfrac{\lambda}{r_i})$$

$$\times\{\int_1^\infty x^{p_i-1}(\ln x)^{p_i(1-\frac{\lambda}{r_i})-1} f_i^{p_i}(x)dx\}^{\frac{1}{p_i}} \quad ;$$

(5.2.13)

$$\int_1^\infty\cdots\int_1^\infty \tfrac{\prod_{i=1}^n f_i(x_i)}{[\ln x_n \ln(x_1\cdots x_{n-1})+1]^\lambda}dx_1\cdots dx_n$$

$$<\tfrac{\Gamma(\frac{\lambda}{2})}{\Gamma(\lambda)}\prod_{i=1}^{n-1}\Gamma(\tfrac{\lambda}{r_i})\{\int_1^\infty x^{p_i-1}(\ln x)^{p_i(1-\frac{\lambda}{r_i})-1} f_i^{p_i}(x)dx\}^{\frac{1}{p_i}}$$

$$\times\{\int_1^\infty x^{p_n-1}(\ln x)^{p_n(1-\frac{\lambda}{2})-1}f_n^{p_n}(x)dx\}^{\frac{1}{p_n}}, \quad (5.2.14)$$

$$\{\int_1^\infty \tfrac{1}{x_n}(\ln x_n)^{\frac{q_n\lambda}{2}}[\int_1^\infty\cdots\int_1^\infty \tfrac{\prod_{i=1}^n f_i(x_i)}{[\ln x_n \ln(x_1\cdots x_{n-1})+1]^\lambda}$$

$$\times dx_1\cdots dx_{n-1}]^{q_n} dx_n\}^{\frac{1}{q_n}}$$

$$<\tfrac{\Gamma(\frac{\lambda}{2})}{\Gamma(\lambda)}\prod_{i=1}^{n-1}\Gamma(\tfrac{\lambda}{r_i})$$

$$\times\{\int_1^\infty x^{p_i-1}(\ln x)^{p_i(1-\frac{\lambda}{r_i})-1} f_i^{p_i}(x)dx\}^{\frac{1}{p_i}} \quad ,$$

(5.2.15)

where the constant factors are the best possible; (2) for $0<p_1<1, p_i<0$ $(i=2,\cdots,n)$, we have the reverses of (5.2.12) -(5.2.15) with the best constant factors (in (5.2.13) , (5.2.14) and the reverses, notice $\sum_{i=1}^{n-1}\frac{1}{r_i}=\frac{1}{2}$).

Example 5.2.3 If $\lambda>0, r_i>1$,

$$k_\lambda(x_1,\cdots,x_n)=\tfrac{1}{(\max\{x_1,\cdots,x_n\})^\lambda},$$

Then by (5.1.2), we set

$$H(n)=\int_0^\infty\cdots\int_0^\infty \tfrac{\prod_{j=1}^{n-1}u_j^{\frac{\lambda}{r_j}-1}}{(\max\{u_1,\cdots u_{n-1},1\})^\lambda}du_1\cdots du_{n-1}. \quad (5.2.16)$$

In the following, we prove that

$$H(n)=\tfrac{1}{\lambda^{n-1}}\prod_{i=1}^n r_i, n\in\mathbf{N}\backslash\{1\}. \quad (5.2.17)$$

For $n=2$,

$$H(2)=\int_0^\infty \tfrac{1}{(\max\{u_1,1\})^\lambda}u_1^{\frac{\lambda}{r_1}-1}du_1$$

$$=\int_0^1 u_1^{\frac{\lambda}{r_1}-1}du_1+\int_1^\infty u_1^{-\frac{\lambda}{r_2}-1}du_1=\tfrac{1}{\lambda}r_1r_2,$$

and (5.2.17) is valid. Assuming that for $n(\geq 2)$, (5.2.17) is valid, then for $n+1$, since $\sum_{i=1}^{n+1}\frac{1}{r_i}=1$,

$$H(n+1)=\int_0^\infty\cdots\int_0^\infty \tfrac{u_1^{\frac{\lambda}{r_1}-1}\cdots u_n^{\frac{\lambda}{r_n}-1}}{(\max_{1\le i\le n}\{u_i,1\})^\lambda}du_1\cdots du_n$$

$$=\int_0^\infty\cdots\int_0^\infty\prod_{i=2}^n u_i^{\frac{\lambda}{r_i}-1}[\int_0^\infty \tfrac{u_1^{\frac{\lambda}{r_1}-1}du_1}{(\max_{1\le i\le n}\{u_i,1\})^\lambda}]du_2\cdots du_n$$

$$=\int_0^\infty\cdots\int_0^\infty\prod_{i=2}^n u_i^{\frac{\lambda}{r_i}-1}[\int_0^{\max\{u_2,\cdots u_n,1\}} \tfrac{u_1^{\frac{\lambda}{r_1}-1}du_1}{(\max_{1\le i\le n}\{u_i,1\})^\lambda}$$

$$+\int_{\max\{u_2,\cdots u_n,1\}}^\infty \tfrac{u_1^{\frac{\lambda}{r_1}-1}du_1}{(\max_{1\le i\le n}\{u_i,1\})^\lambda}]du_2\cdots du_n$$

$$=\int_0^\infty\cdots\int_0^\infty\prod_{i=2}^n u_i^{\frac{\lambda}{r_i}-1}[\int_0^{\max\{u_2,\cdots u_n,1\}} \tfrac{u_1^{\frac{\lambda}{r_1}-1}du_1}{(\max_{2\le i\le n}\{u_i,1\})^\lambda}$$

$$+\int_{\max\{u_2,\cdots u_n,1\}}^\infty \tfrac{u_1^{\frac{\lambda}{r_1}-1}}{u_1^\lambda}du_1]du_2\cdots du_n$$

$$=\tfrac{r_1}{\lambda}\tfrac{r_1}{r_1-1}\int_0^\infty\cdots\int_0^\infty \tfrac{\prod_{i=2}^n u_i^{\frac{\lambda}{r_i}-1}}{(\max_{2\le i\le n}\{u_i,1\})^{\lambda(1-\frac{1}{r_1})}}du_2\cdots du_n. \quad (5.2.18)$$

Setting $v_i=u_i^{1-\frac{1}{r_1}}$ in (5.2.18), then since $u_i=v_i^{\frac{r_1}{r_1-1}}$,

$$du_i=\tfrac{r_1}{r_1-1}v_i^{\frac{1}{r_1-1}}dv_i \ (i=2,\cdots,n),$$

we have

$$H(n+1)=\tfrac{r_1}{\lambda}(\tfrac{r_1}{r_1-1})^n$$

$$\times\int_0^\infty\cdots\int_0^\infty \tfrac{1}{(\max\{v_2,\cdots v_n,1\})^\lambda}\prod_{i=2}^n v_i^{\frac{\lambda}{r_i}(\frac{r_1}{r_1-1})-1}dv_2\cdots dv_n.$$

Since $\sum_{i=2}^{n+1}\frac{1}{r_i}\cdot\frac{r_1}{r_1-1}=1$, by the assumption, we find

$$H(n+1)=\tfrac{r_1}{\lambda}(\tfrac{r_1}{r_1-1})^n\tfrac{1}{\lambda^{n-1}}\prod_{i=2}^{n+1}r_i(\tfrac{r_1-1}{r_1})=\tfrac{1}{\lambda^n}\prod_{i=1}^{n+1}r_i.$$

Then for $n+1$, (5.2.17) is valid. By mathematical induction, we show that for $n\in\mathbf{N}\backslash\{1\}$, (5.2.17) is valid.

By Lemma 5.1.3 and (5.2.17), setting $k_\lambda=\frac{1}{\lambda^{n-1}}\prod_{i=1}^n r_i$, by Theorem 5.1.6 and Theorem 5.1.7, (1) for $p_i>1$ $(i=1,\cdots,n)$, we have the following equivalent inequalities:

$$\int_0^\infty\cdots\int_0^\infty \tfrac{1}{(\max_{1\le i\le n}\{x_i\})^\lambda}\prod_{i=1}^n f_i(x_i)dx_1\cdots dx_n$$

$$<\tfrac{1}{\lambda^{n-1}}\prod_{i=1}^n r_i\{\int_0^\infty x^{p_i(1-\frac{\lambda}{r_i})-1}f_i^{p_i}(x)dx\}^{\frac{1}{p_i}}, \quad (5.2.19)$$

$$\{\int_0^\infty x_n^{\frac{q_n\lambda}{r_n}-1}[\int_0^\infty\cdots\int_0^\infty \tfrac{1}{(\max_{1\le i\le n}\{x_i\})^\lambda}\prod_{i=1}^{n-1}f_i(x_i)$$

$$\times dx_1\cdots dx_{n-1}]^{q_n}dx_n\}^{\frac{1}{q_n}}$$

$$<\frac{r_n}{\lambda^{n-1}}\prod_{i=1}^{n-1} r_i\{\int_0^\infty x^{p_i(1-\frac{\lambda}{r_i})-1} f_i^{p_i}(x)dx\}^{\frac{1}{p_i}}, \quad (5.2.20)$$

where the constant factor $\frac{1}{\lambda^{n-1}}\prod_{i=1}^{n} r_i$ is the best possible; (2) for $0<p_1<1, p_i<0\ (i=2,\cdots,n)$, we have the equivalent reverses of (5.2.19) and (5.2.20) with the same best constant factor.

In particular, for $r_n=2$, (1) if $p_i>1\ (i=1,\cdots,n)$, we have the following equivalent inequalities:

$$\int_0^\infty\cdots\int_0^\infty \frac{1}{(\max_{1\le i\le n-1}\{x_i x_n,1\})^\lambda}\prod_{i=1}^{n} f_i(x_i)dx_1\cdots dx_n$$

$$<\frac{1}{\lambda^{n-1}}\prod_{i=1}^{n} r_i\{\int_0^\infty x^{p_i(1-\frac{\lambda}{r_i})-1} f_i^{p_i}(x)dx\}^{\frac{1}{p_i}}, \quad (5.2.21)$$

$$\{\int_0^\infty x_n^{\frac{q_n\lambda}{2}-1}[\int_0^\infty\cdots\int_0^\infty \frac{1}{(\max_{1\le i\le n-1}\{x_n x_i,1\})^\lambda}\prod_{i=1}^{n-1} f_i(x_i)$$

$$\times dx_1\cdots dx_{n-1}]^{q_n}dx_n\}^{\frac{1}{q_n}}$$

$$<\frac{2}{\lambda^{n-1}}\prod_{i=1}^{n-1} r_i\{\int_0^\infty x^{p_i(1-\frac{\lambda}{r_i})-1} f_i^{p_i}(x)dx\}^{\frac{1}{p_i}}; \quad (5.2.22)$$

(2) if $0<p_1<1, p_i<0\ (i=2,\cdots,n)$, we have the equivalent reverses of (5.2.21) and (5.2.22) with the same best constant factors.

(i) By (5.1.32) - (5.1.35), setting $\alpha_i=\alpha>0$, (1) for $p_i>1\ (i=1,\cdots,n)$, we have the following two pairs of equivalent inequalities:

$$\int_0^\infty\cdots\int_0^\infty \frac{1}{(\max_{1\le i\le n}\{x_i^\alpha\})^\lambda}\prod_{i=1}^{n} f_i(x_i)dx_1\cdots dx_n$$

$$<\frac{1}{(\lambda\alpha)^{n-1}}\prod_{i=1}^{n} r_i\{\int_0^\infty x^{p_i(1-\frac{\alpha\lambda}{r_i})-1} f_i^{p_i}(x)dx\}^{\frac{1}{p_i}}, \quad (5.2.23)$$

$$\{\int_0^\infty x_n^{\frac{q_n\lambda\alpha}{r_n}-1}[\int_0^\infty\cdots\int_0^\infty \frac{1}{(\max_{1\le i\le n}\{x_i^\alpha\})^\lambda}\prod_{i=1}^{n-1} f_i(x_i)$$

$$\times dx_1\cdots dx_{n-1}]^{q_n}dx_n\}^{\frac{1}{q_n}}$$

$$<\frac{r_n}{(\lambda\alpha)^{n-1}}\prod_{i=1}^{n-1} r_i\{\int_0^\infty x^{p_i(1-\frac{\alpha\lambda}{r_i})-1} f_i^{p_i}(x)dx\}^{\frac{1}{p_i}}; \quad (5.2.24)$$

$$\int_0^\infty\cdots\int_0^\infty \frac{1}{(\max_{1\le i\le n-1}\{x_n^\alpha x_i^\alpha,1\})^\lambda}\prod_{i=1}^{n} f_i(x_i)dx_1\cdots dx_n$$

$$<\frac{2}{(\lambda\alpha)^{n-1}}\prod_{i=1}^{n-1} r_i\{\int_0^\infty x^{p_i(1-\frac{\alpha\lambda}{r_i})-1} f_i^{p_i}(x)dx\}^{\frac{1}{p_i}}$$

$$\times\{\int_0^\infty x^{p_n(1-\frac{\alpha\lambda}{2})-1} f_n^{p_n}(x)dx\}^{\frac{1}{p_n}}, \quad (5.2.25)$$

$$\{\int_0^\infty x_n^{\frac{q_n\lambda\alpha}{2}-1}[\int_0^\infty\cdots\int_0^\infty \frac{1}{(\max_{1\le i\le n-1}\{x_n^\alpha x_i^\alpha,1\})^\lambda}\prod_{i=1}^{n-1} f_i(x_i)$$

$$\times dx_1\cdots dx_{n-1}]^{q_n}dx_n\}^{\frac{1}{q_n}}$$

$$<\frac{2}{(\lambda\alpha)^{n-1}}\prod_{i=1}^{n-1} r_i\{\int_0^\infty x^{p_i(1-\frac{\alpha\lambda}{r_i})-1} f_i^{p_i}(x)dx\}^{\frac{1}{p_i}}, (5.2.26)$$

where the constant factors are the best possible; (2) for $0<p_1<1, p_i<0\ (i=2,\cdots,n)$, we have the reverses of (5.2.23) - (5.2.26) with the best constant factors (in (5.2.25), (5.2.26) and the reverses, notice $\sum_{i=1}^{n-1}\frac{1}{r_i}=\frac{1}{2}$).

(ii) By (5.1.36) - (5.1.39), (1) for $p_i>1\ (i=1,\cdots,n)$, we have the following two pairs of equivalent inequalities:

$$\int_1^\infty\cdots\int_1^\infty \frac{1}{(\max_{1\le i\le n}\{\ln x_i\})^\lambda}\prod_{i=1}^{n} f_i(x_i)dx_1\cdots dx_n$$

$$<\frac{1}{\lambda^{n-1}}\prod_{i=1}^{n} r_i\{\int_1^\infty x^{p_i-1}(\ln x)^{p_i(1-\frac{\lambda}{r_i})-1} f_i^{p_i}(x)dx\}^{\frac{1}{p_i}}, \quad (5.2.27)$$

$$\{\int_1^\infty \frac{1}{x_n}(\ln x_n)^{\frac{q_n\lambda}{r_n}}[\int_1^\infty\cdots\int_1^\infty \frac{\prod_{i=1}^{n-1} f_i(x_i)}{(\max_{1\le i\le n}\{\ln x_i\})^\lambda}$$

$$\times dx_1\cdots dx_{n-1}]^{q_n}dx_n\}^{\frac{1}{q_n}}$$

$$<\frac{r_n}{\lambda^{n-1}}\prod_{i=1}^{n-1} r_i\{\int_1^\infty x^{p_i-1}(\ln x)^{p_i(1-\frac{\lambda}{r_i})-1} f_i^{p_i}(x)dx\}^{\frac{1}{p_i}}; \quad (5.2.28)$$

$$\int_1^\infty\cdots\int_1^\infty \frac{1}{(\max_{1\le i\le n-1}\{\ln x_n\ln x_i,1\})^\lambda}\prod_{i=1}^{n} f_i(x_i)dx_1\cdots dx_n$$

$$<\frac{2}{\lambda^{n-1}}\prod_{i=1}^{n-1} r_i\{\int_1^\infty x^{p_i-1}(\ln x)^{p_i(1-\frac{\lambda}{r_i})-1} f_i^{p_i}(x)dx\}^{\frac{1}{p_i}}$$

$$\times\{\int_1^\infty x^{p_n-1}(\ln x)^{p_n(1-\frac{\lambda}{2})-1} f_n^{p_n}(x)dx\}^{\frac{1}{p_n}}, \quad (5.2.29)$$

$$\{\int_1^\infty \frac{1}{x_n}(\ln x_n)^{\frac{q_n\lambda}{2}}[\int_1^\infty\cdots\int_1^\infty \frac{\prod_{i=1}^{n-1} f_i(x_i)}{(\max_{1\le i\le n-1}\{\ln x_n\ln x_i,1\})^\lambda}$$

$$\times dx_1\cdots dx_{n-1}]^{q_n}dx_n\}^{\frac{1}{q_n}}$$

$$<\frac{2}{\lambda^{n-1}}\prod_{i=1}^{n-1} r_i\{\int_1^\infty x^{p_i-1}(\ln x)^{p_i(1-\frac{\lambda}{r_i})-1} f_i^{p_i}(x)dx\}^{\frac{1}{p_i}}, \quad (5.2.30)$$

where the constant factors are the best possible; (2) for $0<p_1<1, p_i<0\ (i=2,\cdots,n)$, we have the reverses of (5.2.27)-(5.2.30) with the best constant factors (in (5.2.29), (5.2.30) and the reverses, notice $\sum_{i=1}^{n-1}\frac{1}{r_i}=\frac{1}{2}$).

5.2.2 A MULTIPLE INTEGRAL INEQUALITY WITH THE HOMOGENEOUS KERNEL OF POSITIVE NUMBER-DEGREE

Example 5.2.4 If $\lambda > 0, r_i > 1$,

$$k_{-\lambda}(x_1,\cdots,x_n) = (\min\{x_1,\cdots,x_n\})^{\lambda},$$

by (5.1.2), we set

$$H(n) = \int_0^\infty \cdots \int_0^\infty (\min\{u_1,\cdots u_{n-1},1\})^{\lambda} \times \prod_{j=1}^{n-1} u_j^{\frac{-\lambda}{r_j}-1} du_1 \cdots du_{n-1}. \quad (5.2.31)$$

In the following, we prove that

$$H(n) = \frac{1}{\lambda^{n-1}} \prod_{i=1}^{n} r_i, n \in \mathbf{N}\backslash\{1\}. \quad (5.2.32)$$

For $n = 2$,

$$H(2) = \int_0^\infty (\min\{u_1,1\})^{\lambda} u_1^{\frac{-\lambda}{r_1}-1} du_1$$

$$= \int_0^1 u_1^{\frac{\lambda}{r_2}-1} du_1 + \int_1^\infty u_1^{\frac{-\lambda}{r_1}-1} du_1 = \frac{1}{\lambda} r_1 r_2,$$

and (5.2.32) is valid. Assuming that for $n(\geq 2)$, (5.2.32) is valid, then for $n+1$, since $\sum_{i=1}^{n+1} \frac{1}{r_i} = 1$,

$$H(n+1) = \int_0^\infty \cdots \int_0^\infty (\min_{1\leq i\leq n}\{u_i,1\})^{\lambda} \times u_1^{\frac{-\lambda}{r_1}-1} \cdots u_n^{\frac{-\lambda}{r_n}-1} du_1 \cdots du_n$$

$$= \int_0^\infty \cdots \int_0^\infty \prod_{i=2}^{n} u_i^{\frac{-\lambda}{r_i}-1} \times [\int_0^\infty (\min_{1\leq i\leq n}\{u_i,1\})^{\lambda} u_1^{\frac{-\lambda}{r_1}-1} du_1] du_2 \cdots du_n$$

$$= \int_0^\infty \cdots \int_0^\infty \prod_{i=2}^{n} u_i^{\frac{-\lambda}{r_i}-1} \times [\int_0^{\min\{u_2,\cdots u_n,1\}} (\min_{1\leq i\leq n}\{u_i,1\})^{\lambda} u_1^{\frac{-\lambda}{r_1}-1} du_1 + \int_{\min\{u_2,\cdots u_n,1\}}^\infty (\min_{1\leq i\leq n}\{u_i,1\})^{\lambda} u_1^{\frac{-\lambda}{r_1}-1} du_1] \times du_2 \cdots du_n$$

$$= \int_0^\infty \cdots \int_0^\infty \prod_{i=2}^{n} u_i^{\frac{-\lambda}{r_i}-1} [\int_0^{\min\{u_2,\cdots u_n,1\}} u_1^{\lambda} u_1^{\frac{-\lambda}{r_1}-1} du_1 + (\min\{u_2,\cdots u_n,1\})^{\lambda} \int_{\min\{u_2,\cdots u_n,1\}}^\infty u_1^{\frac{-\lambda}{r_1}-1} du_1] \times du_2 \cdots du_n$$

$$= \frac{r_1}{\lambda} \frac{r_1}{r_1-1} \int_0^\infty \cdots \int_0^\infty (\min\{u_2,\cdots u_n,1\})^{\lambda(1-\frac{1}{r_1})} \times \prod_{i=2}^{n} u_i^{\frac{-\lambda}{r_i}-1} du_2 \cdots du_n. \quad (5.2.33)$$

Setting $v_i = u_i^{1-\frac{1}{r_1}}$ in (5.2.33), then since $u_i = v_i^{\frac{r_1}{r_1-1}}$,

$$du_i = \frac{r_1}{r_1-1} v_i^{\frac{1}{r_1-1}} dv_i \ (i = 2,\cdots,n),$$

we have

$$H(n+1) = \frac{r_1}{\lambda}(\frac{r_1}{r_1-1})^n \int_0^\infty \cdots \int_0^\infty (\min\{v_2,\cdots v_n,1\})^{\lambda} \times \prod_{i=2}^{n} v_i^{\frac{-\lambda}{r_i}(\frac{r_1}{r_1-1})-1} dv_2 \cdots dv_n.$$

Since $\sum_{i=2}^{n+1} \frac{1}{r_i} \cdot \frac{r_1}{r_1-1} = 1$, by the assumption, we find

$$H(n+1) = (\frac{r_1}{r_1-1})^n \frac{r_1}{\lambda^n} \prod_{i=2}^{n+1} r_i (\frac{r_1-1}{r_1}) = \frac{1}{\lambda^n} \prod_{i=1}^{n+1} r_i.$$

Then for $n+1$, (5.2.32) is valid. By mathematical induction, we show that for $n \in \mathbf{N}\backslash\{1\}$, (5.2.32) is valid.

Setting $k_\lambda = \frac{1}{\lambda^{n-1}} \prod_{i=1}^{n} r_i$, by Theorem 5.1.6 and Theorem 5.1.7, (1) for $p_i > 1 \ (i = 1,\cdots,n)$, we have the following equivalent inequalities:

$$\int_0^\infty \cdots \int_0^\infty (\min_{1\leq i\leq n}\{x_i\})^{\lambda} \prod_{i=1}^{n} f_i(x_i) dx_1 \cdots dx_n < \frac{1}{\lambda^{n-1}} \prod_{i=1}^{n} r_i \{\int_0^\infty x^{p_i(1+\frac{\lambda}{r_i})-1} f_i^{p_i}(x) dx\}^{\frac{1}{p_i}}, \quad (5.2.34)$$

$$\{\int_0^\infty x_n^{\frac{-q_n\lambda}{r_n}-1} [\int_0^\infty \cdots \int_0^\infty (\min_{1\leq i\leq n}\{x_i\})^{\lambda} \prod_{i=1}^{n-1} f_i(x_i) \times dx_1 \cdots dx_{n-1}]^{q_n} dx_n\}^{\frac{1}{q_n}} < \frac{r_n}{\lambda^{n-1}} \prod_{i=1}^{n-1} r_i \{\int_0^\infty x^{p_i(1+\frac{\lambda}{r_i})-1} f_i^{p_i}(x) dx\}^{\frac{1}{p_i}}, \quad (5.2.35)$$

where the constant factor $\frac{1}{\lambda^{n-1}} \prod_{i=1}^{n} r_i$ is the best possible; (2) for $0 < p_1 < 1, p_i < 0 \ (i = 2,\cdots,n)$, we have the equivalent reverses of (5.2.34) and (5.2.35) with the best constant factors.

In particular, for $r_n = 2$, (1) if $p_i > 1$ $(i = 1,2,\cdots,n)$, we have the following equivalent inequalities:

$$\int_0^\infty \cdots \int_0^\infty (\min_{1\leq i\leq n-1}\{x_i x_n,1\})^{\lambda} \prod_{i=1}^{n} f_i(x_i) dx_1 \cdots dx_n < \frac{1}{\lambda^{n-1}} \prod_{i=1}^{n} r_i \{\int_0^\infty x^{p_i(1+\frac{\lambda}{r_i})-1} f_i^{p_i}(x) dx\}^{\frac{1}{p_i}}, \quad (5.2.36)$$

$$\{\int_0^\infty x_n^{\frac{-q_n\lambda}{2}-1}[\int_0^\infty\cdots\int_0^\infty(\min_{1\le i\le n-1}\{x_ix_n,1\})^\lambda\prod_{i=1}^{n-1}f_i(x_i)$$

$$\times dx_1\cdots dx_{n-1}]^{q_n}dx_n\}^{\frac{1}{q_n}}$$

$$<\frac{2}{\lambda^{n-1}}\prod_{i=1}^{n-1}r_i\{\int_0^\infty x^{p_i(1+\frac{\lambda}{r_i})-1}f_i^{p_i}(x)dx\}^{\frac{1}{p_i}};\quad(5.2.37)$$

(2) if $0<p_1<1,p_i<0\ (i=2,\cdots,n)$, we have the equivalent reverses of (5.2.36) and (5.2.37).

(i) By (5.1.32) - (5.1.35), setting $\alpha_i=\alpha>0$, (1) for $p_i>1\ (i=1,\cdots,n)$, we have the following two pairs of equivalent inequalities:

$$\int_0^\infty\cdots\int_0^\infty(\min_{1\le i\le n}\{x_i^\alpha\})^\lambda\prod_{i=1}^{n}f_i(x_i)dx_1\cdots dx_n$$

$$<\frac{1}{(\lambda\alpha)^{n-1}}\prod_{i=1}^{n}r_i\{\int_0^\infty x^{p_i(1+\frac{\alpha\lambda}{r_i})-1}f_i^{p_i}(x)dx\}^{\frac{1}{p_i}},\quad(5.2.38)$$

$$\{\int_0^\infty x_n^{\frac{-q_n\lambda\alpha}{r_n}-1}[\int_0^\infty\cdots\int_0^\infty(\min_{1\le i\le n}\{x_i^\alpha\})^\lambda\prod_{i=1}^{n-1}f_i(x_i)$$

$$\times dx_1\cdots dx_{n-1}]^{q_n}dx_n\}^{\frac{1}{q_n}}$$

$$<\frac{r_n}{(\lambda\alpha)^{n-1}}\prod_{i=1}^{n-1}r_i\{\int_0^\infty x^{p_i(1+\frac{\alpha\lambda}{r_i})-1}f_i^{p_i}(x)dx\}^{\frac{1}{p_i}};\quad(5.2.39)$$

$$\int_0^\infty\cdots\int_0^\infty(\min_{1\le i\le n-1}\{x_n^\alpha x_i^\alpha,1\})^\lambda\prod_{i=1}^{n}f_i(x_i)dx_1\cdots dx_n$$

$$<\frac{2}{(\lambda\alpha)^{n-1}}\prod_{i=1}^{n-1}r_i\{\int_0^\infty x^{p_i(1+\frac{\alpha\lambda}{r_i})-1}f_i^{p_i}(x)dx\}^{\frac{1}{p_i}}$$

$$\times\{\int_0^\infty x^{p_n(1+\frac{\alpha\lambda}{2})-1}f_n^{p_n}(x)dx\}^{\frac{1}{p_n}},\quad(5.2.40)$$

$$\{\int_0^\infty x_n^{\frac{-q_n\lambda\alpha}{2}-1}[\int_0^\infty\cdots\int_0^\infty(\min_{1\le i\le n-1}\{x_n^\alpha x_i^\alpha,1\})^\lambda$$

$$\times\prod_{i=1}^{n-1}f_i(x_i)dx_1\cdots dx_{n-1}]^{q_n}dx_n\}^{\frac{1}{q_n}}$$

$$<\frac{2}{(\lambda\alpha)^{n-1}}\prod_{i=1}^{n-1}r_i\{\int_0^\infty x^{p_i(1+\frac{\alpha\lambda}{r_i})-1}f_i^{p_i}(x)dx\}^{\frac{1}{p_i}},\quad(5.2.41)$$

where the constant factors are the best possible; (2) for $0<p_1<1,p_i<0\ (i=2,\cdots,n)$, we have the reverses of (5.2.38) - (5.2.41) with the best constant factors (in (5.2.40) , (5.2.41) and the reverses, notice $\sum_{i=1}^{n-1}\frac{1}{r_i}=\frac{1}{2}$).

(ii) By (5.1.36)- (5.1.39), (1) for $p_i>1$ $(i=1,\cdots,n)$, we have the following two pair s of equivalent inequalities:

$$\int_1^\infty\cdots\int_1^\infty(\min_{1\le i\le n}\{\ln x_i\})^\lambda\prod_{i=1}^{n}f_i(x_i)dx_1\cdots dx_n$$

$$<\frac{1}{\lambda^{n-1}}\prod_{i=1}^{n}r_i\{\int_1^\infty x^{p_i-1}(\ln x)^{p_i(1+\frac{\lambda}{r_i})-1}f_i^{p_i}(x)dx\}^{\frac{1}{p_i}},\quad(5.2.42)$$

$$\{\int_1^\infty\frac{1}{x_n}(\ln x_n)^{\frac{-q_n\lambda}{r_n}}[\int_1^\infty\cdots\int_1^\infty(\min_{1\le i\le n}\{\ln x_i\})^\lambda$$

$$\times\prod_{i=1}^{n}f_i(x_i)dx_1\cdots dx_{n-1}]^{q_n}dx_n\}^{\frac{1}{q_n}}$$

$$<\frac{r_n}{\lambda^{n-1}}\prod_{i=1}^{n-1}r_i\{\int_1^\infty x^{p_i-1}(\ln x)^{p_i(1+\frac{\lambda}{r_i})-1}f_i^{p_i}(x)dx\}^{\frac{1}{p_i}};\quad(5.2.43)$$

$$\int_1^\infty\cdots\int_1^\infty(\min_{1\le i\le n-1}\{\ln x_n\ln x_i,1\})^\lambda$$

$$\times\prod_{i=1}^{n}f_i(x_i)dx_1\cdots dx_n$$

$$<\frac{2}{\lambda^{n-1}}\prod_{i=1}^{n-1}r_i\{\int_1^\infty x^{p_i-1}(\ln x)^{p_i(1+\frac{\lambda}{r_i})-1}f_i^{p_i}(x)dx\}^{\frac{1}{p_i}}$$

$$\times\{\int_1^\infty x^{p_n-1}(\ln x)^{p_n(1+\frac{\lambda}{2})-1}f_n^{p_n}(x)dx\}^{\frac{1}{p_n}},\quad(5.2.44)$$

$$\{\int_1^\infty\frac{1}{x_n}(\ln x_n)^{\frac{-q_n\lambda}{2}}[\int_1^\infty\cdots\int_1^\infty(\min_{1\le i\le n-1}\{\ln x_n\ln x_i,1\})^\lambda$$

$$\times\prod_{i=1}^{n}f_i(x_i)dx_1\cdots dx_{n-1}]^{q_n}dx_n\}^{\frac{1}{q_n}}$$

$$<\frac{2}{\lambda^{n-1}}\prod_{i=1}^{n-1}r_i\{\int_1^\infty x^{p_i-1}(\ln x)^{p_i(1+\frac{\lambda}{r_i})-1}f_i^{p_i}(x)dx\}^{\frac{1}{p_i}},\quad(5.2.45)$$

where the constant factors are the best possible; (2) for $0<p_1<1,p_i<0\ (i=2,\cdots,n)$, we have the reverses of (5.2.42) - (5.2.45) with the best constant factors (in (5.2.44) , (5.2.45) and the reverses, notice $\sum_{i=1}^{n-1}\frac{1}{r_i}=\frac{1}{2}$).

5.2.3 A MULTIPLE INTEGRAL INEQUALITY WITH THE HOMOGENEOUS KERNEL OF 0-DEGREE

Example 5.2.5 If $n\in\mathrm{N}\setminus\{1\}$,

$$k_0(x_1,\cdots,x_n)=\frac{\min\{x_1,\cdots,x_n\}}{\sqrt[n]{x_1\cdots x_n}},$$

Then we set

$$H(n)=\int_0^\infty\cdots\int_0^\infty\frac{\min\{u_1,\cdots,u_{n-1},1\}}{\sqrt[n]{u_1\cdots u_{n-1}}}$$

$$\times\prod_{j=1}^{n-1} u_j^{-1} du_1 \cdots du_{n-1} .$$

(5.2.46)

In the following, we prove by mathematical induction that

$$H(n) = \prod_{k=1}^{n-1} \frac{(k+1)^{k+1}}{k^k}, n \in \mathbf{N}\setminus\{1\}. \quad (5.2.47)$$

For $n = 2$, we find

$$H(2) = \int_0^\infty \frac{\min\{u_1,1\}}{\sqrt{u_1}} u_1^{-1} du_1$$

$$= \int_0^1 \frac{u_1}{\sqrt{u_1}} u_1^{-1} du_1 + \int_1^\infty \frac{1}{\sqrt{u_1}} u_1^{-1} du_1 = 4,$$

and (5.2.47) is valid. Assuming that for $n (\geq 2)$, (5.2.47) is valid, then for $n+1$,

$$H(n+1) = \int_0^\infty \cdots \int_0^\infty \frac{\min\{u_1,\cdots,u_n,1\}}{\sqrt[n+1]{u_1\cdots u_n}}$$

$$\times u_1^{-1} \cdots u_n^{-1} du_1 \cdots du_n$$

$$= \int_0^\infty \cdots \int_0^\infty \prod_{i=1}^{n-1} u_i^{-1}$$

$$\times[\int_0^\infty \frac{\min\{u_1,\cdots,u_n,1\}}{\sqrt[n+1]{u_1\cdots u_n}} u_n^{-1} du_n] du_1 \cdots du_{n-1}$$

$$= \int_0^\infty \cdots \int_0^\infty \prod_{i=1}^{n-1} u_i^{-1} [\int_0^{\min\{u_1,\cdots,u_{n-1},1\}} \frac{u_n u_n^{-1}}{\sqrt[n+1]{u_1\cdots u_n}} du_n$$

$$+ \int_{\min\{u_1,\cdots,u_{n-1},1\}}^\infty \frac{\min\{u_1,\cdots,u_{n-1},1\}}{u_n \sqrt[n+1]{u_1\cdots u_n}} du_n] du_1 \cdots du_{n-1}$$

$$= \frac{(n+1)^2}{n} \int_0^\infty \cdots \int_0^\infty \prod_{i=1}^{n-1} u_i^{-1}$$

$$\times \frac{(\min\{u_1,\cdots,u_{n-1},1\})^{n/(n+1)}}{\sqrt[n+1]{u_1\cdots u_{n-1}}} du_1 \cdots du_{n-1}.$$

Setting $v_i = u_i^{\frac{n}{n+1}} (i = 1,\cdots,n-1)$ in the above integral, by the assumption of induction, we find

$$H(n+1) = \frac{(n+1)^{n+1}}{n^n} \int_0^\infty \cdots \int_0^\infty \frac{\min\{v_1,\cdots,v_{n-1},1\}}{\sqrt[n]{v_1\cdots v_{n-1}}}$$

$$\times \prod_{j=1}^{n-1} v_j^{-1} dv_1 \cdots dv_{n-1} = \prod_{k=1}^{n} \frac{(k+1)^{k+1}}{k^k}.$$

Then for $n+1$, (5.2.47) is valid. By mathematical induction, we show that for $n \in \mathbf{N}\setminus\{1\}$, (5.2.47) is valid.

Hence by (5.1.2), we set

$$k_0 = \prod_{k=1}^{n-1} \frac{(k+1)^{k+1}}{k^k}.$$

In the following, we show by mathematical induction that there exists $\delta = \frac{1}{2(1+n)} > 0$, such that for $|\eta| < \delta, n \in \mathbf{N}\setminus\{1\}$,

$$0 < H_\eta(n) := \int_0^\infty \cdots \int_0^\infty \frac{\min\{u_1,\cdots,u_{n-1},1\}}{\sqrt[n]{u_1\cdots u_{n-1}}}$$

$$\times \prod_{j=1}^{n-1} u_j^{\eta-1} du_1 \cdots du_{n-1} < \infty. \quad (5.2.48)$$

Without lose of generality, suppose $\eta > 0$. For $n = 2$, we find

$$0 < K_\eta(2) = \int_0^\infty \frac{\min\{u_1,1\}}{\sqrt{u_1}} u_1^{\eta-1} du_1$$

$$= \int_0^1 \frac{u_1}{\sqrt{u_1}} u_1^{\eta-1} du_1 + \int_1^\infty \frac{1}{\sqrt{u_1}} u_1^{\eta-1} du_1$$

$$= \frac{2}{1+2\eta} + \frac{2}{1-2\eta} < \infty.$$

Assuming that for $n \geq 2$, (5.2.48) is valid. Then for $n+1$,

$$0 < K_\eta(n+1) = \int_0^\infty \cdots \int_0^\infty \prod_{j=1}^{n-1} u_j^{\eta-1}$$

$$\times[\int_0^\infty \frac{\min\{u_1,\cdots,u_n,1\}}{\sqrt[n+1]{u_1\cdots u_n}} u_n^{\eta-1} du_n] du_1 \cdots du_{n-1}$$

$$\leq \int_0^\infty \cdots \int_0^\infty \prod_{j=1}^{n-1} u_j^{\eta-1}$$

$$\times[\int_0^{\min\{u_1,\cdots,u_{n-1},1\}} \frac{u_n}{\sqrt[n+1]{u_1\cdots u_{n-1}} u_n^{\frac{1}{n+1}}} u_n^{-1} du_n$$

$$+ \int_{\min\{u_1,\cdots,u_{n-1},1\}}^1 \frac{\min\{u_1,\cdots,u_{n-1},1\}}{\sqrt[n+1]{u_1\cdots u_{n-1}} u_n^{\frac{1}{n+1}}} u_n^{-1} du_n$$

$$+ \int_1^\infty \frac{(\min\{u_1,\cdots,u_{n-1},1\})^{1-\frac{1}{n+1}}}{\sqrt[n+1]{u_1\cdots u_{n-1}} u_n^{\frac{1}{n+1}}} u_n^{\eta-1} du_n] du_1 \cdots du_{n-1}$$

$$\leq (\frac{n+1}{n} + \frac{1}{\frac{1}{n+1} - \eta}) \int_0^\infty \cdots \int_0^\infty \prod_{j=1}^{n-1} u_j^{\eta-1}$$

$$\times \frac{(\min\{u_1,\cdots,u_{n-1},1\})^{\frac{n}{n+1}}}{\sqrt[n+1]{u_1\cdots u_{n-1}}} du_1 \cdots du_{n-1}.$$

Setting $v_i = u_i^{\frac{n}{n+1}} (i = 1,\cdots,n-1)$ in the above integral, by the assumption of mathematical induction, we find

$$0 < K_\eta(n+1) \leq (\frac{n+1}{n} + \frac{1}{\frac{1}{n+1} - \eta})(\frac{n+1}{n})^{n-1}$$

$$\times \int_0^\infty \cdots \int_0^\infty \frac{\min\{v_1,\cdots,v_{n-1},1\}}{\sqrt[n]{v_1\cdots v_{n-1}}} \prod_{j=1}^{n-1} v_j^{\eta(1+\frac{1}{n})-1} dv_1 \cdots dv_{n-1}$$

$$\leq (\frac{n+1}{n} + \frac{1}{\frac{1}{n+1} - \eta})(\frac{n+1}{n})^{n-1} O(1) < \infty.$$

Then for $n+1$, (5.2.48) is valid. By mathematical induction, we show for $n \in \mathbf{N}\setminus\{1\}$, (5.2.48) is valid.

By Theorem 5.1.6 and Theorem 5.1.7, (1) for $p_i > 1$ $(i = 1,\cdots,n)$, we have the following equivalent inequalities:

$$\int_0^\infty \cdots \int_0^\infty \frac{\min\{x_1,\cdots,x_n\}}{\sqrt[n]{x_1\cdots x_n}} \prod_{i=1}^n f_i(x_i)dx_1\cdots dx_n$$
$$< k_0 \prod_{i=1}^n \{\int_0^\infty x^{p_i-1} f_i^{p_i}(x)dx\}^{\frac{1}{p_i}}, \quad (5.2.49)$$
$$\{\int_0^\infty x_n^{-1}[\int_0^\infty \cdots \int_0^\infty \frac{\min\{x_1,\cdots,x_n\}}{\sqrt[n]{x_1\cdots x_n}} \prod_{i=1}^{n-1} f_i(x_i) \times dx_1\cdots dx_{n-1}]^{q_n} dx_n\}^{\frac{1}{q_n}}$$
$$< \prod_{i=1}^{n-1} \frac{(i+1)^{i+1}}{i^i} \{\int_0^\infty x^{p_i-1} f_i^{p_i}(x)dx\}^{\frac{1}{p_i}}, \quad (5.2.50)$$

where the constant factor $k_0 = \prod_{i=1}^{n-1} \frac{(i+1)^{i+1}}{i^i}$ is the best possible . We still have the following equivalent inequalities with the same best constant factor:

$$\int_0^\infty \cdots \int_0^\infty \frac{\min_{1\le i\le n-1}\{x_i x_n, 1\}}{\sqrt[n]{x_n^{n-1} x_1\cdots x_{n-1}}} \prod_{i=1}^n f_i(x_i)dx_1\cdots dx_n$$
$$< k_0 \prod_{i=1}^n \{\int_0^\infty x^{p_i-1} f_i^{p_i}(x)dx\}^{\frac{1}{p_i}}, \quad (5.2.51)$$
$$\{\int_0^\infty x_n^{-1}[\int_0^\infty \cdots \int_0^\infty \frac{\min_{1\le i\le n-1}\{x_i x_n, 1\}}{\sqrt[n]{x_n^{n-1} x_1\cdots x_{n-1}}} \prod_{i=1}^{n-1} f_i(x_i) \times dx_1\cdots dx_{n-1}]^{q_n} dx_n\}^{\frac{1}{q_n}}$$
$$< \prod_{i=1}^{n-1} \frac{(i+1)^{i+1}}{i^i} \{\int_0^\infty x^{p_i-1} f_i^{p_i}(x)dx\}^{\frac{1}{p_i}}; \quad (5.2.52)$$

(2) if $0 < p_1 < 1, p_i < 0$ $(i = 2,\cdots,n)$, we have the equivalent reverses of (5.2.51) - (5.2.52) with the best constant factors.

(i) By (5.1.32) - (5.1.35), setting $\alpha_i = \alpha > 0$, (1) for $p_i > 1$ $(i = 1,\cdots,n)$, we have the following two pairs of equivalent inequalities:

$$\int_0^\infty \cdots \int_0^\infty \left(\frac{\min\{x_1,\cdots,x_n\}}{\sqrt[n]{x_1\cdots x_n}}\right)^\alpha \prod_{i=1}^n f_i(x_i)dx_1\cdots dx_n$$
$$< \frac{k_0}{\alpha^{n-1}} \prod_{i=1}^n \{\int_0^\infty x^{p_i-1} f_i^{p_i}(x)dx\}^{\frac{1}{p_i}}, \quad (5.2.53)$$
$$\{\int_0^\infty x_n^{-1}[\int_0^\infty \cdots \int_0^\infty \left(\frac{\min\{x_1,\cdots,x_n\}}{\sqrt[n]{x_1\cdots x_n}}\right)^\alpha \prod_{i=1}^{n-1} f_i(x_i) \times dx_1\cdots dx_{n-1}]^{q_n} dx_n\}^{\frac{1}{q_n}}$$
$$< \frac{1}{\alpha^{n-1}} \prod_{i=1}^{n-1} \frac{(i+1)^{i+1}}{i^i} \{\int_0^\infty x^{p_i-1} f_i^{p_i}(x)dx\}^{\frac{1}{p_i}}; \quad (5.2.54)$$

$$\int_0^\infty \cdots \int_0^\infty \frac{\min_{1\le i\le n-1}\{x_n^\alpha x_i^\alpha, 1\}}{\sqrt[n]{(x_n^{n-1} x_1\cdots x_{n-1})^\alpha}} \prod_{i=1}^n f_i(x_i)dx_1\cdots dx_n$$
$$< \frac{k_0}{\alpha^{n-1}} \prod_{i=1}^n \{\int_0^\infty x^{p_i-1} f_i^{p_i}(x)dx\}^{\frac{1}{p_i}}, \quad (5.2.55)$$
$$\{\int_0^\infty x_n^{-1}[\int_0^\infty \cdots \int_0^\infty \frac{\min_{1\le i\le n-1}\{x_n^\alpha x_i^\alpha, 1\}}{\sqrt[n]{(x_n^{n-1} x_1\cdots x_{n-1})^\alpha}} \prod_{i=1}^{n-1} f_i(x_i) \times dx_1\cdots dx_{n-1}]^{q_n} dx_n\}^{\frac{1}{q_n}}$$
$$< \frac{1}{\alpha^{n-1}} \prod_{i=1}^{n-1} \frac{(i+1)^{i+1}}{i^i} \{\int_0^\infty x^{p_i-1} f_i^{p_i}(x)dx\}^{\frac{1}{p_i}}, \quad (5.2.56)$$

(2) for $0 < p_1 < 1, p_i < 0$ $(i = 2,\cdots,n)$, we have the reverses of (5.2.53) - (5.2.36) with the best constant factors,

(ii) By (5.1.36) -(5.1.39), (1) for $p_i > 1$, $i = 1,\cdots,n$, we have the following two pairs of equivalent inequalities:

$$\int_1^\infty \cdots \int_1^\infty \frac{\min_{1\le i\le n}\{\ln x_i\}}{\sqrt[n]{\ln x_1\cdots \ln x_n}} \prod_{i=1}^n f_i(x_i)dx_1\cdots dx_n$$
$$< k_0 \prod_{i=1}^n \{\int_1^\infty (x\ln x)^{p_i-1} f_i^{p_i}(x)dx\}^{\frac{1}{p_i}}, \quad (5.2.7)$$
$$\{\int_1^\infty \frac{1}{x_n}[\int_1^\infty \cdots \int_1^\infty \frac{\min_{1\le i\le n}\{\ln x_i\}}{\sqrt[n]{\ln x_1\cdots \ln x_n}} \prod_{i=1}^n f_i(x_i) \times dx_1\cdots dx_{n-1}]^{q_n} dx_n\}^{\frac{1}{q_n}}$$
$$< \prod_{i=1}^{n-1} \frac{(i+1)^{i+1}}{i^i} \{\int_1^\infty (x\ln x)^{p_i-1} f_i^{p_i}(x)dx\}^{\frac{1}{p_i}}; \quad (5.2.58)$$

$$\int_1^\infty \cdots \int_1^\infty \frac{\min_{1\le i\le n-1}\{\ln x_i \ln x_n, 1\}}{\sqrt[n]{(\ln x_n)^{n-1}\ln x_1\cdots \ln x_{n-1}}} \prod_{i=1}^n f_i(x_i)dx_1\cdots dx_n$$
$$< k_0 \prod_{i=1}^n \{\int_1^\infty (x\ln x)^{p_i-1} f_i^{p_i}(x)dx\}^{\frac{1}{p_i}}, \quad (5.2.59)$$
$$\{\int_1^\infty \frac{1}{x_n}[\int_1^\infty \cdots \int_1^\infty \frac{\min_{1\le i\le n-1}\{\ln x_i \ln x_n, 1\}}{\sqrt[n]{(\ln x_n)^{n-1}\ln x_1\cdots \ln x_{n-1}}} \prod_{i=1}^n f_i(x_i) \times dx_1\cdots dx_{n-1}]^{q_n} dx_n\}^{\frac{1}{q_n}}$$
$$< \prod_{i=1}^{n-1} \frac{(i+1)^{i+1}}{i^i} \{\int_1^\infty (x\ln x)^{p_i-1} f_i^{p_i}(x)dx\}^{\frac{1}{p_i}}, \quad (5.2.60)$$

where the constant factor k_0 is the best possible; (2) for $0 < p_1 < 1, p_i < 0$ $(i = 2,\cdots,n)$, we have the reverses of (5.2.56) -(5.2.60) with the best constant factors.

5.3 TWO CLASSES OF MULTIPLE HARDY-TYPE INTEGRAL INEQUALITIES

5.3.1 THE FIRST CLASS OF MULTIPLE HARDY-TYPE INTEGRAL INEQUALITIES

Corollary 5.3.1 Suppose that $m \in \mathbf{N}, n \in \mathbf{N}\setminus\{1\}$, $m \le n-1$, $p_i \ne 0,1, r_i \ne 0\ (i=1,\cdots,n)$, $\sum_{i=1}^{n}\frac{1}{p_i}=1$, $\frac{1}{q_n}=1-\frac{1}{p_n}=\sum_{i=1}^{n-1}\frac{1}{p_i}$, $\sum_{i=1}^{n}\frac{1}{r_i}=1$, $\lambda \in \mathbf{R}$, $\tilde{k}_\lambda(x_1,\cdots,x_n)(\ge 0)$ is a homogeneous function of $-\lambda$-degree in R_+^n, for $i=1,\cdots,n$, there exists $\delta>0$, such that for $|\eta_j|<\delta\ (j=1,\cdots,n-1)$,

$$L_\lambda(\eta_1,\cdots,\eta_{n-1}):=\int_0^\infty\cdots\int_0^\infty[\int_0^1\cdots\int_0^1\tilde{k}_\lambda(u_1,\cdots,u_{n-1},1)\times\prod_{j=1}^{n-1}u_j^{\frac{\lambda}{r_j}-\frac{\eta_j}{p_j}-1}du_1\cdots du_m]du_{m+1}\cdots du_{n-1}$$

is a positive constant, $\tilde{k}_\lambda=L_\lambda(0,\cdots,0)$. If $f_i\ge 0$, $0<\int_0^\infty x^{p_i(1-\frac{\lambda}{r_i})-1}f_i^{p_i}(x)dx<\infty\ (i=1,\cdots,n-1)$, then (1) for $p_i>1\ (i=1,\cdots,n)$, we have the following first class of multiple Hardy-type integral inequality:

$$\{\int_0^\infty x_n^{\frac{q_n\lambda}{r_n}-1}[\int_0^\infty\cdots\int_0^\infty(\int_0^{x_n}\cdots\int_0^{x_n}\tilde{k}_\lambda(x_1,\cdots,x_n)\times\prod_{i=1}^{n-1}f_i(x_i)dx_1\cdots dx_m)dx_{m+1}\cdots dx_{n-1}]^{q_n}dx_n\}^{\frac{1}{q_n}}$$
$$<\tilde{k}_\lambda\prod_{i=1}^{n-1}\{\int_0^\infty x^{p_i(1-\frac{\lambda}{r_i})-1}f_i^{p_i}(x)dx\}^{\frac{1}{p_i}}, \qquad (5.3.1)$$

where the constant factor $\tilde{k}_\lambda$ is the best possible; (2) for $0<p_1<1, p_i<0\ (i=2,\cdots,n)$, we have the reverse of (5.3.1) with the best constant factor $\tilde{k}_\lambda$.

In particular, for $r_n=2$, if $p_i>1(i=1,\cdots,n)$, then we have the following inequality with the best constant factor:

$$\{\int_0^\infty x_n^{\frac{q_n\lambda}{2}-1}[\int_0^\infty\cdots\int_0^\infty(\int_0^{x_n}\cdots\int_0^{x_n}\tilde{k}_\lambda(x_nx_1,\cdots,x_nx_{n-1},1)\times\prod_{i=1}^{n-1}f_i(x_i)dx_1\cdots dx_m)dx_{m+1}\cdots dx_{n-1}]^{q_n}dx_n\}^{\frac{1}{q_n}}$$
$$<\tilde{k}_\lambda\prod_{i=1}^{n-1}\{\int_0^\infty x^{p_i(1-\frac{\lambda}{r_i})-1}f_i^{p_i}(x)dx\}^{\frac{1}{p_i}}; \qquad (5.3.2)$$

if $0<p_1<1, p_i<0\ (i=2,\cdots,n)$, then we have the reverse of (5.3.2) with the best constant factor.

Proof Setting $E:=\{(x_1,\cdots,x_n\}\in R_+^n; 0<x_i\le x_n, i=1,\cdots,m\}$,

$$k_\lambda(x_1,\cdots,x_n):=\begin{cases}\tilde{k}_\lambda(x_1,\cdots,x_n),(x_1,\cdots,x_n)\in E,\\ 0,\ (x_1,\cdots,x_n)\in R_+^n\setminus E,\end{cases}$$

then due to (5.1.3), we find

$$\omega_n(x_n)=x_n^{\frac{\lambda}{r_n}}\int_0^\infty\cdots\int_0^\infty k_\lambda(x_1,\cdots,x_n)\times\prod_{j=1}^{n-1}x_j^{\frac{\lambda}{r_j}-1}dx_1\cdots dx_{n-1}$$
$$=x_n^{\frac{\lambda}{r_n}}\int_0^\infty\cdots\int_0^\infty(\int_0^{x_n}\cdots\int_0^{x_n}\tilde{k}_\lambda(x_1,\cdots,x_n)\times\prod_{j=1}^{n-1}x_j^{\frac{\lambda}{r_j}-1}dx_1\cdots dx_m)dx_{m+1}\cdots dx_{n-1}.$$

Setting $u_i=x_i/x_n(i=1,\cdots,n-1)$ in the above expression, by Lemma 5.1.3, we find

$$\omega_n(x_n)=\int_0^\infty\cdots\int_0^\infty[\int_0^1\cdots\int_0^1\tilde{k}_\lambda(u_1,\cdots,u_{n-1},1)\times\prod_{j=1}^{n-1}u_j^{\frac{\lambda}{r_j}-1}du_1\cdots du_m]du_{m+1}\cdots du_{n-1}=\tilde{k}_\lambda.$$

By Theorem 5.1.7, we have all the results of Corollary5.3.1.□

By Theorem 5.1.9, we still have

Corollary 5.3.2 As the assumption of Corollary 5.3.1, if $v_i(x)$ are strict increasing deliverable functions in (a,b) with $v_i(a^+)=0, v_i(b^-)=\infty, f_i\ge 0$, and

$$0<\int_a^b\frac{(v_i(x))^{p_i(1-\frac{\lambda}{r_i})-1}}{(v_i'(x))^{p_i-1}}f_i^{p_i}(x)dx<\infty\ (i=1,\cdots,n-1),$$

then (1) For $p_i>1\ (i=1,\cdots,n)$, we have the following inequality:

$$\{\int_a^b \frac{v_n'(x_n)}{[v_n(x_n)]^{1-\frac{q_n\lambda}{r_n}}}$$
$$\times[\int_a^b \cdots \int_a^b (\int_a^{v_m^{-1}(x_n)} \cdots \int_a^{v_1^{-1}(x_n)} \tilde{k}_\lambda(v_1(x_1),\cdots,v_n(x_n))$$
$$\times\prod_{i=1}^{n-1} f_i(x_i)dx_1\cdots dx_m)dx_{m+1}\cdots dx_{n-1}]^{q_n} dx_n\}^{\frac{1}{q_n}}$$
$$< \tilde{k}_\lambda \prod_{i=1}^{n-1}\{\int_a^b \frac{[v_i(x)]^{p_i(1-\frac{\lambda}{r_i})-1}}{[v_i'(x)]^{p_i-1}} f_i^{p_i}(x)dx\}^{\frac{1}{p_i}}, \quad (5.3.3)$$

where the constant factor $\tilde{k}_\lambda$ is the best possible; (2) for $0<p_1<1, p_i<0(i=2,\cdots,n)$, we have the reverse of (5.3.3) with the same best constant factor.

In particular, for $r_n=2$, setting

$$K_\lambda(x_1,\cdots,x_n):$$
$$=\tilde{k}_\lambda(v_n(x_n)v_1(x_1),\cdots,v_n(x_n)v_{n-1}(x_{n-1}),1),$$

(1) for $p_i>1\ (i=1,\cdots,n)$, we have the following inequality with the best constant factor:

$$\{\int_a^b \frac{v_n'(x_n)}{[v_n(x_n)]^{1-\frac{q_n\lambda}{2}}}$$
$$\times[\int_a^b \cdots \int_a^b (\int_a^{v_m^{-1}(x_n)} \cdots \int_a^{v_1^{-1}(x_n)} K_\lambda(x_1,\cdots,x_n)$$
$$\times\prod_{i=1}^{n-1} f_i(x_i)dx_1\cdots dx_m)dx_{m+1}\cdots dx_{n-1}]^{q_n} dx_n\}^{\frac{1}{q_n}}$$
$$< \tilde{k}_\lambda \prod_{i=1}^{n-1}\{\int_a^b \frac{[v_i(x)]^{p_i(1-\frac{\lambda}{r_i})-1}}{[v_i'(x)]^{p_i-1}} f_i^{p_i}(x)dx\}^{\frac{1}{p_i}}; \quad (5.3.4)$$

(2) for $0<p_1<1, p_i<0\ (i=2,\cdots,n)$, we have the reverse of (5.3.4) with the best constant factor.

Example 5.3.3 If $\lambda>0, r_i>1$,

$$\tilde{k}_\lambda(x_1,\cdots,x_n)=\frac{1}{(\max\{x_1,\cdots,x_n\})^\lambda},$$

setting $m=n-2\geq 1(n\in \mathbf{N}\setminus\{1,2\})$, then we find

$$\tilde{k}_\lambda = \int_0^\infty (\int_0^1 \cdots \int_0^1 \frac{\prod_{j=1}^{n-1} u_j^{\frac{\lambda}{r_j}-1}}{(\max_{1\leq i\leq n-1}\{u_i,1\})^\lambda} du_1\cdots du_{n-2})du_{n-1}$$
$$= \int_0^\infty (\int_0^1 \cdots \int_0^1 \frac{\prod_{j=1}^{n-1} u_j^{\frac{\lambda}{r_j}-1}}{(\max\{u_{n-1},1\})^\lambda} du_1\cdots du_{n-2})du_{n-1}$$
$$= \frac{\prod_{i=1}^{n-2} r_i}{\lambda^{n-2}}[\int_0^1 u_{n-1}^{\frac{\lambda}{r_{n-1}}-1} du_{n-1} + \int_1^\infty \frac{u_{n-1}^{\frac{\lambda}{r_{n-1}}-1}}{u_{n-1}^\lambda} du_{n-1}]$$
$$= \frac{r_{n-1}}{\lambda^{n-1}(r_{n-1}-1)}\prod_{i=1}^{n-1} r_i .$$

By Corollary 5.3.2, we have the following inequality:

$$\{\int_a^b \frac{v_n'(x_n)}{[v_n(x_n)]^{1-\frac{q_n\lambda}{r_n}}}$$
$$\times[\int_a^b (\int_a^{v_m^{-1}(x_n)} \cdots \int_a^{v_1^{-1}(x_n)} \frac{1}{(\max_{1\leq i\leq n}\{v_i(x_i)\})^\lambda}$$
$$\times\prod_{i=1}^{n-1} f_i(x_i)dx_1\cdots dx_{n-2})dx_{n-1}]^{q_n} dx_n\}^{\frac{1}{q_n}}$$
$$< \frac{r_{n-1}}{\lambda^{n-1}(r_{n-1}-1)} \prod_{i=1}^{n-1} r_i\{\int_a^b \frac{[v_i(x)]^{p_i(1-\frac{\lambda}{r_i})-1}}{[v_i'(x)]^{p_i-1}} f_i^{p_i}(x)dx\}^{\frac{1}{p_i}},$$
(5.3.5)

where the constant factor is the best possible; (2) for $0<p_1<1, p_i<0(i=2,\cdots,n)$, we have the reverse of (5.3.5) with the same best constant factor.

In particular, for $r_n=2$, if $p_i>1(i=1,\cdots,n)$, then we have the following inequality with the best constant factor:

$$\{\int_a^b \frac{v_n'(x_n)}{[v_n(x_n)]^{1-\frac{q_n\lambda}{2}}}$$
$$\times[\int_a^b (\int_a^{v_m^{-1}(x_n)} \cdots \int_a^{v_1^{-1}(x_n)} \frac{1}{(\max_{1\leq i\leq n-1}\{v_n(x_n)v_i(x_i),1\})^\lambda}$$
$$\times\prod_{i=1}^{n-1} f_i(x_i)dx_1\cdots dx_{n-2})dx_{n-1}]^{q_n} dx_n\}^{\frac{1}{q_n}}$$
$$< \frac{r_{n-1}}{\lambda^{n-1}(r_{n-1}-1)} \prod_{i=1}^{n-1} r_i\{\int_a^b \frac{[v_i(x)]^{p_i(1-\frac{\lambda}{r_i})-1}}{[v_i'(x)]^{p_i-1}} f_i^{p_i}(x)dx\}^{\frac{1}{p_i}};$$
(5.3.6)

if $0<p_1<1, p_i<0\ (i=2,\cdots,n)$, we have the reverse of (5.3.6) with the best constant factor.

Example 5.3.4 If $\lambda>0, r_i>1$,

$$\tilde{k}_\lambda(x_1,\cdots,x_n)=\frac{1}{(\max\{x_1,\cdots,x_n\})^\lambda},$$

setting $m=n-1$, then we find

$$\tilde{k}_\lambda = \int_0^1 \cdots \int_0^1 \frac{\prod_{j=1}^{n-1} u_j^{\frac{\lambda}{r_j}-1}}{(\max\{u_1,\cdots u_{n-1},1\})^\lambda} du_1\cdots du_{n-1}$$
$$= \int_0^1 \cdots \int_0^1 \prod_{j=1}^{n-1} u_j^{\frac{\lambda}{r_j}-1} du_1\cdots du_{n-1} = \frac{\prod_{j=1}^{n-1} r_j}{\lambda^{n-1}}.$$

By Corollary 5.3.2, we have the following inequality:

$$\{\int_a^b \frac{v_n'(x_n)}{[v_n(x_n)]^{1-\frac{q_n\lambda}{r_n}}}[\int_a^{v_m^{-1}(x_n)} \cdots \int_a^{v_1^{-1}(x_n)} \frac{1}{(\max_{1\leq i\leq n}\{v_i(x_i)\})^\lambda}$$
$$\times\prod_{i=1}^{n-1} f_i(x_i)dx_1\cdots dx_{n-1}]^{q_n} dx_n\}^{\frac{1}{q_n}}$$
$$< \frac{1}{\lambda^{n-1}} \prod_{i=1}^{n-1} r_i\{\int_a^b \frac{[v_i(x)]^{p_i(1-\frac{\lambda}{r_i})-1}}{[v_i'(x)]^{p_i-1}} f_i^{p_i}(x)dx\}^{\frac{1}{p_i}}, \quad (5.3.7)$$

where the constant factor is the best possible; (2) for $0<p_1<1, p_i<0(i=2,\cdots,n)$, we have the reverse of (5.3.7) with the best constant factor.

In particular, for $r_n=2$, if $p_i>1(i=1,\cdots,n)$, we have the following inequality with the best constant factor:

$$\{\int_a^b \frac{v_n'(x_n)}{[v_n(x_n)]^{1-\frac{q_n\lambda}{2}}} \times[\int_a^{v_m^{-1}(x_n)}\cdots\int_a^{v_1^{-1}(x_n)} \frac{1}{(\max_{1\le i\le n-1}\{v_n(x_n)v_i(x_i),1\})^{\lambda}} \times\prod_{i=1}^{n-1} f_i(x_i)dx_1\cdots dx_{n-1}]^{q_n} dx_n\}^{\frac{1}{q_n}}$$

$$<\frac{1}{\lambda^{n-1}}\prod_{i=1}^{n-1} r_i\{\int_a^b \frac{[v_i(x)]^{p_i(1-\frac{\lambda}{r_i})-1}}{[v_i'(x)]^{p_i-1}} f_i^{p_i}(x)dx\}^{\frac{1}{p_i}}; \quad (5.3.8)$$

if $0<p_1<1, p_i<0\ (i=2,\cdots,n)$, we have the reverse of (5.3.8) with the same best constant factor.

5.3.2 THE SECOND CLASS OF MULTIPLE HARDY-TYPE INTEGRAL INEQUALITIES

Corollary 5.3.5 Suppose that $m\in\mathbf{N}, n\in\mathbf{N}\setminus\{1\}$, $m<n$, $p_i\ne 0,1, r_i\ne 0\ (i=1,\cdots,n)$, $\sum_{i=1}^n \frac{1}{p_i}=1$, $\frac{1}{q_n}=1-\frac{1}{p_n}$, $\sum_{i=1}^n \frac{1}{r_i}=1$, $\lambda\in\mathbf{R}$, $\tilde{k}_\lambda(x_1,\cdots,x_n)(\ge 0)$ is a homogeneous function of $-\lambda$-degree in R_+^n, for $i=1,2,\cdots,n$, there exists $\delta>0$, such that for $|\eta_j|<\delta\ (j=1,\cdots,n-1)$,

$$H_\lambda(\eta_1,\cdots,\eta_{n-1}): =\int_0^\infty\cdots\int_0^\infty[\int_1^\infty\cdots\int_1^\infty \tilde{k}_\lambda(u_1,\cdots,u_{n-1},1) \times\prod_{j=1}^{n-1} u_j^{\frac{\lambda}{r_j}-\frac{\eta_j}{p_j}-1} du_1\cdots du_m]du_{m+1}\cdots du_{n-1}$$

is a positive constant, $\tilde{K}_\lambda=H_\lambda(0,\cdots,0)$. If

$$f_i\ge 0,\ 0<\int_0^\infty x^{p_i(1-\frac{\lambda}{r_i})-1} f_i^{p_i}(x)dx<\infty$$

$(i=1,\cdots,n-1)$, then (1) for $p_i>1\ (i=1,\cdots,n)$, we have the following second class of multiple Hardy-type integral inequality:

$$\{\int_0^\infty x_n^{\frac{q_n\lambda}{r_n}-1}[\int_0^\infty\cdots\int_0^\infty(\int_{x_n}^\infty\cdots\int_{x_n}^\infty \tilde{k}_\lambda(x_1,\cdots,x_n) \times\prod_{i=1}^{n-1} f_i(x_i)dx_1\cdots dx_m)dx_{m+1}\cdots dx_{n-1}]^{q_n} dx_n\}^{\frac{1}{q_n}}$$

$$<\tilde{K}_\lambda\prod_{i=1}^{n-1}\{\int_0^\infty x^{p_i(1-\frac{\lambda}{r_i})-1} f_i^{p_i}(x)dx\}^{\frac{1}{p_i}}, \quad (5.3.9)$$

where the constant factor $\tilde{K}_\lambda$ is the best possible; (2) for $0<p_1<1, p_i<0\quad(i=2,\cdots,n)$, we have the reverse of (5.3.6) with the best constant factor $\tilde{K}_\lambda$.

In particular, for $r_n=2$, if $p_i>1\ (i=1,\cdots,n)$, then we have the following inequality with the best constant factor:

$$\{\int_0^\infty x_n^{\frac{q_n\lambda}{2}-1}[\int_0^\infty\cdots\int_0^\infty(\int_{x_n}^\infty\cdots\int_{x_n}^\infty \tilde{k}_\lambda(x_nx_1,\cdots,x_nx_{n-1},1) \times\prod_{i=1}^{n-1} f_i(x_i)dx_1\cdots dx_m)dx_{m+1}\cdots dx_{n-1}]^{q_n} dx_n\}^{\frac{1}{q_n}}$$

$$<\tilde{K}_\lambda\prod_{i=1}^{n-1}\{\int_0^\infty x^{p_i(1-\frac{\lambda}{r_i})-1} f_i^{p_i}(x)dx\}^{\frac{1}{p_i}}; \quad (5.3.10)$$

if $0<p_1<1, p_i<0\ (i=2,\cdots,n)$, then we have the reverse of (5.3.10) with the best constant factor.

Proof Setting $\tilde{E}:=\{(x_1,\cdots,x_n\}\in R_+^n; x_n<x_i, i=1,\cdots,m\}$,

$$k_\lambda(x_1,\cdots,x_n):=\begin{cases}\tilde{k}_\lambda(x_1,\cdots,x_n),(x_1,\cdots,x_n)\in E,\\ 0,\ (x_1,\cdots,x_n)\in R_+^n\setminus E,\end{cases}$$

then due to (5.1.3), we find

$$\omega_n(x_n)=x_n^{\frac{\lambda}{r_n}}\int_0^\infty\cdots\int_0^\infty k_\lambda(x_1,\cdots,x_n) \times\prod_{j=1}^{n-1} x_j^{\frac{\lambda}{r_j}-1} dx_1\cdots dx_{n-1}$$

$$=x_n^{\frac{\lambda}{r_n}}\int_0^\infty\cdots\int_0^\infty(\int_{x_n}^\infty\cdots\int_{x_n}^\infty \tilde{k}_\lambda(x_1,\cdots,x_n) \times\prod_{j=1}^{n-1} x_j^{\frac{\lambda}{r_j}-1} dx_1\cdots dx_m)dx_{m+1}\cdots dx_{n-1}.$$

Setting $u_i=x_i/x_n(i=1,\cdots,n-1)$ in the above expression, we find

$$\omega_n(x_n)=\int_0^\infty\cdots\int_0^\infty[\int_1^\infty\cdots\int_1^\infty \tilde{k}_\lambda(u_1,\cdots,u_{n-1},1) \times\prod_{j=1}^{n-1} u_j^{\frac{\lambda}{r_j}-\frac{\eta_j}{p_j}-1} du_1\cdots du_m]du_{m+1}\cdots du_{n-1}=\tilde{K}_\lambda.$$

By Theorem 5.1.7, we have all the results of Corollary 5.3.5.□

By Theorem 5.1.9, we still have

Corollary 5.3.6 As the assumption of Corollary 5.3.5, if $v_i(x)$ are strict increasing deliverable functions in (a,b) with $v_i(a^+)=0, v_i(b^-)=\infty$, $f_i\ge 0$, and

$$0<\int_a^b \frac{(v_i(x))^{p_i(1-\frac{\lambda}{r_i})-1}}{(v_i'(x))^{p_i-1}} f_i^{p_i}(x)dx<\infty \ (i=1,\cdots,n-1)$$

,

then (1) For $p_i>1\ (i=1,2,\cdots,n)$, we have the following inequality:

$$\{\int_a^b \frac{v_n'(x_n)}{[v_n(x_n)]^{1-\frac{q_n\lambda}{r_n}}} \times[\int_a^b\cdots\int_a^b(\int_{v_m^{-1}(x_n)}\cdots\int_{v_1^{-1}(x_n)}\tilde{k}_\lambda(v_1(x_1),\cdots,v_n(x_n)) \times\prod_{i=1}^{n-1}f_i(x_i)dx_1\cdots dx_m)dx_{m+1}\cdots dx_{n-1}]^{q_n}dx_n\}^{\frac{1}{q_n}}$$

$$<\tilde{K}_\lambda\prod_{i=1}^{n-1}\{\int_a^b\frac{[v_i(x)]^{p_i(1-\frac{\lambda}{r_i})-1}}{[v_i'(x)]^{p_i-1}}f_i^{p_i}(x)dx\}^{\frac{1}{p_i}},\quad(5.3.11)$$

where the constant factor $\tilde{K}_\lambda$ is the best possible; (2) for $0<p_1<1, p_i<0(i=2,\cdots,n)$, we have the reverse of (5.3.8) with the same best constant factor.

In particular, for $r_n=2$, setting

$$\tilde{K}_\lambda(x_1,\cdots,x_n):=\tilde{k}_\lambda(v_n(x_n)v_1(x_1),\cdots,v_n(x_n)v_{n-1}(x_{n-1}),1),$$

if $p_i>1\ (i=1,2,\cdots,n)$, then we have the following inequality with the best constant factor:

$$\{\int_a^b \frac{v_n'(x_n)}{[v_n(x_n)]^{1-\frac{q_n\lambda}{2}}} \times[\int_a^b\cdots\int_a^b(\int_{v_m^{-1}(x_n)}\cdots\int_{v_1^{-1}(x_n)}\tilde{K}_\lambda(x_1,\cdots,x_n) \times\prod_{i=1}^{n-1}f_i(x_i)dx_1\cdots dx_m)dx_{m+1}\cdots dx_{n-1}]^{q_n}dx_n\}^{\frac{1}{q_n}}$$

$$<\tilde{K}_\lambda\prod_{i=1}^{n-1}\{\int_a^b\frac{[v_i(x)]^{p_i(1-\frac{\lambda}{r_i})-1}}{[v_i'(x)]^{p_i-1}}f_i^{p_i}(x)dx\}^{\frac{1}{p_i}};\ (5.3.12)$$

if $0<p_1<1, p_i<0\ (i=2,\cdots,n)$, then we have the reverse of (5.3.12) with the same best constant factor.

Example 5.3.7 If $\lambda>0, r_i>1$,

$$\tilde{k}_{-\lambda}(x_1,\cdots,x_n)=(\min\{x_1,\cdots,x_n\})^\lambda,$$

setting $m=n-2\geq1$, then we find

$$\tilde{K}_{-\lambda}=\int_0^\infty(\int_1^\infty\cdots\int_1^\infty(\min\{u_{n-1},1\})^\lambda \times\prod_{j=1}^{n-1}u_j^{\frac{-\lambda}{r_j}-1}du_1\cdots du_{n-2})du_{n-1}$$

$$=\frac{\prod_{i=1}^{n-2}r_i}{\lambda^{n-2}}\int_0^\infty(u_{n-1})^{\frac{-\lambda}{r_{n-1}}-1}(\min\{u_{n-1},1\})^\lambda du_{n-1}$$

$$=\frac{r_{n-1}}{\lambda^{n-1}(r_{n-1}-1)}\prod_{i=1}^{n-1}r_i.$$

By Corollary 5.3.6, we have the following inequality:

$$\{\int_a^b \frac{v_n'(x_n)}{[v_n(x_n)]^{1+\frac{q_n\lambda}{r_n}}} \times[\int_a^b(\int_{v_m^{-1}(x_n)}\cdots\int_{v_1^{-1}(x_n)}(\min_{1\le i\le n}\{v_i(x_i)\})^\lambda \times\prod_{i=1}^{n-1}f_i(x_i)dx_1\cdots dx_{n-2})dx_{n-1}]^{q_n}dx_n\}^{\frac{1}{q_n}}$$

$$<\frac{r_{n-1}}{\lambda^{n-1}(r_{n-1}-1)}\prod_{i=1}^{n-1}r_i\{\int_a^b\frac{[v_i(x)]^{p_i(1+\frac{\lambda}{r_i})-1}}{[v_i'(x)]^{p_i-1}}f_i^{p_i}(x)dx\}^{\frac{1}{p_i}},\quad(5.3.13)$$

where the constant factor is the best possible; (2) for $0<p_1<1, p_i<0(i=2,\cdots,n)$, we have the reverse of (5.3.13) with the same best constant factor.

In particular, for $r_n=2$, if $p_i>1(i=1,\cdots,n)$, then we have the following inequality with the best constant factor:

$$\{\int_a^b \frac{v_n'(x_n)}{[v_n(x_n)]^{1+\frac{q_n\lambda}{2}}} \times[\int_a^b(\int_{v_m^{-1}(x_n)}\cdots\int_{v_1^{-1}(x_n)}(\min_{1\le i\le n-1}\{v_n(x_n)v_i(x_i),1\})^\lambda \times\prod_{i=1}^{n-1}f_i(x_i)dx_1\cdots dx_{n-2})dx_{n-1}]^{q_n}dx_n\}^{\frac{1}{q_n}}$$

$$<\frac{r_{n-1}}{\lambda^{n-1}(r_{n-1}-1)}\prod_{i=1}^{n-1}r_i\{\int_a^b\frac{[v_i(x)]^{p_i(1+\frac{\lambda}{r_i})-1}}{[v_i'(x)]^{p_i-1}}f_i^{p_i}(x)dx\}^{\frac{1}{p_i}};\quad(5.3.14)$$

(2) for $0<p_1<1, p_i<0\ (i=2,\cdots,n)$, we have the reverse of (5.3.14) with the best constant factor.

Example 5.3.8 If $\lambda>0, r_i>1$,

$$\tilde{k}_{-\lambda}(x_1,\cdots,x_n)=(\min\{x_1,\cdots,x_n\})^\lambda,$$

setting $m=n-1$, then we find

$$\tilde{K}_{-\lambda}=\int_1^\infty\cdots\int_1^\infty\prod_{j=1}^{n-1}u_j^{\frac{-\lambda}{r_j}-1}du_1\cdots du_{n-1}=\frac{1}{\lambda^{n-1}}\prod_{j=1}^{n-1}r_j.$$

By Corollary 5.3.6, we have the following inequality:

$$\{\int_a^b \frac{v_n'(x_n)}{[v_n(x_n)]^{1+\frac{q_n\lambda}{r_n}}}[\int_{v_m^{-1}(x_n)}^b \cdots \int_{v_1^{-1}(x_n)}^b (\min_{1\le i\le n}\{v_i(x_i)\})^\lambda$$

$$\times\prod_{i=1}^{n-1} f_i(x_i)dx_1\cdots dx_{n-1}]^{q_n} dx_n\}^{\frac{1}{q_n}}$$

$$<\frac{1}{\lambda^{n-1}}\prod_{i=1}^{n-1} r_i\{\int_a^b \frac{[v_i(x)]^{p_i(1+\frac{\lambda}{r_i})-1}}{[v_i'(x)]^{p_i-1}} f_i^{p_i}(x)dx\}^{\frac{1}{p_i}},$$

(5.3.15)

where the constant factor is the best possible; for $0<p_1<1, p_i<0(i=2,\cdots,n)$, we have the reverse of (5.3.15) with the best constant factor.

In particular, for $r_n=2$, if $p_i>1(i=1,\cdots,n)$, then we have the following inequality with the best constant factor:

$$\{\int_a^b \frac{v_n'(x_n)}{[v_n(x_n)]^{1+\frac{q_n\lambda}{2}}}$$

$$\times[\int_{v_m^{-1}(x_n)}^b \cdots \int_{v_1^{-1}(x_n)}^b (\min_{1\le i\le n-1}\{v_n(x_n)v_i(x_i),1\})^\lambda$$

$$\times\prod_{i=1}^{n-1} f_i(x_i)dx_1\cdots dx_{n-1}]^{q_n} dx_n\}^{\frac{1}{q_n}}$$

$$<\frac{1}{\lambda^{n-1}}\prod_{i=1}^{n-1} r_i\{\int_a^b \frac{[v_i(x)]^{p_i(1+\frac{\lambda}{r_i})-1}}{[v_i'(x)]^{p_i-1}} f_i^{p_i}(x)dx\}^{\frac{1}{p_i}};$$

(5.3.16)

if $0<p_1<1, p_i<0\ (i=2,\cdots,n)$, then we have the reverse of (5.3.16) with the best constant factor.

Example 5.3.9 If $n\in\mathbf{N}\setminus\{1,2\}$,

$$\tilde{k}_0(x_1,\cdots,x_n)=\frac{\min\{x_1,\cdots,x_n\}}{\sqrt[n]{x_1\cdots x_n}},$$

setting $m=n-2\ge 1$, then we find

$$\tilde{K}_0=\int_0^\infty(\int_1^\infty\cdots\int_1^\infty \frac{\min\{u_1,\cdots,u_{n-1},1\}}{\sqrt[n]{u_1\cdots u_{n-1}}}$$

$$\times\prod_{j=1}^{n-1} u_j^{-1}du_1\cdots du_{n-2})du_{n-1}$$

$$=\int_0^\infty(\int_1^\infty\cdots\int_1^\infty\prod_{j=1}^{n-1} u_j^{-1-\frac{1}{n}}$$

$$\times du_1\cdots du_{n-2})\min\{u_{n-1},1\}du_{n-1}$$

$$=n^{n-2}\int_0^\infty u_{n-1}^{-1-\frac{1}{n}}\min\{u_{n-1},1\}du_{n-1}=\frac{n^n}{n-1}.$$

By Corollary 5.3.6, we have the following inequality:

$$\{\int_a^b \frac{v_n'(x_n)}{v_n(x_n)}[\int_a^b(\int_{v_m^{-1}(x_n)}^b\cdots\int_{v_1^{-1}(x_n)}^b \frac{\min_{1\le i\le n}\{v_i(x_i)\}}{\sqrt[n]{\prod_{i=1}^n v_i(x_i)}}$$

$$\times\prod_{i=1}^{n-1} f_i(x_i)dx_1\cdots dx_{n-2})dx_{n-1}]^{q_n} dx_n\}^{\frac{1}{q_n}}$$

$$<\frac{n^n}{n-1}\prod_{i=1}^{n-1}\{\int_a^b (\frac{v_i(x)}{v_i'(x)})^{p_i-1} f_i^{p_i}(x)dx\}^{\frac{1}{p_i}}, \quad (5.3.17)$$

where the constant factor is the best possible; (2) for $0<p_1<1, p_i<0(i=2,\cdots,n)$, we have the reverse of (5.3.17) with the same best constant factor.

In particular, for $r_n=2$, if $p_i>1(i=1,\cdots,n)$, then we have the following inequality with the best constant factor:

$$\{\int_a^b \frac{v_n'(x_n)}{v_n(x_n)}[\int_a^b(\int_{v_m^{-1}(x_n)}^b\cdots\int_{v_1^{-1}(x_n)}^b \frac{\min_{1\le i\le n-1}\{v_n(x_n)v_i(x_i),1\}}{\sqrt[n]{v_n^{n-1}(x_n)\prod_{i=1}^{n-1} v_i(x_i)}}$$

$$\times\prod_{i=1}^{n-1} f_i(x_i)dx_1\cdots dx_{n-2})dx_{n-1}]^{q_n} dx_n\}^{\frac{1}{q_n}}$$

$$<\frac{n^n}{n-1}\prod_{i=1}^{n-1}\{\int_a^b (\frac{v_i(x)}{v_i'(x)})^{p_i-1} f_i^{p_i}(x)dx\}^{\frac{1}{p_i}}; \quad (5.3.18)$$

if $0<p_1<1, p_i<0\ (i=2,\cdots,n)$, then we have the reverse of (5.3.18) with the same best constant factor.

Example 5.3.10 If $n\in\mathbf{N}\setminus\{1\}$,

$$\tilde{k}_0(x_1,\cdots,x_n)=\frac{\min\{x_1,\cdots,x_n\}}{\sqrt[n]{x_1\cdots x_n}},$$

setting $m=n-1$, then we find

$$\tilde{K}_0=\int_1^\infty\cdots\int_1^\infty \frac{\min\{u_1,\cdots,u_{n-1},1\}}{\sqrt[n]{u_1\cdots u_{n-1}}}$$

$$\times\prod_{j=1}^{n-1} u_j^{-1}du_1\cdots du_{n-1}$$

$$=\int_1^\infty\cdots\int_1^\infty\prod_{j=1}^{n-1} u_j^{-1-\frac{1}{n}}du_1\cdots du_{n-1}=n^{n-1}.$$

By Corollary 5.3.6, we have the following inequality:

$$\{\int_a^b \frac{v_n'(x_n)}{v_n(x_n)}[\int_{v_m^{-1}(x_n)}^b\cdots\int_{v_1^{-1}(x_n)}^b \frac{\min_{1\le i\le n}\{v_i(x_i)\}}{\sqrt[n]{\prod_{i=1}^n v_i(x_i)}}$$

$$\times\prod_{i=1}^{n-1} f_i(x_i)dx_1\cdots dx_{n-1}]^{q_n} dx_n\}^{\frac{1}{q_n}}$$

$$<n^{n-1}\prod_{i=1}^{n-1}\{\int_a^b (\frac{v_i(x)}{v_i'(x)})^{p_i-1} f_i^{p_i}(x)dx\}^{\frac{1}{p_i}}, \quad (5.3.19)$$

where the constant factor is the best possible; (2) for $0<p_1<1, p_i<0(i=2,\cdots,n)$, we have the reverse of (5.3.19) with the same best constant factor.

In particular, for $r_n = 2$, if $p_i > 1(i = 1,\cdots,n)$, then we have the following inequality with the best constant factor:

$$\{\int_a^b \frac{v_n'(x_n)}{v_n(x_n)}[\int_{v_m^{-1}(x_n)}^b \cdots \int_{v_1^{-1}(x_n)}^b \frac{\min_{1\le i\le n-1}\{v_n(x_n)v_i(x_i),1\}}{\sqrt[n]{v_n^{n-1}(x_n)\prod_{i=1}^{n-1} v_i(x_i)}}$$

$$\times\prod_{i=1}^{n-1} f_i(x_i)dx_1\cdots dx_{n-1}]^{q_n} dx_n\}^{\frac{1}{q_n}}$$

$$< n^{n-1}\prod_{i=1}^{n-1}\{\int_a^b (\frac{v_i(x)}{v_i'(x)})^{p_i-1} f_i^{p_i}(x)dx\}^{\frac{1}{p_i}}; \qquad (5.3.20)$$

if $0 < p_1 < 1, p_i < 0\ (i = 2,\cdots,n)$, then we have the reverse of (5.3.20) with the best constant factor.

5.4 REFERENCES

1. Wang DQ, Guo DR. Introduction to special functions. Science Press, 1979.
2. Yang BC, Brnetc I, Krnic M, Pecaric J. Generalization of Hilbert and Hardy-Hilbert integral inequalities . Math. Ineq. & Appl., 2005;8(2):259-272.

6. Multivariable Hilbert-Type Integral Inequalities

Bicheng Yang

Department of Mathematics, Guangdong Education Institute, Guangzhou, Guangdong 510303, P. R. China; E-mail: bcyang@pub.guangzhou.gd.cn

Abstract: In this chapter, we consider a class of multivariable Hilbert-type integral inequalities and the reverses, which are the best extensions of the corresponding results in Chapter 2. We also consider some equivalent integral inequalities with the non-homogeneous kernels and the Hardy-type integral inequalities.

6.1. MULTIVARIABLE HILBERT-TYPE INTEGRAL INEQUALITIES AND THE OPERATOR EXPRESSIONS

6.1.1. SOME LEMMAS

Lemma 6.1.1 If $\alpha > 0, n \in \mathbf{N}$,

$$R_+^n = \{(x_1,\cdots,x_n) \mid x_i > 0 \ (i=1,\cdots,n)\},$$

$x \in R_+^n, \| x \|_\alpha = (\sum_{i=1}^n x_i^\alpha)^{\frac{1}{\alpha}}$ $r, s \neq 0, \frac{1}{r}+\frac{1}{s}=1$, $\lambda \in \mathbf{R}, k_\lambda(x,y)(\geq 0)$ is a homogeneous function of $-\lambda$-degree in R_+^2, and $k_\lambda(r) := \int_0^\infty k_\lambda(u,1)u^{\frac{\lambda}{r}-1}du$ is a finite number, define the weight functions $\omega_\lambda(r,y)$ and $\tilde{\omega}_\lambda(s,x)$ as follows:

$$\omega_\lambda(r,y) := \int_{R_+^n} k_\lambda(\| x\|_\alpha, \| y\|_\alpha) \frac{\| y\|_\alpha^{\frac{\lambda}{s}}}{\| x\|_\alpha^{n-\frac{\lambda}{r}}} dx,$$

$$\tilde{\omega}_\lambda(s,x) := \int_{R_+^n} k_\lambda(\| x\|_\alpha, \| y\|_\alpha) \frac{\| x\|_\alpha^{\frac{\lambda}{r}}}{\| y\|_\alpha^{n-\frac{\lambda}{s}}} dy,$$

$$x, y \in R_+^n. \quad (6.1.1)$$

Then we have

$$\omega_\lambda(r,y) = \tilde{\omega}_\lambda(s,x) = \frac{\Gamma^n(\frac{1}{\alpha})}{\alpha^{n-1}\Gamma(\frac{n}{\alpha})} k_\lambda(r),$$

$$x, y \in R_+^n. \quad (6.1.2)$$

Proof Replacing $n-1$ by n in (5.2.1), setting $a_i = R, \alpha_i = \alpha, P_i = 1 \ (i=1,\cdots,n)$, then we find

$$\int\cdots\int_{D_R} \psi(\sum_{i=1}^n (\tfrac{x_i}{R})^\alpha) dx_1 \cdots dx_n$$

$$= \frac{R^n\Gamma^n(\frac{1}{\alpha})}{\alpha^n\Gamma(\frac{n}{\alpha})} \int_0^1 \psi(u) u^{\frac{n}{\alpha}-1} du, \quad (6.1.3)$$

$$D_R := \{(x_1,\cdots,x_n) \mid x_i > 0, i=1,\cdots,n, \sum_{i=1}^n x_i^\alpha \leq R^\alpha\}.$$

By (6.1.3), we obtain

$$\omega_\lambda(r,y) = \| y\|_\alpha^{\frac{\lambda}{s}} \times \int_{R_+^n} k_\lambda(\| x\|_\alpha, \| y\|_\alpha) \frac{1}{\| x\|_\alpha^{n-\frac{\lambda}{r}}} dx$$

$$= \| y\|_\alpha^{\frac{\lambda}{s}} \lim_{R\to+\infty} \int\cdots\int_{D_R} k_\lambda(R[\sum_{i=1}^n (\tfrac{x_i}{R})^\alpha]^{\frac{1}{\alpha}}, \| y\|_\alpha)$$

$$\times \{R[\sum_{i=1}^n (\tfrac{x_i}{R})^\alpha]^{\frac{1}{\alpha}}\}^{\frac{\lambda}{r}-n} dx_1 \cdots dx_n$$

$$= \| y\|_\alpha^{\frac{\lambda}{s}} \lim_{R\to+\infty} \frac{R^n\Gamma^n(\frac{1}{\alpha})}{\alpha^n\Gamma(\frac{n}{\alpha})} \times \int_0^1 k_\lambda(Ru^{\frac{1}{\alpha}}, \| y\|_\alpha)(Ru^{\frac{1}{\alpha}})^{\frac{\lambda}{r}-n} u^{\frac{n}{\alpha}-1} du$$

$$= \frac{\Gamma^n(\frac{1}{\alpha})}{\alpha^{n-1}\Gamma(\frac{n}{\alpha})} \int_0^\infty k_\lambda(v,1) v^{\frac{\lambda}{r}-1} dv = \frac{\Gamma^n(\frac{1}{\alpha})k_\lambda(r)}{\alpha^{n-1}\Gamma(\frac{n}{\alpha})}.$$

Due to

$$k_\lambda(r) = \tilde{k}_\lambda(s) = \int_0^\infty k_\lambda(1,v) v^{\frac{\lambda}{s}-1} dv,$$

we have

$$\tilde{\omega}_\lambda(s,x) = \frac{\Gamma^n(\frac{1}{\alpha})\tilde{k}_\lambda(s)}{\alpha^{n-1}\Gamma(\frac{n}{\alpha})} = \omega_\lambda(r,y),$$

and (6.1.2) is valid.□

Lemma 6.1.2 If $\alpha, \varepsilon > 0, n \in \mathbf{N}$, then the integrals

$$I(\varepsilon) := \int_{\{x\in R_+^n; \| x\|_\alpha \geq 1\}} \| x\|_\alpha^{-n-\varepsilon} dx$$

and

$$\tilde{I}(\varepsilon) := \int_{\{x\in R_+^n; \| x\|_\alpha \leq 1\}} \| x\|_\alpha^{-n+\varepsilon} dx$$

are all convergent, satisfying $\tilde{I}(\varepsilon) = I(\varepsilon)$ and

$$\lim_{\varepsilon\to 0^+} I(\varepsilon) = \lim_{\varepsilon\to 0^+} \tilde{I}(\varepsilon) = \infty. \quad (6.1.4)$$

Proof We Set $R > 1$ and

$$\psi(u) := \begin{cases} 0, & 0 < u \leq \frac{1}{R^\alpha}, \\ \frac{1}{(Ru^{1/\alpha})^{n+\varepsilon}}, & \frac{1}{R^\alpha} < u \leq 1. \end{cases}$$

By (6.1.3), we find

$$I(\varepsilon)=\lim_{R\to\infty}\frac{R^n\Gamma^n(\frac{1}{\alpha})}{\alpha^n\Gamma(\frac{n}{\alpha})}\int_{\frac{1}{R^\alpha}}^1 (Ru^{\frac{1}{\alpha}})^{-n-\varepsilon}u^{\frac{n}{\alpha}-1}du$$

$$=\frac{\Gamma^n(\frac{1}{\alpha})}{\alpha^n\Gamma(\frac{n}{\alpha})}\lim_{R\to\infty}\frac{1}{R^\varepsilon}\int_{\frac{1}{R^\alpha}}^1 u^{\frac{-\varepsilon}{\alpha}-1}du$$

$$=\frac{\Gamma^n(\frac{1}{\alpha})}{\alpha^{n-1}\Gamma(\frac{n}{\alpha})}\lim_{R\to\infty}\frac{1}{\varepsilon R^\varepsilon}(R^\varepsilon-1)=\frac{\Gamma^n(\frac{1}{\alpha})}{\alpha^{n-1}\Gamma(\frac{n}{\alpha})}\frac{1}{\varepsilon}<\infty.$$

If we Set

$$\psi(u):=\begin{cases}(Ru^{1/\alpha})^{-n+\varepsilon}, 0<u\le\frac{1}{R^\alpha},\\ 0, \quad \frac{1}{R^\alpha}<u\le 1,\end{cases}$$

then by (6.1.3), we find

$$\tilde{I}(\varepsilon)=\lim_{R\to\infty}\frac{R^n\Gamma^n(\frac{1}{\alpha})}{\alpha^n\Gamma(\frac{n}{\alpha})}\int_0^{\frac{1}{R^\alpha}}(Ru^{\frac{1}{\alpha}})^{-n+\varepsilon}u^{\frac{n}{\alpha}-1}du$$

$$=\frac{\Gamma^n(\frac{1}{\alpha})}{\alpha^n\Gamma(\frac{n}{\alpha})}\lim_{R\to\infty}R^\varepsilon\int_0^{\frac{1}{R^\alpha}}u^{\frac{\varepsilon}{\alpha}-1}du$$

$$=\frac{\Gamma^n(\frac{1}{\alpha})}{\alpha^{n-1}\Gamma(\frac{n}{\alpha})}\frac{1}{\varepsilon}=I(\varepsilon).$$

It is obvious that $I(\varepsilon)=\tilde{I}(\varepsilon)\to\infty(\varepsilon\to 0^+)$ and (6.1.4) is valid.

Lemma 6.1.3 As the assumption of Lemma 6.1.1, if

$$K_\lambda(x):=\int_0^\infty k_\lambda(u,1)u^{x-1}du$$

is finite in a neighborhood I_λ of $\frac{\lambda}{r}$, $p>0\ (p\ne 1)$, $\varepsilon_0>0$, such that $\frac{\lambda}{r}-\frac{\varepsilon}{p}\in I_\lambda(0<\varepsilon\le\varepsilon_0)$, then

$$J_\varepsilon(r):=\int_{\{y\in R_+^n;\|y\|_\alpha\ge 1\}}\int_{\{x\in R_+^n;\|x\|_\alpha\ge 1\}}k_\lambda(\|x\|_\alpha,\|y\|_\alpha)$$

$$\times\|x\|_\alpha^{\frac{\lambda}{r}-\frac{\varepsilon}{p}-n}\|y\|_\alpha^{\frac{\lambda}{s}-\frac{\varepsilon}{q}-n}dxdy=[\frac{\Gamma^n(\frac{1}{\alpha})k_\lambda(r)}{\alpha^{n-1}\Gamma(\frac{n}{\alpha})}+o(1)]$$

$$\times\int_{\{y\in R_+^n;\|y\|_\alpha\ge 1\}}\|y\|_\alpha^{-n-\varepsilon}dy(\varepsilon\to 0^+). \quad (6.1.5)$$

Proof We set

$$\tilde{k}_\lambda(x,y):=\begin{cases}0, \quad 0<x<1, y>0,\\ k_\lambda(x,y),\ x\ge 1, y>0,\end{cases}$$

Then by (6.1.2), we find

$$J_\varepsilon(r)=\int_{\{y\in R_+^n;\|y\|_\alpha\ge 1\}}\|y\|_\alpha^{-n-\varepsilon}[\|y\|_\alpha^{\frac{\lambda}{s}+\frac{\varepsilon}{p}}$$

$$\times\int_{\{x\in R_+^n;\|x\|_\alpha\ge 1\}}k_\lambda(\|x\|_\alpha,\|y\|_\alpha)$$

$$\times\|x\|_\alpha^{\frac{\lambda}{r}-\frac{\varepsilon}{p}-n}dx]dy$$

$$=\int_{\{y\in R_+^n;\|y\|_\alpha\ge 1\}}\|y\|_\alpha^{-n-\varepsilon}[\|y\|_\alpha^{(\frac{\lambda}{s}+\frac{\varepsilon}{p})}$$

$$\times\int_{R_+^n}\tilde{k}_\lambda(\|x\|_\alpha,\|y\|_\alpha)\|x\|_\alpha^{(\frac{\lambda}{r}-\frac{\varepsilon}{p})-n}dx]dy$$

$$=\int_{\{y\in R_+^n;\|y\|_\alpha\ge 1\}}\|y\|_\alpha^{-n-\varepsilon}[\frac{\Gamma^n(\frac{1}{\alpha})}{\alpha^{n-1}\Gamma(\frac{n}{\alpha})}$$

$$\times\int_0^\infty\tilde{k}_\lambda(v,1)v^{\frac{\lambda}{r}-\frac{\varepsilon}{p}-1}dv]dy$$

$$=\int_{\{y\in R_+^n;\|y\|_\alpha\ge 1\}}\|y\|_\alpha^{-n-\varepsilon}[\frac{\Gamma^n(\frac{1}{\alpha})}{\alpha^{n-1}\Gamma(\frac{n}{\alpha})}$$

$$\times\int_{\|y\|_\alpha^{-1}}^\infty k_\lambda(v,1)v^{\frac{\lambda}{r}-\frac{\varepsilon}{p}-1}dv]dy$$

$$=\frac{\Gamma^n(\frac{1}{\alpha})}{\alpha^{n-1}\Gamma(\frac{n}{\alpha})}[\int_{\{y\in R_+^n;\|y\|_\alpha\ge 1\}}\|y\|_\alpha^{-n-\varepsilon}$$

$$\times\int_{\|y\|_\alpha^{-1}}^1 k_\lambda(v,1)v^{\frac{\lambda}{r}-\frac{\varepsilon}{p}-1}dvdy$$

$$+\int_{\{y\in R_+^n;\|y\|\ge 1\}}\|y\|_\alpha^{-n-\varepsilon}\int_1^\infty k_\lambda(v,1)v^{\frac{\lambda}{r}-\frac{\varepsilon}{p}-1}dvdy]. \quad (6.1.6)$$

Since by Fubini Theorem, we find

$$\int_{\{y\in R_+^n;\|y\|_\alpha\ge 1\}}\|y\|_\alpha^{-n-\varepsilon}\int_{\|y\|_\alpha^{-1}}^1 k_\lambda(v,1)v^{\frac{\lambda}{r}-\frac{\varepsilon}{p}-1}dvdy$$

$$=\int_0^1 k_\lambda(v,1)v^{\frac{\lambda}{r}-\frac{\varepsilon}{p}-1}\int_{\{y\in R_+^n;\|y\|_\alpha\ge v^{-1}\}}\|y\|_\alpha^{-n-\varepsilon}dy$$

$$\overset{z=vy}{=}\int_0^1 k_\lambda(v,1)v^{\frac{\lambda}{r}+\frac{\varepsilon}{q}-1}\int_{\{z\in R_+^n;\|z\|_\alpha\ge 1\}}\|z\|_\alpha^{-n-\varepsilon}dz,$$

then by (6.1.6), it follows

$$J_\varepsilon(r)=\frac{\Gamma^n(\frac{1}{\alpha})}{\alpha^{n-1}\Gamma(\frac{n}{\alpha})}\int_{\{y\in R_+^n;\|y\|_\alpha\ge 1\}}\|y\|_\alpha^{-n-\varepsilon}dy$$

$$\times[\int_0^1 k_\lambda(v,1)v^{\frac{\lambda}{r}+\frac{\varepsilon}{q}-1}dv+\int_1^\infty k_\lambda(v,1)v^{\frac{\lambda}{r}-\frac{\varepsilon}{p}-1}dv]. \quad (6.1.7)$$

Since $p>0$, then we find $\frac{\varepsilon}{q}>-\frac{\varepsilon_0}{p}$ and

$$\int_0^1 k_\lambda(v,1)v^{\frac{\lambda}{r}+\frac{\varepsilon}{q}-1}dv+\int_1^\infty k_\lambda(v,1)v^{\frac{\lambda}{r}-\frac{\varepsilon}{p}-1}dv$$

$$\le\int_0^1 k_\lambda(v,1)v^{\frac{\lambda}{r}-\frac{\varepsilon_0}{p}-1}dv+\int_1^\infty k_\lambda(v,1)v^{\frac{\lambda}{r}-1}dv$$

$$\le K_\lambda(\frac{\lambda}{r}-\frac{\varepsilon_0}{p})+k_\lambda(r)<\infty.$$

By Lebesgue Control Convergence Theorem, we have

$$\int_0^1 k_\lambda(v,1)v^{\frac{\lambda}{r}+\frac{\varepsilon}{q}-1}dv+\int_1^\infty k_\lambda(v,1)v^{\frac{\lambda}{r}-\frac{\varepsilon}{p}-1}dv$$

$$=\int_0^\infty k_\lambda(v,1)v^{\frac{\lambda}{r}-1}dv+o_1(1)$$

$$=k_\lambda(r)(1+o(1))(\varepsilon\to 0^+).$$

Hence in view of (6.1.7), we have (6.1.5). □

6.1.2. BASIC RESULTS

Theorem 6.1.4 Suppose that (p,q) and (r,s) are two pairs of conjugate exponents, $p>1$, $r,s\ne 0$, and

$\lambda \in \mathbf{R}$, $k_\lambda(x,y)(\geq 0)$ is a homogeneous function of $-\lambda$-degree in R_+^2, if

$$K_\lambda(x) := \int_0^\infty k_\lambda(u,1)u^{x-1}du$$

is finite in a neighborhood I_λ of $\frac{\lambda}{r}$,

$$k_\lambda(r) := \int_0^\infty k_\lambda(u,1)u^{\frac{\lambda}{r}-1}du,$$

$\alpha > 0, n \in \mathbf{N}$, $\| x \|_\alpha = (\sum_{i=1}^n x_i^\alpha)^{\frac{1}{\alpha}}$ $f(x)$ and $g(x)$ are non-negative measurable functions in R_+^n, then we have the following equivalent inequalities:

$$J := \int_{R_+^n} \| y \|_\alpha^{\frac{p\lambda}{s}-n} [\int_{R_+^n} k_\lambda(\| x \|_\alpha, \| y \|_\alpha) \times f(x)dx]^p dy \leq [\frac{\Gamma^n(\frac{1}{\alpha})k_\lambda(r)}{\alpha^{n-1}\Gamma(\frac{n}{\alpha})}]^p \int_{R_+^n} \| x \|_\alpha^{p(n-\frac{\lambda}{r})-n} f^p(x)dx; \quad (6.1.8)$$

$$I := \int_{R_+^n} \int_{R_+^n} k_\lambda(\| x \|_\alpha, \| y \|_\alpha) f(x)g(y)dxdy \leq \frac{\Gamma^n(\frac{1}{\alpha})k_\lambda(r)}{\alpha^{n-1}\Gamma(\frac{n}{\alpha})} (\int_{R_+^n} \| x \|_\alpha^{p(n-\frac{\lambda}{r})-n} f^p(x)dx)^{\frac{1}{p}} \times (\int_{R_+^n} \| x \|_\alpha^{q(n-\frac{\lambda}{s})-n} g^q(x)dx)^{\frac{1}{q}}. \quad (6.1.9)$$

Proof By Hölder's inequality with weight and Lemma 6.1.1, we find

$$\{\int_{R_+^n} k_\lambda(\| x \|_\alpha, \| y \|_\alpha) \times [\frac{\| x \|_\alpha^{(n-\frac{\lambda}{r})/q}}{\| y \|_\alpha^{(n-\frac{\lambda}{s})/p}} f(x)][\frac{\| y \|_\alpha^{(n-\frac{\lambda}{s})/p}}{\| x \|_\alpha^{(n-\frac{\lambda}{r})/q}}]dx\}^p$$

$$\leq \int_{R_+^n} k_\lambda(\| x \|_\alpha, \| y \|_\alpha) \frac{\| x \|_\alpha^{(n-\frac{\lambda}{r})(p-1)}}{\| y \|_\alpha^{n-\frac{\lambda}{s}}} f^p(x)dx \times \{\int_{R_+^n} k_\lambda(\| x \|_\alpha, \| y \|_\alpha) \frac{\| y \|_\alpha^{(n-\frac{\lambda}{s})(q-1)}}{\| x \|_\alpha^{n-\frac{\lambda}{r}}} dx\}^{p-1}$$

$$= [\omega_\lambda(r,y)]^{p-1} \| y \|_\alpha^{[q(n-\frac{\lambda}{s})-n](p-1)} \times \int_{R_+^n} k_\lambda(\| x \|_\alpha, \| y \|_\alpha) \frac{\| x \|_\alpha^{(n-\frac{\lambda}{r})(p-1)}}{\| y \|_\alpha^{n-\frac{\lambda}{s}}} f^p(x)dx$$

$$= [\frac{\Gamma^n(\frac{1}{\alpha})k_\lambda(r)}{\alpha^{n-1}\Gamma(\frac{n}{\alpha})}]^{p-1} \| y \|_\alpha^{(1-p)\frac{\lambda}{s}} \times \int_{R_+^n} k_\lambda(\| x \|_\alpha, \| y \|_\alpha) \| x \|_\alpha^{(n-\frac{\lambda}{r})(p-1)} f^p(x)dx. \quad (6.1.10)$$

Then by (6.1.2), we find

$$J \leq [\frac{\Gamma^n(\frac{1}{\alpha})k_\lambda(r)}{\alpha^{n-1}\Gamma(\frac{n}{\alpha})}]^{p-1} \int_{R_+^n} \| y \|_\alpha^{\frac{\lambda}{s}-n} \times \int_{R_+^n} k_\lambda(\| x \|_\alpha, \| y \|_\alpha) \times \| x \|_\alpha^{(n-\frac{\lambda}{r})(p-1)} f^p(x)dxdy$$

$$= [\frac{\Gamma^n(\frac{1}{\alpha})k_\lambda(r)}{\alpha^{n-1}\Gamma(\frac{n}{\alpha})}]^{p-1} \times \int_{R_+^n} [\int_{R_+^n} k_\lambda(\| x \|_\alpha, \| y \|_\alpha) \frac{\| x \|_\alpha^{\frac{\lambda}{r}}}{\| y \|_\alpha^{n-\frac{\lambda}{s}}} dy] \times \| x \|_\alpha^{p(n-\frac{\lambda}{r})-n} f^p(x)dx$$

$$= [\frac{\Gamma^n(\frac{1}{\alpha})k_\lambda(r)}{\alpha^{n-1}\Gamma(\frac{n}{\alpha})}]^{p-1} \times \int_{R_+^n} \tilde{\omega}_\lambda(s,x) \times \| x \|_\alpha^{p(n-\frac{\lambda}{r})-n} f^p(x)dx$$

$$= [\frac{\Gamma^n(\frac{1}{\alpha})k_\lambda(r)}{\alpha^{n-1}\Gamma(\frac{n}{\alpha})}]^p \int_{R_+^n} \| x \|_\alpha^{p(n-\frac{\lambda}{r})-n} f^p(x)dx, \quad (6.1.11)$$

and (6.1.8) is valid.

By Hölder's inequality, we find

$$I = \int_{R_+^n} [\| y \|_\alpha^{\frac{\lambda}{s}-\frac{n}{p}} \int_{R_+^n} k_\lambda(\| x \|_\alpha, \| y \|_\alpha) f(x)dx] \times [\| y \|_\alpha^{\frac{n}{p}-\frac{\lambda}{s}} g(y)]dy \leq J^{\frac{1}{p}} \{\int_{R_+^n} \| y \|_\alpha^{q(n-\frac{\lambda}{s})-n} g^q(y)dy\}^{\frac{1}{q}}. \quad (6.1.12)$$

Then by (6.1.8), we have (6.1.9).

Setting

$$g(y) := \| y \|_\alpha^{\frac{p\lambda}{s}-n} \times [\int_{R_+^n} k_\lambda(\| x \|_\alpha, \| y \|_\alpha) f(x)dx]^{p-1} dy, y \in R_+^n,$$

then by (6.1.9), we have

$$\int_{R_+^n} \| y \|_\alpha^{q(n-\frac{\lambda}{s})-n} g^q(y)dy = I = J \leq \frac{\Gamma^n(\frac{1}{\alpha})k_\lambda(r)}{\alpha^{n-1}\Gamma(\frac{n}{\alpha})} (\int_{R_+^n} \| x \|_\alpha^{p(n-\frac{\lambda}{r})-n} f^p(x)dx)^{\frac{1}{p}} \times (\int_{R_+^n} \| y \|_\alpha^{q(n-\frac{\lambda}{s})-n} g^q(y)dy)^{\frac{1}{q}}. \quad (6.1.13)$$

If $J = \infty$, then by (6.1.11), we conclude that (6.1.8) takes the form of equality; if $J = 0$, then (6.1.8) is naturally valid. Assuming that $0 < J < \infty$, dividing by $J^{\frac{1}{q}}$ in (6.1.13), we have

$$J^{\frac{1}{p}} \leq \frac{\Gamma^n(\frac{1}{\alpha})k_\lambda(r)}{\alpha^{n-1}\Gamma(\frac{n}{\alpha})} (\int_{R_+^n} \| x \|_\alpha^{p(n-\frac{\lambda}{r})-n} f^p(x)dx)^{\frac{1}{p}}.$$

Hence (6.1.8) is valid, which is equivalent to (6.1.9).

As the assumption of Theorem 6.1.4, setting

$$\tilde{\phi}(x) = \| x \|_\alpha^{p(n-\frac{\lambda}{r})-n}, \tilde{\psi}(x) = \| x \|_\alpha^{q(n-\frac{\lambda}{s})-n}, x \in R_+^n$$

and then $\tilde{\psi}^{1-p}(x) = \| x \|_\alpha^{\frac{p\lambda}{s}-n}$, define the following real function spaces as:

$$L^p_{\tilde{\phi}}(R^n_+)=\{f;\|f\|_{p,\tilde{\phi}}:=\{\int_{R^n_+}\tilde{\phi}(x)|f(x)|^p dx\}^{\frac{1}{p}}<\infty\},$$

$$L^q_{\tilde{\psi}}(R^n_+)=\{g;\|g\|_{q,\tilde{\psi}}:=\{\int_{R^n_+}\tilde{\psi}(x)|g(x)|^q dx\}^{\frac{1}{q}}<\infty\}.$$

For $f(\geq 0)\in L^p_{\tilde{\phi}}(R^n_+)$, define a multivariable Hilbert-type integral operator

$$T: L^p_{\tilde{\phi}}(R^n_+)\to L^p_{\tilde{\psi}^{1-p}}(R^n_+)$$

as

$$(Tf)(y)=\int_{R^n_+}k_\lambda(\|x\|_\alpha,\|y\|_\alpha)f(x)dx,$$
$$y\in R^n_+. \qquad (6.1.14)$$

Then by (6.1.9), it follows $Tf\in L^p_{\tilde{\psi}^{1-p}}(R^n_+)$. If $g(\geq 0)\in L^q_{\tilde{\psi}}(R^n_+)$, setting the formal inner product of Tf and g as

$$(Tf,g):=\int_{R^n_+}\int_{R^n_+}k_\lambda(\|x\|_\alpha,\|y\|_\alpha)$$
$$\times f(x)g(y)dxdy,$$

then we may rewrite (6.1.8) and (6.1.9) to the following equivalent forms:

$$\|Tf\|_{p,\tilde{\psi}^{1-p}}\leq\frac{\Gamma^n(\frac{1}{\alpha})k_\lambda(r)}{\alpha^{n-1}\Gamma(\frac{n}{\alpha})}\|f\|_{p,\tilde{\phi}}; \qquad (6.1.15)$$

$$(Tf,g)\leq\frac{\Gamma^n(\frac{1}{\alpha})k_\lambda(r)}{\alpha^{n-1}\Gamma(\frac{n}{\alpha})}\|f\|_{p,\tilde{\phi}}\|g\|_{q,\tilde{\psi}}. \qquad (6.1.16)$$

By (6.1.15), it follows that T is a bounded operator and

$$\|T\|\leq\frac{\Gamma^n(1/\alpha)k_\lambda(r)}{\alpha^{n-1}\Gamma(n/\alpha)}.$$

Theorem 6.1.5 As the assumption of Theorem 6.1.4, if the operator T is defined by (6.1.13), then we have

$$\|T\|=\frac{\Gamma^n(\frac{1}{\alpha})}{\alpha^{n-1}\Gamma(\frac{n}{\alpha})}k_\lambda(r). \qquad (6.1.17)$$

Proof For $\varepsilon>0$ small enough, setting

$$\tilde{f}(x)=\begin{cases}0, & x\in\{x;\|x\|_\alpha<1,x\in R^n_+\},\\ \|x\|_\alpha^{\frac{\lambda}{r}-\frac{\varepsilon}{p}-1}, & x\in\{x;\|x\|_\alpha\geq 1,x\in R^n_+\},\end{cases}$$

$$\tilde{g}(x)=\begin{cases}0, & x\in\{x;\|x\|_\alpha<1,x\in R^n_+\},\\ \|x\|_\alpha^{\frac{\lambda}{s}-\frac{\varepsilon}{q}-1}, & x\in\{x;\|x\|_\alpha\geq 1,x\in R^n_+\},\end{cases}$$

we find

$$\|\tilde{f}\|_{p,\tilde{\phi}}\|\tilde{g}\|_{q,\tilde{\psi}}=\int_{\{x\in R^n_+;\|x\|_\alpha\geq 1\}}\|x\|_\alpha^{-n-\varepsilon}dx.$$

If there exists

$$0\leq k\leq\frac{\Gamma^n(\frac{1}{\alpha})}{\alpha^{n-1}\Gamma(\frac{n}{\alpha})}k_\lambda(r),$$

such that (6.1.16) is still valid as we replace $\frac{\Gamma^n(1/\alpha)k_\lambda(r)}{\alpha^{n-1}\Gamma(n/\alpha)}$ by k, then in particular, by (6.1.5), we have

$$[\frac{\Gamma^n(\frac{1}{\alpha})k_\lambda(r)}{\alpha^{n-1}\Gamma(\frac{n}{\alpha})}+o(1)]\int_{\{y\in R^n_+;\|y\|_\alpha\geq 1\}}\|y\|_\alpha^{-n-\varepsilon}dy$$
$$=I_\varepsilon=(T\tilde{f},\tilde{g})\leq k\|\tilde{f}\|_{p,\tilde{\phi}}\|\tilde{g}\|_{q,\tilde{\psi}}$$
$$=k\int_{\{y\in R^n_+;\|y\|_\alpha>1\}}\|y\|_\alpha^{-n-\varepsilon}dy,$$

and $\frac{\Gamma^n(1/\alpha)k_\lambda(r)}{\alpha^{n-1}\Gamma(n/\alpha)}+o(1)\leq k$. For $\varepsilon\to 0^+$, it follows $\frac{\Gamma^n(1/\alpha)k_\lambda(r)}{\alpha^{n-1}\Gamma(n/\alpha)}\leq k$. Hence $k=\frac{\Gamma^n(1/\alpha)k_\lambda(r)}{\alpha^{n-1}\Gamma(n/\alpha)}$ is the best value of (6.1.16). By (6.1.12), we can show that the constant factor $k=\frac{\Gamma^n(1/\alpha)k_\lambda(r)}{\alpha^{n-1}\Gamma(n/\alpha)}$ in (6.1.15) is the best possible. And we have (6.1.17).□

Theorem 6.1.6 As the assumption of Theorem 6.1.4, if $p>0(p\neq 1)$,

$$\tilde{\phi}(x)=\|x\|_\alpha^{p(n-\frac{\lambda}{r})-n},\tilde{\psi}(x)=\|x\|_\alpha^{q(n-\frac{\lambda}{s})-n},x\in R^n_+,$$

$f(\geq 0)\in L^p_{\tilde{\phi}}(R^n_+)$, $g(\geq 0)\in L^q_{\tilde{\psi}}(R^n_+)$, and

$$\|f\|_{p,\tilde{\phi}}=\{\int_{R^n_+}\|x\|_\alpha^{p(n-\frac{\lambda}{r})-n}f^p(x)dx\}^{\frac{1}{p}}>0,$$

$$\|g\|_{q,\tilde{\psi}}=\{\int_{R^n_+}\|x\|_\alpha^{q(n-\frac{\lambda}{s})-n}g^q(x)dx\}^{\frac{1}{q}}>0,$$

then (1) for $p>1$, we have the following equivalent inequalities:

$$\int_{R^n_+}\int_{R^n_+}k_\lambda(\|x\|_\alpha,\|y\|_\alpha)f(x)g(y)dxdy$$
$$<\frac{\Gamma^n(\frac{1}{\alpha})k_\lambda(r)}{\alpha^{n-1}\Gamma(\frac{n}{\alpha})}\|f\|_{p,\tilde{\phi}}\|g\|_{q,\tilde{\psi}}, \qquad (6.1.18)$$

$$\int_{R^n_+}\|y\|_\alpha^{\frac{p\lambda}{s}-n}[\int_{R^n_+}k_\lambda(\|x\|_\alpha,\|y\|_\alpha)$$
$$\times f(x)dx]^p dy$$
$$<[\frac{\Gamma^n(\frac{1}{\alpha})k_\lambda(r)}{\alpha^{n-1}\Gamma(\frac{n}{\alpha})}]^p\|f\|^p_{p,\tilde{\phi}}, \qquad (6.1.19)$$

where the constant factors are the best possible; (2) for $0<p<1$, we have the equivalent reverses of (6.1.18) and (6.1.19) with the best constant factors.

Proof (1) We conform that (6.1.10) takes the strict sign-inequality, otherwise, since $k_\lambda(r)>0$, then $k_\lambda(x,y)$ is not zero almost everywhere, there exist $y\in R^n_+$, constants A and B, which are not all zero and

$$A\frac{\|x\|_\alpha^{(n-\frac{\lambda}{r})(p-1)}}{\|y\|_\alpha^{n-\frac{\lambda}{s}}}f^p(x)=B\frac{\|y\|_\alpha^{(n-\frac{\lambda}{s})(q-1)}}{\|x\|_\alpha^{n-\frac{\lambda}{r}}}\ a.e.\ \text{in } R^n_+.$$

It follows

$$A\|x\|_\alpha^{p(n-\frac{\lambda}{r})}f^p(x)=B\|y\|_\alpha^{q(n-\frac{\lambda}{s})}\ a.e.\ \text{in } R^n_+.$$

It is obvious that $A \neq 0$,otherwise, $A = B = 0$. Then we have

$$\| x \|_{\alpha}^{p(n-\frac{\lambda}{r})-n} f^{p}(x) = B \| y \|_{\alpha}^{q(n-\frac{\lambda}{s})} /(A \| x \|_{\alpha}^{n})$$

a.e. in R_{+}^{n},

which contradicts $0 < \| f \|_{p,\tilde{\phi}} < \infty$ due to (6.1.4). Then (6.1.8) still takes the strict sign-inequality and we have (6.1.19). By (6.1.12), we have (6.1.18), which is still equivalent to (6.1.19). And the constant factors are all the best possible by the proof of Theorem 6.1.5.

(2) By the same way of the case in $p > 1$ and applying the reverse Hölder's inequality, we have the equivalent reverses of (6.1.18) and (6.1.19). If there exists a constant

$$K \geq \frac{\Gamma^{n}(\frac{1}{\alpha})}{\alpha^{n-1}\Gamma(\frac{n}{\alpha})} k_{\lambda}(r),$$

such that the reverse of (6.1.18) is valid as we replace $\frac{\Gamma^{n}(1/\alpha) k_{\lambda}(r)}{\alpha^{n-1}\Gamma(n/\alpha)}$ by K, then in particular, by (6.1.5), we find

$$[\frac{\Gamma^{n}(\frac{1}{\alpha}) k_{\lambda}(r)}{\alpha^{n-1}\Gamma(\frac{n}{\alpha})} + o(1)] \int_{\{y \in R_{+}^{n}; \|y\|_{\alpha} \geq 1\}} \| y \|_{\alpha}^{-n-\varepsilon} dy$$

$$= I_{\varepsilon} > K \| \tilde{f} \|_{p,\tilde{\phi}} \| \tilde{g} \|_{q,\tilde{\psi}}$$

$$= K \int_{\{y \in R_{+}^{n}; \|y\|_{\alpha} > 1\}} \| y \|_{\alpha}^{-n-\varepsilon} dy,$$

and $\frac{\Gamma^{n}(1/\alpha) k_{\lambda}(r)}{\alpha^{n-1}\Gamma(n/\alpha)} + o(1) > K$. For $\varepsilon \to 0^{+}$, it follows $\frac{\Gamma^{n}(1/\alpha) k_{\lambda}(r)}{\alpha^{n-1}\Gamma(n/\alpha)} \geq K$. Hence $K = \frac{\Gamma^{n}(1/\alpha) k_{\lambda}(r)}{\alpha^{n-1}\Gamma(n/\alpha)}$ is the best value of the reverse of (6.1.18). By using the reverse of (6.1.12), we can show that the constant factor in the reverse of (6.1.19) is the best possible.

6.1.3. SOME EXAMPLES

In the following examples, we omit to write the assumption of f and g in Theorem 6.1.6 and the conclusions that the constant factors are the best possible.

Example 6.1.7 If $\beta, \lambda > 0, r, s > 1$,

$$k_{\lambda}(x,y) = \frac{1}{(x^{\beta} + y^{\beta})^{\lambda/\beta}},$$

then we find $k_{\lambda}(r) = \frac{1}{\beta} B(\frac{\lambda}{\beta r}, \frac{\lambda}{\beta s})$. By Theorem 6.1.6, (1) for $p > 1$, we have the following equivalent inequalities:

$$\int_{R_{+}^{n}} \int_{R_{+}^{n}} \frac{f(x) g(y)}{(\|x\|_{\alpha}^{\beta} + \|y\|_{\alpha}^{\beta})^{\lambda/\beta}} dx dy$$

$$< \frac{\Gamma^{n}(\frac{1}{\alpha})}{\alpha^{n-1}\Gamma(\frac{n}{\alpha})\beta} B(\frac{\lambda}{\beta r}, \frac{\lambda}{\beta s}) \| f \|_{p,\tilde{\phi}} \| g \|_{q,\tilde{\psi}}, \quad (6.1.20)$$

$$\int_{R_{+}^{n}} \| y \|_{\alpha}^{\frac{p\lambda}{s}-n} [\int_{R_{+}^{n}} \frac{f(x)}{(\|x\|_{\alpha}^{\beta} + \|y\|_{\alpha}^{\beta})^{\lambda/\beta}} dx]^{p} dy$$

$$< [\frac{\Gamma^{n}(\frac{1}{\alpha})}{\alpha^{n-1}\Gamma(\frac{n}{\alpha})\beta} B(\frac{\lambda}{\beta r}, \frac{\lambda}{\beta s})]^{p} \| f \|_{p,\tilde{\phi}}^{p}; \quad (6.1.21)$$

(2) for $0 < p < 1$, we have the equivalent reverses of (6.1.20) and (6.1.21). In particular,

(i) if $\beta = 1, \lambda > 0$, then (1) for $p > 1$, we have the following equivalent inequalities (Hong JIA, 2006) [1]:

$$\int_{R_{+}^{n}} \int_{R_{+}^{n}} \frac{1}{(\|x\|_{\alpha} + \|y\|_{\alpha})^{\lambda}} f(x) g(y) dx dy$$

$$< \frac{\Gamma^{n}(\frac{1}{\alpha})}{\alpha^{n-1}\Gamma(\frac{n}{\alpha})} B(\frac{\lambda}{r}, \frac{\lambda}{s}) \| f \|_{p,\tilde{\phi}} \| g \|_{q,\tilde{\psi}}, \quad (6.1.22)$$

$$\int_{R_{+}^{n}} \| y \|_{\alpha}^{\frac{p\lambda}{s}-n} [\int_{R_{+}^{n}} \frac{f(x)}{(\|x\|_{\alpha} + \|y\|_{\alpha})^{\lambda}} dx]^{p} dy$$

$$< [\frac{\Gamma^{n}(\frac{1}{\alpha})}{\alpha^{n-1}\Gamma(\frac{n}{\alpha})} B(\frac{\lambda}{r}, \frac{\lambda}{s})]^{p} \| f \|_{p,\tilde{\phi}}^{p}; \quad (6.1.23)$$

(2) for $0 < p < 1$, we have the equivalent reverses of (6.1.22) and (6.1.23).

(ii) If $\beta = \lambda > 0$, then (1) for $p > 1$, we have the following equivalent inequalities

$$\int_{R_{+}^{n}} \int_{R_{+}^{n}} \frac{1}{\|x\|_{\alpha}^{\lambda} + \|y\|_{\alpha}^{\lambda}} f(x) g(y) dx dy$$

$$< \frac{\Gamma^{n}(\frac{1}{\alpha})\pi}{\alpha^{n-1}\Gamma(\frac{n}{\alpha})\lambda \sin(\frac{\pi}{r})} \| f \|_{p,\tilde{\phi}} \| g \|_{q,\tilde{\psi}}, \quad (6.1.24)$$

$$\int_{R_{+}^{n}} \| y \|_{\alpha}^{\frac{p\lambda}{s}-n} [\int_{R_{+}^{n}} \frac{f(x)}{\|x\|_{\alpha}^{\lambda} + \|y\|_{\alpha}^{\lambda}} dx]^{p} dy$$

$$< [\frac{\Gamma^{n}(\frac{1}{\alpha})\pi}{\alpha^{n-1}\Gamma(\frac{n}{\alpha})\lambda \sin(\frac{\pi}{r})}]^{p} \| f \|_{p,\tilde{\phi}}^{p}. \quad (6.1.25)$$

(2) for $0 < p < 1$, we have the equivalent reverses of (6.1.24) and (6.1.25).

Example 6.1.8 If $\lambda > 0, r, s > 1$,

$$k_{\lambda}(x,y) = \frac{1}{(\max\{x,y\})^{\lambda}},$$

then we find $k_{\lambda}(r) = \frac{rs}{\lambda}$. By Theorem 6.1.6, (1) for $p > 1$, we have the following equivalent inequalities:

$$\int_{R_{+}^{n}} \int_{R_{+}^{n}} \frac{f(x) g(y)}{(\max\{\|x\|_{\alpha}, \|y\|_{\alpha}\})^{\lambda}} dx dy$$

$$< \frac{\Gamma^{n}(\frac{1}{\alpha}) rs}{\alpha^{n-1}\Gamma(\frac{n}{\alpha})\lambda} \| f \|_{p,\tilde{\phi}} \| g \|_{q,\tilde{\psi}}, \quad (6.1.26)$$

$$\int_{R_{+}^{n}} \| y \|_{\alpha}^{\frac{p\lambda}{s}-n} [\int_{R_{+}^{n}} \frac{f(x)}{(\max\{\|x\|_{\alpha}, \|y\|_{\alpha}\})^{\lambda}} dx]^{p} dy$$

$$< [\frac{\Gamma^{n}(\frac{1}{\alpha}) rs}{\alpha^{n-1}\Gamma(\frac{n}{\alpha})\lambda}]^{p} \| f \|_{p,\tilde{\phi}}^{p}; \quad (6.1.27)$$

(2) for $0 < p < 1$, we have the equivalent reverses of (6.1.26) and (6.1.27).

Example 6.1.9 If $\lambda > 0, r, s > 1, k_\lambda(x,y) = \frac{\ln(x/y)}{x^\lambda - y^\lambda}$, then we find $k_\lambda(r) = [\frac{\pi}{\lambda\sin(\frac{\pi}{r})}]^2$. By Theorem 6.1.6, (1) for $p > 1$, we have the following equivalent inequalities (Zhong MIA 2007) [2]:

$$\int_{R_+^n}\int_{R_+^n} \frac{\ln(\|x\|_\alpha/\|y\|_\alpha)}{\|x\|_\alpha^\lambda - \|y\|_\alpha^\lambda} f(x)g(y)dxdy$$
$$< \frac{\Gamma^n(\frac{1}{\alpha})}{\alpha^{n-1}\Gamma(\frac{n}{\alpha})}[\frac{\pi}{\lambda\sin(\frac{\pi}{r})}]^2 \| f \|_{p,\tilde{\phi}} \| g \|_{q,\tilde{\psi}}, \quad (6.1.28)$$

$$\int_{R_+^n} \| y \|_\alpha^{\frac{p\lambda}{s}-n} [\int_{R_+^n} \frac{\ln(\|x\|_\alpha/\|y\|_\alpha)}{\|x\|_\alpha^\lambda - \|y\|_\alpha^\lambda} f(x)dx]^p dy$$
$$< [\frac{\Gamma^n(\frac{1}{\alpha})}{\alpha^{n-1}\Gamma(\frac{n}{\alpha})}(\frac{\pi}{\lambda\sin(\frac{\pi}{r})})^2]^p \| f \|_{p,\tilde{\phi}}^p; \quad (6.1.29)$$

(2) for $0 < p < 1$, we have the equivalent reverses of (6.1.28) and (6.1.29).

Example 6.1.10 If $\lambda > 0, r, s > 1$,

$$k_\lambda(x,y) = \frac{|\ln(x/y)|}{(\max\{x,y\})^\lambda},$$

then we find $k_\lambda(r) = \frac{r^2+s^2}{\lambda^2}$. By Theorem 6.1.6, (1) for $p > 1$, we have the following equivalent inequalities

$$\int_{R_+^n}\int_{R_+^n} \frac{|\ln(\|x\|_\alpha/\|y\|_\alpha)|}{(\max\{\|x\|_\alpha,\|y\|_\alpha\})^\lambda} f(x)g(y)dxdy$$
$$< \frac{\Gamma^n(\frac{1}{\alpha})(r^2+s^2)}{\alpha^{n-1}\Gamma(\frac{n}{\alpha})\lambda^2} \| f \|_{p,\tilde{\phi}} \| g \|_{q,\tilde{\psi}}, \quad (6.1.30)$$

$$\int_{R_+^n} \| y \|_\alpha^{\frac{p\lambda}{s}-n} [\int_{R_+^n} \frac{|\ln(\|x\|_\alpha/\|y\|_\alpha)|}{(\max\{\|x\|_\alpha,\|y\|_\alpha\})^\lambda} f(x)dx]^p dy$$
$$< [\frac{\Gamma^n(\frac{1}{\alpha})(r^2+s^2)}{\alpha^{n-1}\Gamma(\frac{n}{\alpha})\lambda^2}]^p \| f \|_{p,\tilde{\phi}}^p; \quad (6.1.31)$$

(2) for $0 < p < 1$, we have the equivalent reverses of (6.1.30) and (6.1.31).

Example 6.1.11 If $0 < \lambda < 1, r, s > 1, k_\lambda(x,y) = \frac{1}{|x-y|^\lambda}$, then

$$k_\lambda(r) = B(1-\lambda, \tfrac{\lambda}{r}) + B(1-\lambda, \tfrac{\lambda}{s}).$$

By Theorem 6.1.6, (1) for $p > 1$, we have the following equivalent inequalities:

$$\int_{R_+^n}\int_{R_+^n} \frac{f(x)g(y)}{|\|x\|_\alpha - \|y\|_\alpha|^\lambda}dxdy < \frac{\Gamma^n(\frac{1}{\alpha})}{\alpha^{n-1}\Gamma(\frac{n}{\alpha})}$$
$$\times[B(1-\lambda,\tfrac{\lambda}{r}) + B(1-\lambda,\tfrac{\lambda}{s})] \| f \|_{p,\tilde{\phi}} \| g \|_{q,\tilde{\psi}}, \quad (6.1.32)$$

$$\int_{R_+^n} \| y \|_\alpha^{\frac{p\lambda}{s}-n} [\int_{R_+^n} \frac{f(x)}{|\|x\|_\alpha - \|y\|_\alpha|^\lambda}dx]^p dy$$
$$< [\frac{\Gamma^n(\frac{1}{\alpha})}{\alpha^{n-1}\Gamma(\frac{n}{\alpha})}(B(1-\lambda,\tfrac{\lambda}{r}) + B(1-\lambda,\tfrac{\lambda}{s}))]^p \| f \|_{p,\tilde{\phi}}^p; \quad (6.1.33)$$

(2) for $0 < p < 1$, we have the equivalent reverses of (6.1.32) and (6.1.33).

Example 6.1.12 If $\lambda \in \mathbf{R}, r, s \neq 0$,

$$\beta > \max\{0, \tfrac{\lambda}{r}, \tfrac{\lambda}{s}\}, |\eta| < \beta - \max\{\tfrac{\lambda}{r}, \tfrac{\lambda}{s}\},$$

$k_\lambda(x,y) = \frac{(\min\{x,y\})^{\beta-\lambda}}{x^\beta + y^\beta}$, then by Example 2.2.14, we obtain $0 < K_\lambda(\frac{\lambda}{r} + \eta) < \infty$ and

$$k_{\lambda,\beta}(r) = \sum_{k=1}^{\infty}(-1)^{k-1}[\frac{1}{\beta k - \frac{\lambda}{s}} + \frac{1}{\beta k - \frac{\lambda}{r}}].$$

By Theorem 6.1.6, (1) for $p > 1$, we have the following equivalent inequalities:

$$\int_{R_+^n}\int_{R_+^n} \frac{(\min\{\|x\|_\alpha,\|y\|_\alpha\})^{\beta-\lambda}}{\|x\|_\alpha^\beta + \|y\|_\alpha^\beta} f(x)g(y)dxdy$$
$$< \frac{\Gamma^n(\frac{1}{\alpha})}{\alpha^{n-1}\Gamma(\frac{n}{\alpha})} k_{\lambda,\beta}(r) \| f \|_{p,\tilde{\phi}} \| g \|_{q,\tilde{\psi}}, \quad (6.1.34)$$

$$\int_{R_+^n} \| y \|_\alpha^{\frac{p\lambda}{s}-n} [\int_{R_+^n} \frac{(\min\{\|x\|_\alpha,\|y\|_\alpha\})^{\beta-\lambda}}{\|x\|_\alpha^\beta + \|y\|_\alpha^\beta} f(x)dx]^p dy$$
$$< [\frac{\Gamma^n(\frac{1}{\alpha})}{\alpha^{n-1}\Gamma(\frac{n}{\alpha})} k_{\lambda,\beta}(r)]^p \| f \|_{p,\tilde{\phi}}^p; \quad (6.1.35)$$

(2) for $0 < p < 1$, we have the equivalent reverses of (6.1.34) and (6.1.35).

Example 6.1.13 If $\lambda \in \mathbf{R}$, $r, s \neq 0$,

$$\beta > \max\{\tfrac{\lambda}{r}, \tfrac{\lambda}{s}\}, |\eta| < \beta - \max\{\tfrac{\lambda}{r}, \tfrac{\lambda}{s}\},$$
$$k_\lambda(x,y) = \frac{(\min\{x,y\})^{\beta-\lambda}}{(\max\{x,y\})^\beta},$$

then by Example 2.2.15, we obtain

$$0 < K_\lambda(\tfrac{\lambda}{r} + \eta) < \infty$$

and

$$k_\lambda(r) = \frac{1}{\beta - \frac{\lambda}{s}} + \frac{1}{\beta - \frac{\lambda}{r}}.$$

By Theorem 6.1.6, (1) for $p > 1$, we have the following equivalent inequalities:

$$\int_{R_+^n}\int_{R_+^n} \frac{(\min\{\|x\|_\alpha,\|y\|_\alpha\})^{\beta-\lambda}}{(\max\{\|x\|_\alpha,\|y\|_\alpha\})^\beta} f(x)g(y)dxdy$$
$$< \frac{\Gamma^n(\frac{1}{\alpha})}{\alpha^{n-1}\Gamma(\frac{n}{\alpha})}[\frac{1}{\beta - \frac{\lambda}{r}} + \frac{1}{\beta - \frac{\lambda}{s}}] \| f \|_{p,\tilde{\phi}} \| g \|_{q,\tilde{\psi}}, \quad (6.1.36)$$

$$\int_{R_+^n} \| y \|_\alpha^{\frac{p\lambda}{s}-n} [\int_{R_+^n} \frac{(\min\{\|x\|_\alpha,\|y\|_\alpha\})^{\beta-\lambda}}{(\max\{\|x\|_\alpha,\|y\|_\alpha\})^\beta} f(x)dx]^p dy$$
$$< [\frac{\Gamma^n(\frac{1}{\alpha})}{\alpha^{n-1}\Gamma(\frac{n}{\alpha})}(\frac{1}{\beta - \frac{\lambda}{r}} + \frac{1}{\beta - \frac{\lambda}{s}})]^p \| f \|_{p,\tilde{\phi}}^p; \quad (6.1.37)$$

(2) for $0 < p < 1$, we have the equivalent reverses of (6.1.36) and (6.1.37).

Example 6.1.14 If $\lambda \in \mathbf{R}, r, s \neq 0$,

$$\max\{\tfrac{\lambda}{r}, \tfrac{\lambda}{s}\} < \beta < 1, |\eta| < \beta - \max\{\tfrac{\lambda}{r}, \tfrac{\lambda}{s}\},$$
$$k_\lambda(x,y) = \frac{(\min\{x,y\})^{\beta-\lambda}}{|x-y|^\beta},$$

then by Example 2.2.16, we obtain

$$0 < K_\lambda(\tfrac{\lambda}{r}+\eta) < \infty$$

and

$$k_\lambda(r) = B(1-\beta,\beta-\tfrac{\lambda}{r}) + B(1-\beta,\beta-\tfrac{\lambda}{s}).$$

By Theorem6.1.6, (1) for $p>1$, we have the following equivalent inequalities:

$$\int_{R_+^n}\int_{R_+^n}\frac{(\min\{\|x\|_\alpha,\|y\|_\alpha\})^{\beta-\lambda}}{|\|x\|_\alpha-\|y\|_\alpha|^\beta}f(x)g(y)dxdy$$
$$<\frac{\Gamma^n(\frac{1}{\alpha})}{\alpha^{n-1}\Gamma(\frac{n}{\alpha})}[B(1-\beta,\beta-\tfrac{\lambda}{r})+B(1-\beta,\beta-\tfrac{\lambda}{s})]$$
$$\times\|f\|_{p,\tilde\phi}\|g\|_{q,\tilde\psi}, \quad (6.1.38)$$

$$\int_{R_+^n}\|y\|_\alpha^{\frac{p\lambda}{s}-n}[\int_{R_+^n}\frac{(\min\{\|x\|_\alpha,\|y\|_\alpha\})^{\beta-\lambda}}{|\|x\|_\alpha-\|y\|_\alpha|^\beta}f(x)dx]^p dy$$
$$<\{\frac{\Gamma^n(\frac{1}{\alpha})}{\alpha^{n-1}\Gamma(\frac{n}{\alpha})}[B(1-\beta,\beta-\tfrac{\lambda}{r})$$
$$+B(1-\beta,\beta-\tfrac{\lambda}{s})]\}^p\|f\|_{p,\tilde\phi}^p; \quad (6.1.39)$$

(2) for $0<p<1$, we have the equivalent reverses of (6.1.38) and (6.1.39).

Example 6.1.15 If $\lambda\in\mathbf{R}$, $r,s\neq 0$,

$$\beta>\max\{\tfrac{\lambda}{r},\tfrac{\lambda}{s}\},\ |\eta|<\beta-\max\{\tfrac{\lambda}{r},\tfrac{\lambda}{s}\},$$
$$k_\lambda(x,y)=\frac{(\min\{x,y\})^{\beta-\lambda}}{(x+y)^\beta},$$

then we obtain $0<K_\lambda(\frac{\lambda}{r}+\eta)<\infty$ and

$$k_\lambda(r)=\int_0^1\frac{u^{\beta-\lambda}}{(1+u)^\beta}(u^{\frac{\lambda}{r}-1}+u^{\frac{\lambda}{s}-1})du$$
$$=\sum_{k=0}^\infty\binom{-\beta}{k}\left(\frac{1}{k+\beta-\frac{\lambda}{s}}+\frac{1}{k+\beta-\frac{\lambda}{r}}\right).$$

By Theorem6.1.6, (1) for $p>1$, we have the following equivalent inequalities:

$$\int_{R_+^n}\int_{R_+^n}\frac{(\min\{\|x\|_\alpha,\|y\|_\alpha\})^{\beta-\lambda}}{(\|x\|_\alpha+\|y\|_\alpha)^\beta}f(x)g(y)dxdy$$
$$<\frac{\Gamma^n(\frac{1}{\alpha})}{\alpha^{n-1}\Gamma(\frac{n}{\alpha})}k_\lambda(r)\|f\|_{p,\tilde\phi}\|g\|_{q,\tilde\psi}, \quad (6.1.40)$$

$$\int_{R_+^n}\|y\|_\alpha^{\frac{p\lambda}{s}-n}[\int_{R_+^n}\frac{(\min\{\|x\|_\alpha,\|y\|_\alpha\})^{\beta-\lambda}}{(\|x\|_\alpha+\|y\|_\alpha)^\beta}f(x)dx]^p dy$$
$$<[\frac{\Gamma^n(\frac{1}{\alpha})}{\alpha^{n-1}\Gamma(\frac{n}{\alpha})}k_\lambda(r)]^p\|f\|_{p,\tilde\phi}^p; \quad (6.1.41)$$

(2) for $0<p<1$, we have the equivalent reverses of (6.1.40) and (6.1.41).

6.1.4. SOME MULTIVARIABLE HARDY-TYPE INTEGRAL INEQUALITIES

By Theorem 6.1.6, we have

Corollary 6.1.16 Suppose that (p,q) and (r,s) are two pairs of conjugate exponents with $p>1$, $r,s\neq 0$, $\lambda\in\mathbf{R}$, $\tilde k_\lambda(x,y)$ is a homogeneous function of $-\lambda$ -degree,

$$k_\lambda(x,y)=\begin{cases}\tilde k_\lambda(x,y), & 0<y\le x,\\ 0, & y>x,\end{cases}$$

and

$$\tilde H_\lambda(x):=\int_0^1\tilde k_\lambda(1,u)u^{x-1}du$$

is a positive number in a neighborhood I_λ of $\frac{\lambda}{r}$,

$$0<\tilde k_\lambda(r):=\int_0^1\tilde k_\lambda(1,u)u^{\frac{\lambda}{r}-1}du<\infty.$$

Setting $\tilde\phi(x)=\|x\|_\alpha^{p(n-\frac{\lambda}{r})-n}$, $\tilde\psi(x)=\|x\|_\alpha^{q(n-\frac{\lambda}{s})-n}$, $x\in R_+^n$, if $f(\ge 0)\in L_{\tilde\phi}^p(R_+^n)$, $g(\ge 0)\in L_{\tilde\psi}^q(R_+^n)$,

$$\|f\|_{p,\tilde\phi}=\{\int_{R_+^n}\|x\|_\alpha^{p(n-\frac{\lambda}{r})-n}f^p(x)dx\}^{\frac{1}{p}}>0,$$
$$\|g\|_{q,\tilde\psi}=\{\int_{R_+^n}\|x\|_\alpha^{q(n-\frac{\lambda}{s})-n}g^q(x)dx\}^{\frac{1}{q}}>0,$$

then (1) for $p>1$, we have the following equivalent inequalities:

$$\int_{R_+^n}[\int_{\{y\in R_+^n;\|y\|_\alpha\le\|x\|_\alpha\}}\tilde k_\lambda(\|x\|_\alpha,\|y\|_\alpha)$$
$$\times g(y)dy]f(x)dx$$
$$=\int_{R_+^n}[\int_{\{x\in R_+^n;\|x\|_\alpha\ge\|y\|_\alpha\}}\tilde k_\lambda(\|x\|_\alpha,\|y\|_\alpha)$$
$$\times f(x)dx]g(y)dy$$
$$<\frac{\Gamma^n(\frac{1}{\alpha})\tilde k_\lambda(r)}{\alpha^{n-1}\Gamma(\frac{n}{\alpha})}\|f\|_{p,\tilde\phi}\|g\|_{q,\tilde\psi}, \quad (6.1.42)$$

$$\int_{R_+^n}\|y\|_\alpha^{\frac{p\lambda}{s}-n}[\int_{\{x\in R_+^n;\|x\|_\alpha\ge\|y\|_\alpha\}}k_\lambda(\|x\|_\alpha,\|y\|_\alpha)$$
$$\times f(x)dx]^p dy<[\frac{\Gamma^n(\frac{1}{\alpha})\tilde k_\lambda(r)}{\alpha^{n-1}\Gamma(\frac{n}{\alpha})}]^p\|f\|_{p,\tilde\phi}^p, \quad (6.1.43)$$

$$\int_{R_+^n}\|x\|_\alpha^{\frac{q\lambda}{r}-n}[\int_{\{y\in R_+^n;\|y\|_\alpha\le\|x\|_\alpha\}}\tilde k_\lambda(\|x\|_\alpha,\|y\|_\alpha)$$
$$\times g(y)dy]^q dx<[\frac{\Gamma^n(\frac{1}{\alpha})\tilde k_\lambda(r)}{\alpha^{n-1}\Gamma(\frac{n}{\alpha})}]^q\|g\|_{q,\tilde\psi}^q, \quad (6.1.44)$$

where the constant factors are the best possible; (2) for $0<p<1$, we have the equivalent reverses of (6.1.42), (6.1.43) and (6.1.44) with the best constant factors (but the reverse of (6.1.44) keeps the same form).

In the following examples, the conditions $\tilde\phi(x)=\|x\|_\alpha^{p(n-\frac{\lambda}{r})-n}$, $\tilde\psi(x)=\|x\|_\alpha^{q(n-\frac{\lambda}{s})-n}$, $x\in R_+^n$, $f(\ge 0)\in L_{\tilde\phi}^p(R_+^n)$, $g(\ge 0)\in L_{\tilde\psi}^q(R_+^n)$,

$$\| f \|_{p,\tilde{\phi}} = \{ \int_{R_+^n} \| x \|_\alpha^{p(n-\frac{\lambda}{r})-n} f^p(x)dx \}^{\frac{1}{p}} > 0,$$

$$\| g \|_{q,\tilde{\psi}} = \{ \int_{R_+^n} \| x \|_\alpha^{q(n-\frac{\lambda}{s})-n} g^q(x)dx \}^{\frac{1}{q}} > 0,$$

and the conclusions that the constant factors are the best possible are omitted.

Example 6.1.17 If $\beta > -1, r, s > 1, \lambda > 0$,

$$\tilde{k}_\lambda(x, y) = \frac{|\ln(x/y)|^\beta}{(\max\{x, y\})^\lambda},$$

by Example 2.2.25, we obtain

$$\tilde{k}_\lambda(r) = \Gamma(\beta+1)(\tfrac{r}{\lambda})^{\beta+1}.$$

Then by Corollary 6.1.16, (1) for $p > 1$, we have the following equivalent inequalities:

$$\int_{R_+^n} [\int_{\{y\in R_+^n; \|y\|_\alpha \le \|x\|_\alpha\}} \frac{|\ln(\|x\|_\alpha / \|y\|_\alpha)|^\beta}{(\max\{\|x\|_\alpha, \|y\|_\alpha\})^\lambda} \times g(y)dy] f(x)dx$$

$$= \int_{R_+^n} [\int_{\{x\in R_+^n; \|x\|_\alpha \ge \|y\|_\alpha\}} \frac{|\ln(\|x\|_\alpha / \|y\|_\alpha)|^\beta}{(\max\{\|x\|_\alpha, \|y\|_\alpha\})^\lambda} \times f(x)dx] g(y)dy$$

$$< \frac{\Gamma^n(\frac{1}{\alpha})\Gamma(\beta+1)}{\alpha^{n-1}\Gamma(\frac{n}{\alpha})} (\tfrac{r}{\lambda})^{\beta+1} \| f \|_{p,\tilde{\phi}} \| g \|_{q,\tilde{\psi}}, \quad (6.1.45)$$

$$\int_{R_+^n} \| y \|_\alpha^{\frac{p\lambda}{s}-n} [\int_{\{x\in R_+^n; \|x\|_\alpha \ge \|y\|_\alpha\}} \frac{|\ln(\|x\|_\alpha / \|y\|_\alpha)|^\beta}{(\max\{\|x\|_\alpha, \|y\|_\alpha\})^\lambda} \times f(x)dx]^p dy$$

$$< [\frac{\Gamma^n(\frac{1}{\alpha})\Gamma(\beta+1)}{\alpha^{n-1}\Gamma(\frac{n}{\alpha})} (\tfrac{r}{\lambda})^{\beta+1}]^p \| f \|_{p,\tilde{\phi}}^p, \quad (6.1.46)$$

$$\int_{R_+^n} \| x \|_\alpha^{\frac{q\lambda}{r}-n} [\int_{\{y\in R_+^n; \|y\|_\alpha \le \|x\|_\alpha\}} \frac{|\ln(\|x\|_\alpha / \|y\|_\alpha)|^\beta}{(\max\{\|x\|_\alpha, \|y\|_\alpha\})^\lambda} \times g(y)dy]^q dx$$

$$< [\frac{\Gamma^n(\frac{1}{\alpha})\Gamma(\beta+1)}{\alpha^{n-1}\Gamma(\frac{n}{\alpha})} (\tfrac{r}{\lambda})^{\beta+1}]^q \| g \|_{q,\tilde{\psi}}^q; \quad (6.1.47)$$

(2) for $0 < p < 1$, we have the equivalent reverses of (6.1.45), (6.1.45) and (6.1.47) (but the reverse form of (6.1.47) keeps the same). In particular, for $\beta = 0$, (1) if $p > 1$, we have the following equivalent extended Hardy's integral inequalities:

$$\int_{R_+^n} [\int_{\{y\in R_+^n; \|y\|_\alpha \le \|x\|_\alpha\}} g(y)dy] \frac{1}{\|x\|_\alpha^\lambda} f(x)dx$$

$$= \int_{R_+^n} [\int_{\{x\in R_+^n; \|x\|_\alpha \ge \|y\|_\alpha\}} \frac{f(x)}{\|x\|_\alpha^\lambda} dx] g(y)dy$$

$$< \frac{r\Gamma^n(\frac{1}{\alpha})}{\alpha^{n-1}\lambda\Gamma(\frac{n}{\alpha})} \| f \|_{p,\tilde{\phi}} \| g \|_{q,\tilde{\psi}}, \quad (6.1.48)$$

$$\int_{R_+^n} \| y \|_\alpha^{\frac{p\lambda}{s}-n} [\int_{\{x\in R_+^n; \|x\|_\alpha \ge \|y\|_\alpha\}} \frac{f(x)}{\|x\|_\alpha^\lambda} dx]^p dy$$

$$< [\frac{r\Gamma^n(\frac{1}{\alpha})}{\alpha^{n-1}\lambda\Gamma(\frac{n}{\alpha})}]^p \| f \|_{p,\tilde{\phi}}^p, \quad (6.1.49)$$

$$\int_{R_+^n} \frac{1}{\|x\|_\alpha^{\frac{q\lambda}{s}+n}} [\int_{\{y\in R_+^n; \|y\|_\alpha \le \|x\|_\alpha\}} g(y)dy]^q dx$$

$$< [\frac{r\Gamma^n(\frac{1}{\alpha})}{\alpha^{n-1}\lambda\Gamma(\frac{n}{\alpha})}]^q \| g \|_{q,\tilde{\psi}}^q; \quad (6.1.50)$$

(2) for $0 < p < 1$, we have the equivalent reverses of (6.1.48), (6.1.49) and (6.1.50) (but the reverse of (6.1.50) keeps the same form).

Example 6.1.18 If $\lambda \in \mathbf{R}, r, s \ne 0, \frac{\lambda}{s} < \beta < 1$,

$$\tilde{k}_\lambda(x, y) = \frac{(\min\{x, y\})^{\beta-\lambda}}{|x-y|^\beta},$$

by Example 2.2.26, we obtain

$$\tilde{k}_\lambda(r) = B(1-\beta, \beta-\tfrac{\lambda}{s}).$$

Then by Corollary 6.1.16, (1) for $p > 1$, we have the following equivalent inequalities:

$$\int_{R_+^n} [\int_{\{y\in R_+^n; \|y\|_\alpha \le \|x\|_\alpha\}} \frac{(\min\{\|x\|_\alpha, \|y\|_\alpha\})^{\beta-\lambda}}{|\|x\|_\alpha - \|y\|_\alpha|^\beta} \times g(y)dy] f(x)dx$$

$$= \int_{R_+^n} [\int_{\{x\in R_+^n; \|x\|_\alpha \ge \|y\|_\alpha\}} \frac{(\min\{\|x\|_\alpha, \|y\|_\alpha\})^{\beta-\lambda}}{|\|x\|_\alpha - \|y\|_\alpha|^\beta} \times f(x)dx] g(y)dy$$

$$< \frac{\Gamma^n(\frac{1}{\alpha})}{\alpha^{n-1}\Gamma(\frac{n}{\alpha})} B(1-\beta, \beta-\tfrac{\lambda}{s}) \| f \|_{p,\tilde{\phi}} \| g \|_{q,\tilde{\psi}}, \quad (6.1.51)$$

$$\int_{R_+^n} \| y \|_\alpha^{\frac{p\lambda}{s}-n} [\int_{\{x\in R_+^n; \|x\|_\alpha \ge \|y\|_\alpha\}} \frac{(\min\{\|x\|_\alpha, \|y\|_\alpha\})^{\beta-\lambda}}{|\|x\|_\alpha - \|y\|_\alpha|^\beta} \times f(x)dx]^p dy$$

$$< [\frac{\Gamma^n(\frac{1}{\alpha})}{\alpha^{n-1}\Gamma(\frac{n}{\alpha})} B(1-\beta, \beta-\tfrac{\lambda}{s})]^p \| f \|_{p,\tilde{\phi}}^p, \quad (6.1.52)$$

$$\int_{R_+^n} \| x \|^{\frac{q\lambda}{r}-n} [\int_{\{y\in R_+^n; \|y\|_\alpha \le \|x\|_\alpha\}} \frac{(\min\{\|x\|_\alpha, \|y\|_\alpha\})^{\beta-\lambda}}{|\|x\|_\alpha - \|y\|_\alpha|^\beta} \times g(y)dy]^q dx$$

$$< [\frac{\Gamma^n(\frac{1}{\alpha})}{\alpha^{n-1}\Gamma(\frac{n}{\alpha})} B(1-\beta, \beta-\tfrac{\lambda}{s})]^q \| g \|_{q,\tilde{\psi}}^q; \quad (6.1.53)$$

(2) for $0 < p < 1$, we have the equivalent reverses of (6.1.51), (6.1.52) and (6.1.53) (but the reverse of (6.1.53) keeps the same form).

Example 6.1.19 If $\lambda \in \mathbf{R}, r, s \ne 0, \beta > \frac{\lambda}{s}$,

$$\tilde{k}_\lambda(x, y) = \frac{(\min\{x, y\})^{\beta-\lambda}}{(x+y)^\beta},$$

by Example 6.1.15, we obtain

$$\tilde{k}_\lambda(r) = \sum_{k=0}^{\infty} \binom{-\beta}{k} \frac{1}{k+\beta-\frac{\lambda}{s}}.$$

Then by Corollary 6.1.16, (1) for $p > 1$, we have the following equivalent inequalities:

$$\int_{R_+^n}\Big[\int_{\{y\in R_+^n;\|y\|_\alpha\le\|x\|_\alpha\}}\frac{(\min\{\|x\|_\alpha,\|y\|_\alpha\})^{\beta-\lambda}}{(\|x\|_\alpha+\|y\|_\alpha)^\beta}\times g(y)dy\Big]f(x)dx$$

$$=\int_{R_+^n}\Big[\int_{\{x\in R_+^n;\|x\|_\alpha\ge\|y\|_\alpha\}}\frac{(\min\{\|x\|_\alpha,\|y\|_\alpha\})^{\beta-\lambda}}{(\|x\|_\alpha+\|y\|_\alpha)^\beta}\times f(x)dx\Big]g(y)dy$$

$$<\frac{\Gamma^n(\frac{1}{\alpha})}{\alpha^{n-1}\Gamma(\frac{n}{\alpha})}\tilde{k}_\lambda(r)\,\|f\|_{p,\tilde{\phi}}\,\|g\|_{q,\tilde{\psi}},\qquad(6.1.54)$$

$$\int_{R_+^n}\|y\|_\alpha^{\frac{p\lambda}{s}-n}\Big[\int_{\{x\in R_+^n;\|x\|_\alpha\ge\|y\|_\alpha\}}\frac{(\min\{\|x\|_\alpha,\|y\|_\alpha\})^{\beta-\lambda}}{(\|x\|_\alpha+\|y\|_\alpha)^\beta}\times f(x)dx\Big]^p dy$$

$$<\Big[\frac{\Gamma^n(\frac{1}{\alpha})}{\alpha^{n-1}\Gamma(\frac{n}{\alpha})}\tilde{k}_\lambda(r)\Big]^p\|f\|_{p,\tilde{\phi}}^p,\qquad(6.1.55)$$

$$\int_{R_+^n}\|x\|_\alpha^{\frac{q\lambda}{r}-n}\Big[\int_{\{y\in R_+^n;\|y\|_\alpha\le\|x\|_\alpha\}}\frac{(\min\{\|x\|_\alpha,\|y\|_\alpha\})^{\beta-\lambda}}{(\|x\|_\alpha+\|y\|_\alpha)^\beta}\times g(y)dy\Big]^q dx$$

$$<\Big[\frac{\Gamma^n(\frac{1}{\alpha})}{\alpha^{n-1}\Gamma(\frac{n}{\alpha})}\tilde{k}_\lambda(r)\Big]^q\|g\|_{q,\tilde{\psi}}^q;\qquad(6.1.56)$$

(2) for $0<p<1$, we have the equivalent reverses of (6.1.55), (6.1.55) and (6.1.56) (but the reverse form of (6.1.53) keeps the same).

6.2. MULTIVARIABLE HILBERT-TYPE INTEGRAL INEQUALITIES WITH THE NON-HOMOGENEOUS KERNELS

6.2.1. SOME LEMMAS

Lemma 6.2.1 If $\alpha>0, n\in\mathbf{N}$, $R_+^n=\{(x_1,\cdots,x_n)\,|\,x_i>0\ (i=1,\cdots,n)\}$, $x\in R_+^n$, $\|x\|_\alpha=(\sum_{i=1}^n x_i^\alpha)^{\frac{1}{\alpha}}$, $\lambda\in\mathbf{R}$, $k_\lambda(x,y)(\ge 0)$ is a homogeneous function of $-\lambda$-degree in R_+^2, and

$$k_\lambda:=\int_0^\infty k_\lambda(1,u)u^{\frac{\lambda}{2}-1}du$$

is a finite number, define the following weight functions:

$$\omega_\lambda(y):=\int_{R_+^n}k_\lambda(1,\|x\|_\alpha\|y\|_\alpha)\frac{\|y\|_\alpha^{\frac{\lambda}{2}}}{\|x\|_\alpha^{n-\frac{\lambda}{2}}}dx,$$

$$\tilde{\omega}_\lambda(x):=\int_{R_+^n}k_\lambda(1,\|x\|_\alpha\|y\|_\alpha)\frac{\|x\|_\alpha^{\frac{\lambda}{2}}}{\|y\|_\alpha^{n-\frac{\lambda}{2}}}dy,$$

$$x,y\in R_+^n.\qquad(6.2.1)$$

Then we have

$$\omega_\lambda(r,y)=\tilde{\omega}_\lambda(s,x)=\frac{\Gamma^n(\frac{1}{\alpha})}{\alpha^{n-1}\Gamma(\frac{n}{\alpha})}k_\lambda,$$

$$x,y\in R_+^n.\qquad(6.2.2)$$

Proof Replacing $n-1$ by n in (5.2.1), setting $a_i=R,\alpha_i=\alpha,P_i=1\ (i=1,\cdots,n)$, then

$$\int\cdots\int_{D_R}\psi\Big(\sum_{i=1}^n(\tfrac{x_i}{R})^\alpha\Big)dx_1\cdots dx_n$$

$$=\frac{R^n\Gamma^n(\frac{1}{\alpha})}{\alpha^n\Gamma(\frac{n}{\alpha})}\int_0^1\psi(u)u^{\frac{n}{\alpha}-1}du,\qquad(6.2.3)$$

where,

$$D_R:=\{(x_1,\cdots,x_n)\,|\,x_i>0,\ i=1,\cdots,n,\textstyle\sum_{i=1}^n x_i^\alpha\le R^\alpha\}.$$

By (5.2.3), we have

$$\omega_\lambda(y)=\|y\|_\alpha^{\frac{\lambda}{2}}\int_{R_+^n}k_\lambda(1,\|x\|_\alpha\|y\|_\alpha)\frac{1}{\|x\|_\alpha^{n-\frac{\lambda}{2}}}dx$$

$$=\|y\|_\alpha^{\frac{\lambda}{2}}\lim_{R\to+\infty}\int\cdots\int_{D_R}k_\lambda\Big(1,R\Big[\sum_{i=1}^n(\tfrac{x_i}{R})^\alpha\Big]^{\frac{1}{\alpha}}\|y\|_\alpha\Big)\times\Big\{R\Big[\sum_{i=1}^n(\tfrac{x_i}{R})^\alpha\Big]^{\frac{1}{\alpha}}\Big\}^{\frac{\lambda}{2}-n}dx_1\cdots dx_n$$

$$=\|y\|_\alpha^{\frac{\lambda}{2}}\lim_{R\to+\infty}\frac{R^n\Gamma^n(\frac{1}{\alpha})}{\alpha^n\Gamma(\frac{n}{\alpha})}\times\int_0^1 k_\lambda(1,Ru^{\frac{1}{\alpha}}\|y\|_\alpha)(Ru^{\frac{1}{\alpha}})^{\frac{\lambda}{2}-n}u^{\frac{n}{\alpha}-1}du$$

$$=\|y\|_\alpha^{\frac{\lambda}{2}}\lim_{R\to+\infty}\frac{R^n\Gamma^n(\frac{1}{\alpha})}{\alpha^n\Gamma(\frac{n}{\alpha})}\times\int_0^{R\|y\|_\alpha}k_\lambda(1,v)\Big(\frac{v}{\|y\|_\alpha}\Big)^{\frac{\lambda}{2}-n}\Big(\frac{v}{R\|y\|_\alpha}\Big)^{n-\alpha}d\Big(\frac{v}{R\|y\|_\alpha}\Big)^\alpha$$

$$=\frac{\Gamma^n(\frac{1}{\alpha})}{\alpha^{n-1}\Gamma(\frac{n}{\alpha})}\int_0^\infty k_\lambda(1,v)v^{\frac{\lambda}{2}-1}dv=\frac{\Gamma^n(\frac{1}{\alpha})k_\lambda}{\alpha^{n-1}\Gamma(\frac{n}{\alpha})}.$$

Due to

$$k_\lambda=\tilde{k}_\lambda=\int_0^\infty k_\lambda(u,1)u^{\frac{\lambda}{2}-1}du,$$

we have $\tilde{\omega}_\lambda(x)=\omega_\lambda(y)$ and (6.2.2) is valid.

Lemma 6.2.2 As the assumption of Lemma 6.2.1, if $p>0(p\ne1)$,

$$K_\lambda(x)=\int_0^\infty k_\lambda(u,1)u^{x-1}du$$

is finite in a neighborhood I_λ of $\frac{\lambda}{2},\varepsilon>0$, such that $\frac{\lambda}{2}-\frac{\varepsilon}{p}\in I_\lambda$, then

$$\tilde{I}_\varepsilon:=\int_{\{y\in R_+^n;\|y\|_\alpha>1\}}\int_{\{x\in R_+^n;\|x\|_\alpha<1\}}k_\lambda(1,\|x\|_\alpha\|y\|_\alpha)\times\|x\|_\alpha^{\frac{\lambda}{2}+\frac{\varepsilon}{p}-n}\|y\|_\alpha^{\frac{\lambda}{2}-\frac{\varepsilon}{q}-n}dxdy$$

$$=\Big[\frac{\Gamma^n(\frac{1}{\alpha})k_\lambda}{\alpha^{n-1}\Gamma(\frac{n}{\alpha})}+o(1)\Big]\times\int_{\{y\in R_+^n;\|y\|_\alpha>1\}}\|y\|_\alpha^{-n-\varepsilon}dy\ (\varepsilon\to0^+).\qquad(6.2.4)$$

Proof Setting

$$\tilde{k}_\lambda(1,xy) := \begin{cases} k_\lambda(1,xy), & 0<x<1, y>0, \\ 0 & , x\ge 1, y>0, \end{cases}$$

we have

$$\tilde{I}_\varepsilon = \int_{\{y\in R_+^n;\|y\|_\alpha\ge 1\}} \|y\|_\alpha^{-n-\varepsilon} [\|y\|_\alpha^{\frac{\lambda}{2}+\frac{\varepsilon}{p}} \times \int_{\{x\in R_+^n;\|x\|_\alpha<1\}} k_\lambda(1,\|x\|_\alpha\|y\|_\alpha) \times \|x\|_\alpha^{\frac{\lambda}{2}+\frac{\varepsilon}{p}-n} dx]dy$$

$$= \int_{\{y\in R_+^n;\|y\|_\alpha\ge 1\}} \|y\|_\alpha^{-n-\varepsilon} [\|y\|_\alpha^{\frac{\lambda}{2}+\frac{\varepsilon}{p}} \times \int_{R_+^n} \tilde{k}_\lambda(1,\|x\|_\alpha\|y\|_\alpha)\|x\|_\alpha^{\frac{\lambda}{2}+\frac{\varepsilon}{p}-n} dx]dy$$

$$= \int_{\{y\in R_+^n;\|y\|_\alpha\ge 1\}} \|y\|_\alpha^{-n-\varepsilon} [\frac{\Gamma^n(\frac{1}{\alpha})}{\alpha^{n-1}\Gamma(\frac{n}{\alpha})} \times \int_0^\infty \tilde{k}_\lambda(1,v)v^{\frac{\lambda}{2}+\frac{\varepsilon}{p}-1}dv]dy$$

$$= \int_{\{y\in R_+^n;\|y\|_\alpha\ge 1\}} \|y\|_\alpha^{-n-\varepsilon} [\frac{\Gamma^n(\frac{1}{\alpha})}{\alpha^{n-1}\Gamma(\frac{n}{\alpha})} \times \int_0^{\|y\|_\alpha} k_\lambda(1,v)v^{\frac{\lambda}{2}+\frac{\varepsilon}{p}-1}dv]dy$$

$$= \int_{\{y\in R_+^n;\|y\|_\alpha\ge 1\}} \|y\|_\alpha^{-n-\varepsilon} [\frac{\Gamma^n(\frac{1}{\alpha})}{\alpha^{n-1}\Gamma(\frac{n}{\alpha})} \times \int_{\|y\|_\alpha^{-1}}^\infty k_\lambda(u,1)u^{\frac{\lambda}{2}-\frac{\varepsilon}{p}-1}du]dy = J_\varepsilon(2).$$

Then by (6.1.5) (for $r=s=2$), we have (6.2.4).□

6.2.2. BASIC RESULTS

Theorem 6.2.3 Suppose that (p,q) is one pair of conjugate exponents with $p>1, \frac{1}{p}+\frac{1}{q}=1$, and $\lambda\in\mathbf{R}, k_\lambda(x,y)\ge 0$ is a homogeneous function of $-\lambda$-degree in R_+^2, $k_\lambda = \int_0^\infty k_\lambda(u,1)u^{\frac{\lambda}{2}-1}du$ is finite. If $\alpha>0, n\in\mathbf{N}, \|x\|_\alpha = (\sum_{i=1}^n x_i^\alpha)^{\frac{1}{\alpha}}$, $f(x)$ and $g(x)$ are non-negative measurable functions in R_+^n, then we have the following equivalent inequalities:

$$\tilde{J} := \int_{R_+^n} \|y\|_\alpha^{\frac{p\lambda}{2}-n} [\int_{R_+^n} k_\lambda(1,\|x\|_\alpha\|y\|_\alpha) \times f(x)dx]^p dy \le [\frac{\Gamma^n(\frac{1}{\alpha})k_\lambda}{\alpha^{n-1}\Gamma(\frac{n}{\alpha})}]^p \int_{R_+^n} \|x\|_\alpha^{p(n-\frac{\lambda}{2})-n} f^p(x)dx; \quad (6.2.5)$$

$$\tilde{I} := \int_{R_+^n}\int_{R_+^n} k_\lambda(1,\|x\|_\alpha\|y\|_\alpha)f(x)g(y)dxdy \le \frac{\Gamma^n(\frac{1}{\alpha})k_\lambda}{\alpha^{n-1}\Gamma(\frac{n}{\alpha})} (\int_{R_+^n}\|x\|_\alpha^{p(n-\frac{\lambda}{2})-n} f^p(x)dx)^{\frac{1}{p}} \times (\int_{R_+^n}\|x\|_\alpha^{q(n-\frac{\lambda}{2})-n} g^q(x)dx)^{\frac{1}{q}}. \quad (6.2.6)$$

Proof By Hölder's inequality with weight and Lemma 6.2.1, we have

$$\{\int_{R_+^n} k_\lambda(1,\|x\|_\alpha\|y\|_\alpha)[\frac{\|x\|_\alpha^{(n-\frac{\lambda}{2})/q}}{\|y\|_\alpha^{(n-\frac{\lambda}{2})/p}} f(x)] \times [\frac{\|y\|_\alpha^{(n-\frac{\lambda}{2})/p}}{\|x\|_\alpha^{(n-\frac{\lambda}{2})/q}}]dx\}^p$$

$$\le \int_{R_+^n} k_\lambda(1,\|x\|_\alpha\|y\|_\alpha)\frac{\|x\|_\alpha^{(n-\frac{\lambda}{2})(p-1)}}{\|y\|_\alpha^{n-\frac{\lambda}{2}}} f^p(x)dx \times\{\int_{R_+^n} k_\lambda(1,\|x\|_\alpha\|y\|_\alpha)\frac{\|y\|_\alpha^{(n-\frac{\lambda}{2})(q-1)}}{\|x\|_\alpha^{n-\frac{\lambda}{2}}}dx\}^{p-1}$$

$$= [\omega_\lambda(y)]^{p-1}\|y\|_\alpha^{[q(n-\frac{\lambda}{2})-n](p-1)} \times \int_{R_+^n} k_\lambda(1,\|x\|_\alpha\|y\|_\alpha)\frac{\|x\|_\alpha^{(n-\frac{\lambda}{2})(p-1)}}{\|y\|_\alpha^{n-\frac{\lambda}{2}}} f^p(x)dx$$

$$= [\frac{\Gamma^n(\frac{1}{\alpha})k_\lambda}{\alpha^{n-1}\Gamma(\frac{n}{\alpha})}]^{p-1}\|y\|_\alpha^{(1-p)\frac{\lambda}{2}}\int_{R_+^n} k_\lambda(1,\|x\|_\alpha\|y\|_\alpha) \times\|x\|_\alpha^{(n-\frac{\lambda}{2})(p-1)} f^p(x)dx. \quad (6.2.7)$$

Then by (6.2.2), we find

$$\tilde{J} \le [\frac{\Gamma^n(\frac{1}{\alpha})k_\lambda}{\alpha^{n-1}\Gamma(\frac{n}{\alpha})}]^{p-1} \times \int_{R_+^n}\|y\|_\alpha^{\frac{\lambda}{2}-n}\int_{R_+^n} k_\lambda(1,\|x\|_\alpha\|y\|_\alpha) \times\|x\|_\alpha^{(n-\frac{\lambda}{2})(p-1)} f^p(x)dxdy$$

$$= [\frac{\Gamma^n(\frac{1}{\alpha})k_\lambda}{\alpha^{n-1}\Gamma(\frac{n}{\alpha})}]^{p-1} \times\int_{R_+^n}[\int_{R_+^n} k_\lambda(1,\|x\|_\alpha\|y\|_\alpha)\|y\|_\alpha^{\frac{\lambda}{2}-n}dy] \times\|x\|_\alpha^{(n-\frac{\lambda}{2})(p-1)} f^p(x)dx$$

$$= [\frac{\Gamma^n(\frac{1}{\alpha})k_\lambda}{\alpha^{n-1}\Gamma(\frac{n}{\alpha})}]^{p}\int_{R_+^n}\|x\|_\alpha^{p(n-\frac{\lambda}{2})-n} f^p(x)dx,$$

and (6.2.5) is valid.

By Hölder's inequality, we have

$$\tilde{I} = \int_{R_+^n}[\|y\|_\alpha^{\frac{\lambda}{2}-\frac{n}{p}}\int_{R_+^n} k_\lambda(1,\|x\|_\alpha\|y\|_\alpha)f(x)dx] \times[\|y\|_\alpha^{\frac{n}{p}-\frac{\lambda}{2}} g(y)]dy \le \tilde{J}^{\frac{1}{p}}\{\int_{R_+^n}\|y\|_\alpha^{q(n-\frac{\lambda}{2})-n}g^q(y)dy\}^{\frac{1}{q}}. \quad (6.2.8)$$

Then by (6.2.5), we have (6.2.6).

Setting

$$g(y) := \|y\|_\alpha^{\frac{p\lambda}{2}-n}[\int_{R_+^n} k_\lambda(1,\|x\|_\alpha\|y\|_\alpha)$$

$$\times f(x)dx]^{p-1}dy, y \in R_+^n,$$

by (6.2.8), we have

$$\int_{R_+^n} \| y \|_\alpha^{q(n-\frac{\lambda}{2})-n} g^q(y)dy = \tilde{I} = \tilde{J}$$

$$\leq \frac{\Gamma^n(\frac{1}{\alpha})k_\lambda}{\alpha^{n-1}\Gamma(\frac{n}{\alpha})}\left(\int_{R_+^n} \| x \|_\alpha^{p(n-\frac{\lambda}{2})-n} f^p(x)dx\right)^{\frac{1}{p}}$$

$$\times\left(\int_{R_+^n} \| y \|_\alpha^{q(n-\frac{\lambda}{2})-n} g^q(y)dy\right)^{\frac{1}{q}}. \quad (6.2.9)$$

If $\tilde{J} = \infty$, then by (6.2.7), we conclude that (6.2.5) takes the form of equality; if $\tilde{J} = 0$, then (6.2.5) is naturally valid. Assuming that $0 < \tilde{J} < \infty$, then in (6.2.9), we have

$$\tilde{J}^{\frac{1}{p}} \leq \frac{\Gamma^n(\frac{1}{\alpha})k_\lambda}{\alpha^{n-1}\Gamma(\frac{n}{\alpha})}\left(\int_{R_+^n} \| x \|_\alpha^{p(n-\frac{\lambda}{2})-n} f^p(x)dx\right)^{\frac{1}{p}}.$$

Hence (6.2.5) is valid, which is equivalent to (6.2.6).

As the assumption of Theorem 6.2.3, setting

$$\phi(x) = \| x \|_\alpha^{p(n-\frac{\lambda}{2})-n}, \psi(x) = \| x \|_\alpha^{q(n-\frac{\lambda}{2})-n}, x \in R_+^n$$

and then $\psi^{1-p}(x) = \| x \|_\alpha^{\frac{p\lambda}{2}-n}$, define the following real function spaces as:

$$L_\phi^p(R_+^n) = \{f; \| f \|_{p,\phi} := \{\int_{R_+^n} \phi(x) | f(x) |^p \, dx\}^{\frac{1}{p}} < \infty\},$$

$$L_\psi^q(R_+^n) = \{g; \| g \|_{q,\psi} := \{\int_{R_+^n} \psi(x) | g(x) |^q \, dx\}^{\frac{1}{q}} < \infty\}.$$

For $f(\geq 0) \in L_\phi^p(R_+^n)$, define a multivariable Hilbert-type integral operator

$$\tilde{T} : L_\phi^p(R_+^n) \to L_{\psi^{1-p}}^p(R_+^n)$$

as

$$(\tilde{T}f)(y) = \int_{R_+^n} k_\lambda(1, \| x \|_\alpha \| y \|_\alpha) f(x)dx,$$

$$y \in R_+^n. \quad (6.2.10)$$

Then by (6.2.5), it follows $\tilde{T}f \in L_{\psi^{1-p}}^p(R_+^n)$. If $g(\geq 0) \in L_\psi^q(R_+^n)$, setting the formal inner product of $\tilde{T}f$ and g as

$$(\tilde{T}f, g) := \int_{R_+^n}\int_{R_+^n} k_\lambda(1, \| x \|_\alpha \| y \|_\alpha)$$

$$\times f(x)g(y)dxdy, \quad (6.2.11)$$

then we may rewrite (6.2.5) and (6.2.6) to the following equivalent forms:

$$\| \tilde{T}f \|_{p,\psi^{1-p}} \leq \frac{\Gamma^n(\frac{1}{\alpha})k_\lambda}{\alpha^{n-1}\Gamma(\frac{n}{\alpha})} \| f \|_{p,\phi}; \quad (6.2.12)$$

$$(\tilde{T}f, g) \leq \frac{\Gamma^n(\frac{1}{\alpha})k_\lambda}{\alpha^{n-1}\Gamma(\frac{n}{\alpha})} \| f \|_{p,\phi} \| g \|_{q,\psi}. \quad (6.2.13)$$

By (6.2.12), it follows that $\tilde{T}$ is a bounded operator and

$$\| \tilde{T} \| \leq \frac{\Gamma^n(\frac{1}{\alpha})}{\alpha^{n-1}\Gamma(\frac{n}{\alpha})} k_\lambda.$$

In particular, for $n = 1$, setting

$$\phi_1(x) = x^{p(1-\frac{\lambda}{2})-1}, \psi_1(x) = x^{q(1-\frac{\lambda}{2})-1} \quad (x \in R_+)$$

and then $\psi_1^{1-p}(x) = x^{\frac{p\lambda}{2}-1}$, define the following real function spaces as:

$$L_{\phi_1}^p(R_+) = \{f; \| f \|_{p,\phi_1} := \{\int_{R_+} \phi_1(x) | f(x) |^p \, dx\}^{\frac{1}{p}} < \infty\},$$

$$L_{\psi_1}^q(R_+) = \{g; \| g \|_{q,\psi_1} := \{\int_{R_+} \psi_1(x) | g(x) |^q \, dx\}^{\frac{1}{q}} < \infty\}.$$

For $f(\geq 0) \in L_{\phi_1}^p(R_+)$, define a Hilbert-type integral operator $\tilde{T}_1 : L_{\phi_1}^p(R_+) \to L_{\psi_1^{1-p}}^p(R_+)$ as

$$(\tilde{T}_1 f)(y) = \int_{R_+} k_\lambda(1, xy) f(x)dx, y \in R_+. \quad (6.2.14)$$

If $g(\geq 0) \in L_{\psi_1}^q(R_+)$, setting the formal inner product of $\tilde{T}_1 f$ and g as

$$(\tilde{T}_1 f, g) := \int_{R_+}\int_{R_+} k_\lambda(1, xy) f(x)g(y)dxdy, \quad (6.2.15)$$

Then we have the following equivalent forms:

$$\| \tilde{T}_1 f \|_{p,\psi_1^{1-p}} \leq k_\lambda \| f \|_{p,\phi_1}, \quad (6.2.16)$$

$$(\tilde{T}_1 f, g) \leq k_\lambda \| f \|_{p,\phi_1} \| g \|_{q,\psi_1}. \quad (6.2.17)$$

By (6.2.16), it follows that $\tilde{T}_1$ is a bounded operator and $\| \tilde{T}_1 \| \leq k_\lambda$.

Theorem 6.2.4 As the assumption of Theorem 6.2.3, if the operator $\tilde{T}$ is defined by (6.2.11), then we have

$$\| \tilde{T} \| = \frac{\Gamma^n(\frac{1}{\alpha})}{\alpha^{n-1}\Gamma(\frac{n}{\alpha})} k_\lambda. \quad (6.2.18)$$

In particular, for $n = 1$, it follows $\| \tilde{T}_1 \| = k_\lambda$.

Proof For $\varepsilon > 0$ small enough, setting

$$\tilde{f}(x) = \begin{cases} \| x \|_\alpha^{\frac{\lambda}{2}+\frac{\varepsilon}{p}-1}, x \in \{x; \| x \|_\alpha < 1, x \in R_+^n\}, \\ 0, \quad x \in \{x; \| x \|_\alpha \geq 1, x \in R_+^n\}, \end{cases}$$

$$\tilde{g}(x) = \begin{cases} 0, \quad x \in \{x; \| x \|_\alpha < 1, x \in R_+^n\}, \\ \| x \|_\alpha^{\frac{\lambda}{2}-\frac{\varepsilon}{q}-1}, x \in \{x; \| x \|_\alpha \geq 1, x \in R_+^n\}. \end{cases}$$

Then we find

$$\| \tilde{f} \|_{p,\phi} \| \tilde{g} \|_{q,\psi} = \int_{\{x \in R_+^n; \| x \|_\alpha > 1\}} \| x \|_\alpha^{-n-\varepsilon} \, dx.$$

If there exists a constant

$$0 \leq k \leq \frac{\Gamma^n(\frac{1}{\alpha})}{\alpha^{n-1}\Gamma(\frac{n}{\alpha})} k_\lambda,$$

such that (6.2.13) is still valid as we replace $\frac{\Gamma^n(1/\alpha)k_\lambda}{\alpha^{n-1}\Gamma(n/\alpha)}$ by k, then in particular, by (6.2.4), we have

$$[\frac{\Gamma^n(\frac{1}{\alpha})k_\lambda}{\alpha^{n-1}\Gamma(\frac{n}{\alpha})}+o(1)]\int_{\{y\in R_+^n;\|y\|_\alpha>1\}}\|y\|_\alpha^{-n-\varepsilon}dy$$
$$=I_\varepsilon=(\tilde{T}\tilde{f},\tilde{g})\le k\|\tilde{f}\|_{p,\phi}\|\tilde{g}\|_{q,\psi}$$
$$=k\int_{\{y\in R_+^n;\|y\|_\alpha\ge 1\}}\|y\|_\alpha^{-n-\varepsilon}dy,$$

and $\frac{\Gamma^n(1/\alpha)k_\lambda}{\alpha^{n-1}\Gamma(n/\alpha)}+o(1)\le k$. For $\varepsilon\to 0^+$, it follows $\frac{\Gamma^n(1/\alpha)k_\lambda}{\alpha^{n-1}\Gamma(n/\alpha)}\le k$. Hence $k=\frac{\Gamma^n(1/\alpha)k_\lambda}{\alpha^{n-1}\Gamma(n/\alpha)}$ is the best value of (6.2.13). By (6.2.8), we can show that the constant factor $k=\frac{\Gamma^n(1/\alpha)k_\lambda}{\alpha^{n-1}\Gamma(n/\alpha)}$ in (6.2.12) is the best possible. And (6.2.18) is valid.

Theorem 6.2.5 As the assumption of Theorem 6.2.3, if $p>0(p\ne 1)$,

$$K_\lambda(x)=\int_0^\infty k_\lambda(u,1)u^{x-1}du$$

is a positive number in a neighborhood I_λ of $\frac{\lambda}{2}$,

$$k_\lambda=\int_0^\infty k_\lambda(u,1)u^{\frac{\lambda}{2}-1}du,\ \phi(x)=\|x\|_\alpha^{p(n-\frac{\lambda}{2})-n},$$

$\psi(x)=\|x\|_\alpha^{q(n-\frac{\lambda}{2})-n}$ $(x\in R_+^n)$, $f(\ge 0)\in L_\phi^p(R_+^n)$, $g(\ge 0)\in L_\psi^q(R_+^n)$, and

$$\|f\|_{p,\phi}=\{\int_{R_+^n}\|x\|_\alpha^{p(n-\frac{\lambda}{2})-n}f^p(x)dx\}^{\frac{1}{p}}>0,$$
$$\|g\|_{q,\psi}=\{\int_{R_+^n}\|x\|_\alpha^{q(n-\frac{\lambda}{2})-n}g^q(x)dx\}^{\frac{1}{q}}>0,$$

then (1) for $p>1$, we have the following equivalent inequalities:

$$\int_{R_+^n}\int_{R_+^n}k_\lambda(1,\|x\|_\alpha\|y\|_\alpha)f(x)g(y)dxdy$$
$$<\frac{\Gamma^n(\frac{1}{\alpha})}{\alpha^{n-1}\Gamma(\frac{n}{\alpha})}k_\lambda\|f\|_{p,\phi}\|g\|_{q,\psi},\qquad (6.2.19)$$
$$\int_{R_+^n}\|y\|_\alpha^{\frac{p\lambda}{2}-n}[\int_{R_+^n}k_\lambda(1,\|x\|_\alpha\|y\|_\alpha)$$
$$\times f(x)dx]^p dy$$
$$<[\frac{\Gamma^n(\frac{1}{\alpha})}{\alpha^{n-1}\Gamma(\frac{n}{\alpha})}k_\lambda]^p\|f\|_{p,\phi}^p,\qquad (6.2.20)$$

where the constant factors are the best possible; (2) for $0<p<1$, we have the equivalent reverses of (6.2.19) and (6.2.20) with the best constant factors.

In particular, for $n=1$, (1) for $p>1$, we have the following equivalent inequalities:

$$\int_{R_+}\int_{R_+}k_\lambda(1,xy)f(x)g(y)dxdy$$
$$<k_\lambda\|f\|_{p,\phi_1}\|g\|_{q,\psi_1},\qquad (6.2.21)$$
$$\int_{R_+}y^{\frac{p\lambda}{2}-1}[\int_{R_+}k_\lambda(1,xy)f(x)dx]^p dy$$
$$<k_\lambda^p\|f\|_{p,\phi_1}^p,\qquad (6.2.22)$$

where the constant factors are the best possible; (2) for $0<p<1$, we have the equivalent reverses of (6.2.21) and (6.2.22) with the best constant factors.

Proof (1) We conform that (6.2.7) takes the strict sign-inequality, otherwise, since $k_\lambda>0$, then $k_\lambda(1,xy)$ is not zero almost everywhere, there exist a $y\in R_+^n$, constants A and B, which are not all zero and

$$A\frac{\|x\|_\alpha^{(n-\frac{\lambda}{2})(p-1)}}{\|y\|_\alpha^{\frac{n-\lambda}{2}}}f^p(x)=B\frac{\|y\|_\alpha^{(n-\frac{\lambda}{2})(q-1)}}{\|x\|_\alpha^{\frac{n-\lambda}{2}}}\quad\text{a.e. in } R_+^n.$$

It follows

$$A\|x\|_\alpha^{p(n-\frac{\lambda}{2})}f^p(x)=B\|y\|_\alpha^{q(n-\frac{\lambda}{2})}\text{ a.e. in } R_+^n.$$

It is obvious that $A\ne 0$ (otherwise, $A=B=0$), and

$$\|x\|_\alpha^{p(n-\frac{\lambda}{2})-n}f^p(x)=B\|y\|_\alpha^{q(n-\frac{\lambda}{2})}/(A\|x\|_\alpha^n)$$

a.e. in R_+^n, which contradicts the fact that $0<\|f\|_{p,\phi}<\infty$. Then (6.2.5) still takes the strict sign-inequality and we have (6.2.20). By (6.2.8), we have (6.2.19), which is still equivalent to (6.2.20). And the constant factors are the best possible due to the proof of Theorem 6.2.4.

(2) By the same way of Theorem 6.2.3 and using the reverse Hölder's inequality, we have the equivalent reverses of (6.2.19) and (6.2.20). If there exists a constant $K\ge\frac{\Gamma^n(1/\alpha)k_\lambda}{\alpha^{n-1}\Gamma(n/\alpha)}$, such that the reverse of (6.2.19) is valid as we replace $\frac{\Gamma^n(1/\alpha)k_\lambda}{\alpha^{n-1}\Gamma(n/\alpha)}$ by K, then in particular, by (6.2.4), we find

$$[\frac{\Gamma^n(\frac{1}{\alpha})k_\lambda}{\alpha^{n-1}\Gamma(\frac{n}{\alpha})}+o(1)]\int_{\{y\in R_+^n;\|y\|_\alpha>1\}}\|y\|_\alpha^{-n-\varepsilon}dy$$
$$=I_\varepsilon>K\|\tilde{f}\|_{p,\phi}\|\tilde{g}\|_{q,\psi}$$
$$=K\int_{\{y\in R_+^n;\|y\|_\alpha>1\}}\|y\|_\alpha^{-n-\varepsilon}dy,$$

and $\frac{\Gamma^n(1/\alpha)k_\lambda}{\alpha^{n-1}\Gamma(n/\alpha)}+o(1)>K$. For $\varepsilon\to 0^+$, it follows $\frac{\Gamma^n(1/\alpha)k_\lambda}{\alpha^{n-1}\Gamma(n/\alpha)}\ge K$. Hence $K=\frac{\Gamma^n(1/\alpha)k_\lambda}{\alpha^{n-1}\Gamma(n/\alpha)}$ is the best value of the reverse of (6.2.19). By using the reverse of (6.2.8), we can show that the constant factor in the reverse of (6.2.20) is the best possible.

6.2.3. SOME EXAMPLES

In the following examples, we omit to write the assumption of f and g in Theorem 6.2.5 and the conclusions that the constant factors are the best possible.

Example 6.2.6 If $\beta,\lambda>0, k_\lambda(x,y)=\frac{1}{(x^\beta+y^\beta)^{\lambda/\beta}}$, then we find $k_\lambda=\frac{1}{\beta}B(\frac{\lambda}{2\beta},\frac{\lambda}{2\beta})$. By Theorem 6.2.5, (1) for $p>1$, we have the following equivalent inequalities:

$$\int_{R_+^n}\int_{R_+^n}\frac{f(x)g(y)}{(1+\|x\|_\alpha^\beta\|y\|_\alpha^\beta)^{\lambda/\beta}}dxdy < \frac{\Gamma^n(\frac{1}{\alpha})}{\alpha^{n-1}\Gamma(\frac{n}{\alpha})\beta}B(\frac{\lambda}{2\beta},\frac{\lambda}{2\beta})\|f\|_{p,\phi}\|g\|_{q,\psi}, \quad (6.2.23)$$

$$\int_{R_+^n}\|y\|_\alpha^{\frac{p\lambda}{2}-n}[\int_{R_+^n}\frac{f(x)}{(1+\|x\|_\alpha^\beta\|y\|_\alpha^\beta)^{\lambda/\beta}}dx]^p dy < [\frac{\Gamma^n(\frac{1}{\alpha})}{\alpha^{n-1}\Gamma(\frac{n}{\alpha})\beta}B(\frac{\lambda}{2\beta},\frac{\lambda}{2\beta})]^p\|f\|_{p,\phi}^p; \quad (6.2.24)$$

(2) for $0<p<1$, we have the equivalent reverses of (6.2.23) and (6.2.24).

In particular, for $n=1$, (1) if $p>1$, then we have the following equivalent inequalities:

$$\int_{R_+}\int_{R_+}\frac{1}{(1+x^\beta y^\beta)^{\lambda/\beta}}f(x)g(y)dxdy < \frac{1}{\beta}B(\frac{\lambda}{2\beta},\frac{\lambda}{2\beta})\|f\|_{p,\phi_1}\|g\|_{q,\psi_1}, \quad (6.2.25)$$

$$\int_{R_+}y^{\frac{p\lambda}{2}-1}[\int_{R_+}\frac{f(x)}{(1+x^\beta y^\beta)^{\lambda/\beta}}dx]^p dy < [\frac{1}{\beta}B(\frac{\lambda}{2\beta},\frac{\lambda}{2\beta})]^p\|f\|_{p,\phi_1}^p; \quad (6.2.26)$$

(2) if $0<p<1$, then we have the equivalent reverses of (6.2.25) and (6.2.26).

Example 6.2.7 If $\lambda>0, k_\lambda(x,y)=\frac{1}{(\max\{x,y\})^\lambda}$, then we find $k_\lambda=\frac{4}{\lambda}$. By Theorem 6.2.5, (1) for $p>1$, we have the following equivalent inequalities:

$$\int_{R_+^n}\int_{R_+^n}\frac{f(x)g(y)}{(\max\{1,\|x\|_\alpha\|y\|_\alpha\})^\lambda}dxdy < \frac{4\Gamma^n(\frac{1}{\alpha})}{\alpha^{n-1}\Gamma(\frac{n}{\alpha})\lambda}\|f\|_{p,\phi}\|g\|_{q,\psi}, \quad (6.2.27)$$

$$\int_{R_+^n}\|y\|_\alpha^{\frac{p\lambda}{2}-n}[\int_{R_+^n}\frac{f(x)}{(\max\{1,\|x\|_\alpha\|y\|_\alpha\})^\lambda}dx]^p dy < [\frac{4\Gamma^n(\frac{1}{\alpha})}{\alpha^{n-1}\Gamma(\frac{n}{\alpha})\lambda}]^p\|f\|_{p,\phi}^p; \quad (6.2.28)$$

(2) for $0<p<1$, we have the equivalent reverses of (6.2.27) and (6.2.28).

In particular, for $n=1$, (1) if $p>1$, then we have the following equivalent inequalities:

$$\int_{R_+}\int_{R_+}\frac{f(x)g(y)}{(\max\{1,xy\})^\lambda}dxdy < \frac{4}{\lambda}\|f\|_{p,\phi_1}\|g\|_{q,\psi_1}, \quad (6.2.29)$$

$$\int_{R_+}y^{\frac{p\lambda}{2}-1}[\int_{R_+}\frac{f(x)}{(\max\{1,xy\})^\lambda}dx]^p dy < (\frac{4}{\lambda})^p\|f\|_{p,\phi_1}^p; \quad (6.2.30)$$

(2) if $0<p<1$, then we have the equivalent reverses of (6.2.29) and (6.2.30).

Example 6.2.8 If $\lambda>0, k_\lambda(x,y)=\frac{\ln(x/y)}{x^\lambda-y^\lambda}$, then we find $k_\lambda=(\frac{\pi}{\lambda})^2$. By Theorem 6.2.5, (1) for $p>1$, we have the following equivalent inequalities:

$$\int_{R_+^n}\int_{R_+^n}\frac{\ln(\|x\|_\alpha\|y\|_\alpha)}{\|x\|_\alpha^\lambda\|y\|_\alpha^\lambda-1}f(x)g(y)dxdy < \frac{\Gamma^n(\frac{1}{\alpha})}{\alpha^{n-1}\Gamma(\frac{n}{\alpha})}(\frac{\pi}{\lambda})^2\|f\|_{p,\phi}\|g\|_{q,\psi}, \quad (6.2.31)$$

$$\int_{R_+^n}\|y\|_\alpha^{\frac{p\lambda}{2}-n}[\int_{R_+^n}\frac{\ln(\|x\|_\alpha\|y\|_\alpha)}{\|x\|_\alpha^\lambda\|y\|_\alpha^\lambda-1}f(x)dx]^p dy < [\frac{\Gamma^n(\frac{1}{\alpha})}{\alpha^{n-1}\Gamma(\frac{n}{\alpha})}(\frac{\pi}{\lambda})^2]^p\|f\|_{p,\phi}^p; \quad (6.2.32)$$

(2) for $0<p<1$, we have the equivalent reverses of (6.2.31) and (6.2.32).

In particular, for $n=1$, (1) if $p>1$, then we have the following equivalent inequalities:

$$\int_{R_+}\int_{R_+}\frac{\ln(xy)}{x^\lambda y^\lambda-1}f(x)g(y)dxdy < (\frac{\pi}{\lambda})^2\|f\|_{p,\phi_1}\|g\|_{q,\psi_1}, \quad (6.2.33)$$

$$\int_{R_+}y^{\frac{p\lambda}{2}-1}[\int_{R_+}\frac{\ln(xy)}{x^\lambda y^\lambda-1}f(x)dx]^p dy < (\frac{\pi}{\lambda})^{2p}\|f\|_{p,\phi_1}^p; \quad (6.2.34)$$

(2) if $0<p<1$, then we have the equivalent reverses of (6.2.33) and (6.2.34).

Example 6.2.9 If $\lambda>0, k_\lambda(x,y)=\frac{|\ln(x/y)|}{(\max\{x,y\})^\lambda}$, then we find $k_\lambda=\frac{8}{\lambda^2}$. By Theorem 6.2.5, (1) for $p>1$, we have the following equivalent inequalities

$$\int_{R_+^n}\int_{R_+^n}\frac{|\ln(\|x\|_\alpha\|y\|_\alpha)|}{(\max\{1,\|x\|_\alpha\|y\|_\alpha\})^\lambda}f(x)g(y)dxdy < \frac{8\Gamma^n(\frac{1}{\alpha})}{\alpha^{n-1}\Gamma(\frac{n}{\alpha})\lambda^2}\|f\|_{p,\phi}\|g\|_{q,\psi}, \quad (6.2.35)$$

$$\int_{R_+^n}\|y\|_\alpha^{\frac{p\lambda}{2}-n}[\int_{R_+^n}\frac{|\ln(\|x\|_\alpha\|y\|_\alpha)|}{(\max\{1,\|x\|_\alpha\|y\|_\alpha\})^\lambda}f(x)dx]^p dy < [\frac{8\Gamma^n(\frac{1}{\alpha})}{\alpha^{n-1}\Gamma(\frac{n}{\alpha})\lambda^2}]^p\|f\|_{p,\phi}^p; \quad (6.2.36)$$

(2) for $0<p<1$, we have the equivalent reverses of (6.2.35) and (6.2.36).

In particular, for $n=1$, (1) if $p>1$, then we have the following equivalent inequalities:

$$\int_{R_+}\int_{R_+}\frac{|\ln(xy)|}{(\max\{1,xy\})^{\lambda}}f(x)g(y)dxdy$$
$$<\frac{8}{\lambda^2}\|f\|_{p,\phi_1}\|g\|_{q,\psi_1}, \quad (6.2.37)$$
$$\int_{R_+}y^{\frac{p\lambda}{2}-1}\Big[\int_{R_+}\frac{|\ln(xy)|}{(\max\{1,xy\})^{\lambda}}f(x)dx\Big]^p dy$$
$$<\Big(\frac{8}{\lambda^2}\Big)^p\|f\|_{p,\phi_1}^p; \quad (6.2.38)$$

(2) if $0<p<1$, then we have the equivalent reverses of (6.2.37) and (6.2.38).

Example 6.2.10 If $0<\lambda<1, k_\lambda(x,y)=\frac{1}{|x-y|^{\lambda}}$, then

$$k_\lambda=2B\Big(1-\lambda,\frac{\lambda}{2}\Big).$$

By Theorem 6.2.5, (1) for $p>1$, we have the following equivalent inequalities:

$$\int_{R_+^n}\int_{R_+^n}\frac{1}{|1-\|x\|_\alpha\|y\|_\alpha|^{\lambda}}f(x)g(y)dxdy$$
$$<\frac{2\Gamma^n(\frac{1}{\alpha})}{\alpha^{n-1}\Gamma(\frac{n}{\alpha})}B\Big(1-\lambda,\frac{\lambda}{2}\Big)\|f\|_{p,\phi}\|g\|_{q,\psi}, \quad (6.2.39)$$
$$\int_{R_+^n}\|y\|_\alpha^{\frac{p\lambda}{2}-n}\Big[\int_{R_+^n}\frac{f(x)}{|1-\|x\|_\alpha\|y\|_\alpha|^{\lambda}}dx\Big]^p dy$$
$$<\Big[\frac{2\Gamma^n(\frac{1}{\alpha})}{\alpha^{n-1}\Gamma(\frac{n}{\alpha})}B\Big(1-\lambda,\frac{\lambda}{2}\Big)\Big]^p\|f\|_{p,\phi}^p; \quad (6.2.40)$$

(2) for $0<p<1$, we have the equivalent reverses of (6.2.39) and (6.2.40).

In particular, for $n=1$, (1) if $p>1$, then we have the following equivalent inequalities:

$$\int_{R_+}\int_{R_+}\frac{1}{|1-xy|^{\lambda}}f(x)g(y)dxdy$$
$$<2B\Big(1-\lambda,\frac{\lambda}{2}\Big)\|f\|_{p,\phi_1}\|g\|_{q,\psi_1}, \quad (6.2.41)$$
$$\int_{R_+}y^{\frac{p\lambda}{2}-1}\Big[\int_{R_+}\frac{f(x)}{|1-xy|^{\lambda}}dx\Big]^p dy$$
$$<\Big[2B\Big(1-\lambda,\frac{\lambda}{2}\Big)\Big]^p\|f\|_{p,\phi_1}^p; \quad (6.2.42)$$

(2) if $0<p<1$, then we have the equivalent reverses of (6.2.41) and (6.2.42).

Example 6.2.11 If $\lambda\in$ **R**, $\beta>\max\{0,\frac{\lambda}{2}\}$, $|\eta|<\beta-\frac{\lambda}{2}$,

$$k_\lambda(x,y)=\frac{(\min\{x,y\})^{\beta-\lambda}}{x^\beta+y^\beta},$$

then by Example 2.2.14, we obtain $0<K_\lambda(\frac{\lambda}{2}+\eta)<\infty$ and

$$k_{\lambda,\beta}(2)=2\sum_{k=1}^{\infty}\frac{(-1)^{k-1}}{\beta k-\frac{\lambda}{2}}.$$

By Theorem 6.2.5, (1) for $p>1$, we have the following equivalent inequalities:

$$\int_{R_+^n}\int_{R_+^n}\frac{(\min\{1,\|x\|_\alpha\|y\|_\alpha\})^{\beta-\lambda}}{1+\|x\|_\alpha^\beta\|y\|_\alpha^\beta}f(x)g(y)dxdy$$
$$<\frac{\Gamma^n(\frac{1}{\alpha})}{\alpha^{n-1}\Gamma(\frac{n}{\alpha})}k_{\lambda,\beta}(2)\|f\|_{p,\phi}\|g\|_{q,\psi}, \quad (6.2.43)$$
$$\int_{R_+^n}\|y\|_\alpha^{\frac{p\lambda}{2}-n}\Big[\int_{R_+^n}\frac{(\min\{1,\|x\|_\alpha\|y\|_\alpha\})^{\beta-\lambda}}{1+\|x\|_\alpha^\beta\|y\|_\alpha^\beta}f(x)dx\Big]^p dy$$
$$<\Big[\frac{\Gamma^n(\frac{1}{\alpha})}{\alpha^{n-1}\Gamma(\frac{n}{\alpha})}k_{\lambda,\beta}(2)\Big]^p\|f\|_{p,\phi}^p; \quad (6.2.44)$$

(2) for $0<p<1$, we have the equivalent reverses of (6.2.43) and (6.2.44).

In particular, for $n=1$, (1) if $p>1$, then we have the following equivalent inequalities:

$$\int_{R_+}\int_{R_+}\frac{(\min\{1,xy\})^{\beta-\lambda}}{1+x^\beta y^\beta}f(x)g(y)dxdy$$
$$<k_{\lambda,\beta}(2)\|f\|_{p,\phi_1}\|g\|_{q,\psi_1}, \quad (6.2.45)$$
$$\int_{R_+}y^{\frac{p\lambda}{2}-1}\Big[\int_{R_+}\frac{(\min\{1,xy\})^{\beta-\lambda}}{1+x^\beta y^\beta}f(x)dx\Big]^p dy$$
$$<[k_{\lambda,\beta}(2)]^p\|f\|_{p,\phi_1}^p; \quad (6.2.46)$$

(2) if $0<p<1$, then we have the equivalent reverses of (6.2.45) and (6.2.46).

Example 6.2.12 If $\lambda\in$ **R**, $\beta>\frac{\lambda}{2}, |\eta|<\beta-\frac{\lambda}{2}$,

$$k_\lambda(x,y)=\frac{(\min\{x,y\})^{\beta-\lambda}}{(\max\{x,y\})^{\beta}},$$

then by Example 2.2.15, we obtain $0<K_\lambda(\frac{\lambda}{2}+\eta)<\infty$ and $k_\lambda=\frac{4}{2\beta-\lambda}$. By Theorem 6.2.5, (1) for $p>1$, we have the following equivalent inequalities:

$$\int_{R_+^n}\int_{R_+^n}\frac{(\min\{1,\|x\|_\alpha\|y\|_\alpha\})^{\beta-\lambda}}{(\max\{1,\|x\|_\alpha\|y\|_\alpha\})^{\beta}}f(x)g(y)dxdy$$
$$<\frac{4\Gamma^n(\frac{1}{\alpha})}{\alpha^{n-1}(2\beta-\lambda)\Gamma(\frac{n}{\alpha})}\|f\|_{p,\phi}\|g\|_{q,\psi}, \quad (6.2.47)$$
$$\int_{R_+^n}\|y\|_\alpha^{\frac{p\lambda}{2}-n}\Big[\int_{R_+^n}\frac{(\min\{1,\|x\|_\alpha\|y\|_\alpha\})^{\beta-\lambda}}{(\max\{1,\|x\|_\alpha\|y\|_\alpha\})^{\beta}}f(x)dx\Big]^p dy$$
$$<\Big[\frac{4\Gamma^n(\frac{1}{\alpha})}{\alpha^{n-1}(2\beta-\lambda)\Gamma(\frac{n}{\alpha})}\Big]^p\|f\|_{p,\phi}^p; \quad (6.2.48)$$

(2) for $0<p<1$, we have the equivalent reverses of (6.2.47) and (6.2.48).

In particular, for $n=1$, (1) if $p>1$, then we have the following equivalent inequalities:

$$\int_{R_+}\int_{R_+}\frac{(\min\{1,xy\})^{\beta-\lambda}}{(\max\{1,x\|y\})^{\beta}}f(x)g(y)dxdy$$
$$<\frac{4}{2\beta-\lambda}\|f\|_{p,\phi_1}\|g\|_{q,\psi_1}, \quad (6.2.49)$$

$$\int_{R_+} y^{\frac{p\lambda}{s}-1}[\int_{R_+} \frac{(\min\{1,xy\})^{\beta-\lambda}}{(\max\{1,xy\})^{\beta}} f(x)dx]^p dy$$
$$< (\tfrac{4}{2\beta-\lambda})^p \| f \|_{p,\phi_1}^p ; \qquad (6.2.50)$$

(2) if $0<p<1$, then we have the equivalent reverses of (6.2.49) and (6.2.50).

Example 6.2.13 If $\lambda<2$, $\frac{\lambda}{2}<\beta<1$, $|\eta|<\beta-\frac{\lambda}{2}$,

$$k_\lambda(x,y)=\frac{(\min\{x,y\})^{\beta-\lambda}}{|x-y|^\beta},$$

then by Example 2.2.16, we obtain $0<K_\lambda(\frac{\lambda}{2}+\eta)<\infty$ and

$$k_\lambda = 2B(1-\beta,\beta-\tfrac{\lambda}{2}).$$

By Theorem 6.2.5, (1) for $p>1$, we have the following equivalent inequalities:

$$\int_{R_+^n}\int_{R_+^n} \frac{(\min\{1,\|x\|_\alpha\|y\|_\alpha\})^{\beta-\lambda}}{|1-\|x\|_\alpha\|y\|_\alpha|^\beta} f(x)g(y)dxdy$$
$$< \frac{2\Gamma^n(\frac{1}{\alpha})}{\alpha^{n-1}\Gamma(\frac{n}{\alpha})} B(1-\beta,\beta-\tfrac{\lambda}{2}) \| f \|_{p,\phi} \| g \|_{q,\psi}, \qquad (6.2.51)$$
$$\int_{R_+^n} \| y \|_\alpha^{\frac{p\lambda}{2}-n} [\int_{R_+^n} \frac{(\min\{1,\|x\|_\alpha\|y\|_\alpha\})^{\beta-\lambda}}{|1-\|x\|_\alpha\|y\|_\alpha|^\beta} f(x)dx]^p dy$$
$$< [\frac{2\Gamma^n(\frac{1}{\alpha})}{\alpha^{n-1}\Gamma(\frac{n}{\alpha})} B(1-\beta,\beta-\tfrac{\lambda}{2})]^p \| f \|_{p,\phi}^p; \qquad (6.2.52)$$

(2) for $0<p<1$, we have the equivalent reverses of (6.2.51) and (6.2.52).

In particular, for $n=1$, (1) if $p>1$, then we have the following equivalent inequalities:

$$\int_{R_+}\int_{R_+} \frac{(\min\{1,xy\})^{\beta-\lambda}}{|1-xy|^\beta} f(x)g(y)dxdy$$
$$< 2B(1-\beta,\beta-\tfrac{\lambda}{2}) \| f \|_{p,\phi_1} \| g \|_{q,\psi_1}, \qquad (6.2.53)$$
$$\int_{R_+} y^{\frac{p\lambda}{2}-1}[\int_{R_+} \frac{(\min\{1,xy\})^{\beta-\lambda}}{|1-xy|^\beta} f(x)dx]^p dy$$
$$< [2B(1-\beta,\beta-\tfrac{\lambda}{2})]^p \| f \|_{p,\phi_1}^p ; \qquad (6.2.54)$$

(2) if $0<p<1$, then we have the equivalent reverses of (6.2.53) and (6.2.54).

Example 6.2.14 If $\lambda<2$, $\frac{\lambda}{2}<\beta<1$, $|\eta|<\beta-\frac{\lambda}{2}$,

$$k_\lambda(x,y)=\frac{(\min\{x,y\})^{\beta-\lambda}}{(x+y)^\beta},$$

then we obtain $0<K_\lambda(\frac{\lambda}{2}+\eta)<\infty$ and

$$k_\lambda = 2\int_0^1 \frac{u^{\beta-\lambda}}{(1+u)^\beta} u^{\frac{\lambda}{2}-1} du = 2\sum_{k=0}^{\infty} \binom{-\beta}{k} \frac{1}{k+\beta-\frac{\lambda}{2}}.$$

By Theorem 6.2.5, (1) for $p>1$, we have the following equivalent inequalities:

$$\int_{R_+^n}\int_{R_+^n} \frac{(\min\{1,\|x\|_\alpha\|y\|_\alpha\})^{\beta-\lambda}}{(1+\|x\|_\alpha\|y\|_\alpha)^\beta} f(x)g(y)dxdy$$
$$< \frac{2\Gamma^n(\frac{1}{\alpha})}{\alpha^{n-1}\Gamma(\frac{n}{\alpha})} k_\lambda \| f \|_{p,\phi} \| g \|_{q,\psi}, \qquad (6.2.55)$$
$$\int_{R_+^n} \| y \|_\alpha^{\frac{p\lambda}{2}-n} [\int_{R_+^n} \frac{(\min\{1,\|x\|_\alpha\|y\|_\alpha\})^{\beta-\lambda}}{(1+\|x\|_\alpha\|y\|_\alpha)^\beta} f(x)dx]^p dy$$
$$< [\frac{2\Gamma^n(\frac{1}{\alpha})}{\alpha^{n-1}\Gamma(\frac{n}{\alpha})} k_\lambda]^p \| f \|_{p,\phi}^p ; \qquad (6.2.56)$$

(2) for $0<p<1$, we have the equivalent reverses of (6.2.55) and (6.2.56).

In particular, for $n=1$, (1) if $p>1$, then we have the following equivalent inequalities:

$$\int_{R_+}\int_{R_+} \frac{(\min\{1,xy\})^{\beta-\lambda}}{(1+xy)^\beta} f(x)g(y)dxdy$$
$$< k_\lambda \| f \|_{p,\phi_1} \| g \|_{q,\psi_1}, \qquad (6.2.57)$$
$$\int_{R_+} y^{\frac{p\lambda}{2}-1}[\int_{R_+} \frac{(\min\{1,xy\})^{\beta-\lambda}}{(1+xy)^\beta} f(x)dx]^p dy$$
$$< k_\lambda^p \| f \|_{p,\phi_1}^p ; \qquad (6.2.58)$$

(2) if $0<p<1$, then we have the equivalent reverses of (6.2.57) and (6.2.58).

6.2.4. TWO EQUIVALENT HILBERT-TYPE INTEGRAL INEQUALITIES WITH VARIABLES AS PARAMETERS AND THE REVERSES

Theorem 6.2.15 Suppose that $p>0(p\neq 1)$, $\lambda\in\mathbf{R}$, $k_\lambda(x,y)(\geq 0)$ is a homogeneous function of $-\lambda$ -degree in $(0,\infty)\times(0,\infty)$, such that

$$0<k_\lambda=\int_0^\infty k_\lambda(1,u)u^{\frac{\lambda}{2}-1}du<\infty,$$

$u(x)$ and $v(x)$ are strict increasing deliverable functions in (a,b) with $u(a^+)=v(a^+)=0$, $u(b^-)=v(b^-)=\infty$. If $f(x),g(x)\geq 0$, such that

$$0<\int_a^b \frac{(u(x))^{p(1-\frac{\lambda}{2})-1}}{(u'(x))^{p-1}} f^p(x)dx<\infty$$

an $0<\int_a^b \frac{(v(x))^{q(1-\frac{\lambda}{2})-1}}{(v'(x))^{q-1}} g^q(x)dx<\infty$, then (1) for $p>1$, we have the following equivalent inequalities:

$$\int_a^b\int_a^b k_\lambda(1,u(x)v(y))f(x)g(y)dxdy$$
$$<k_\lambda\{\int_a^b \frac{(u(x))^{p(1-\frac{\lambda}{2})-1}}{(u'(x))^{p-1}} f^p(x)dx\}^{\frac{1}{p}}$$
$$\times\{\int_a^b \frac{(v(x))^{q(1-\frac{\lambda}{2})-1}}{(v'(x))^{q-1}} g^q(x)dx\}^{\frac{1}{q}}, \qquad (6.2.59)$$
$$\int_a^b \frac{v'(y)}{(v(y))^{1-(p\lambda)/2}}[\int_a^b k_\lambda(1,u(x)v(y))f(x)dx]^p dy$$
$$<k_\lambda^p \int_a^b \frac{(u(x))^{p(1-\frac{\lambda}{2})-1}}{(u'(x))^{p-1}} f^p(x)dx, \qquad (6.2.60)$$

where the constant factors k_λ and k_λ^p are all the best possible; (2) for $0<p<1$, we have the equivalent reverse forms of (6.2.59) and (6.2.60) with the best constant factors.

Proof (1) For $p>1$, replacing x and y respectively by $u(x)$ $v(y)$ in (6.2.21) and (6.2.22), by simplifications, then replacing $u'(x)f(u(x))$ and $v'(y)g(v(y))$ respectively by $f(x)$, $g(y)$, we find (6.2.59) and (6.2.60). It is obvious that inequalities (6.2.21) and (6.2.59) are equivalent; so are (6.2.22) and (6.2.60). Hence inequalities (6.2.59) and (6.2.60) are equivalent. We can prove that the constant factors in (6.2.59) and (6.2.60) are the best possible by using the equivalent relationship of them. (2) for $0<p<1$, by using the above way in the reverses of (6.2.21) and (6.2.22), we can obtain the equivalent reverses of (6.2.59) and (6.2.60) with the best constant factors.

Note 6.2.16 For some results of particular $k_\lambda(x,y)$, please see (Yang SJM 2007) [3].

Example 6.2.17 If $u(x)=x^\alpha, v(x)=x^\beta$ $(\alpha,\beta>0; x\in(0,\infty))$, then for $p>1$ in (6.2.59) and (6.2.60), we have the following equivalent inequalities:

$$\int_0^\infty\int_0^\infty k_\lambda(1,x^\alpha y^\beta)f(x)g(y)dxdy$$
$$<\frac{k_\lambda}{\alpha^{1/q}\beta^{1/p}}\{\int_0^\infty x^{p(1-\frac{\alpha\lambda}{2})-1}f^p(x)dx\}^{\frac{1}{p}}$$
$$\times\{\int_0^\infty x^{q(1-\frac{\beta\lambda}{2})-1}g^q(x)dx\}^{\frac{1}{q}}, \qquad (6.2.61)$$

$$\int_0^\infty y^{\frac{p\beta\lambda}{2}-1}[\int_0^\infty k_\lambda(1,x^\alpha y^\beta)f(x)dx]^p dy$$
$$<(\frac{k_\lambda}{\alpha^{1/q}\beta^{1/p}})^p\int_0^\infty x^{p(1-\frac{\alpha\lambda}{2})-1}f^p(x)dx; \qquad (6.2.62)$$

for $0<p<1$, we have the reverses of (6.2.61) and (6.2.62).

In particular, setting

$$k_\lambda(x,y)=\frac{1}{(x+y)^\lambda}\quad(\lambda>0),$$

for $p>1$, we have the following equivalent inequalities (Yang JMAA 2006) [4]:

$$\int_0^\infty\int_0^\infty \frac{1}{(1+x^\alpha y^\beta)^\lambda}f(x)g(y)dxdy$$
$$<\frac{B(\lambda/2,\lambda/2)}{\alpha^{1/q}\beta^{1/p}}\{\int_0^\infty x^{p(1-\frac{\alpha\lambda}{2})-1}f^p(x)dx\}^{\frac{1}{p}}$$
$$\times\{\int_0^\infty x^{q(1-\frac{\beta\lambda}{2})-1}g^q(x)dx\}^{\frac{1}{q}}, \qquad (6.2.63)$$

$$\int_0^\infty y^{\frac{p\beta\lambda}{2}-1}[\int_0^\infty \frac{1}{(1+x^\alpha y^\beta)^\lambda}f(x)dx]^p dy$$
$$<[\frac{B(\lambda/2,\lambda/2)}{\alpha^{1/q}\beta^{1/p}}]^p\int_0^\infty x^{p(1-\frac{\alpha\lambda}{2})-1}f^p(x)dx; \qquad (6.2.64)$$

for $0<p<1$, we have the equivalent reverses of (6.2.63) and (6.2.64).

Example 6.2.18 If $u(x)=v(x)=\ln x$ $(x\in(1,\infty))$, then for $p>1$ in (6.2.59) and (6.2.60), we have the equivalent inequalities:

$$\int_1^\infty\int_1^\infty k_\lambda(1,\ln x\ln y)f(x)g(y)dxdy$$
$$<k_\lambda\ \{\int_1^\infty x^{p-1}(\ln x)^{p(1-\frac{\lambda}{2})-1}f^p(x)dx\}^{\frac{1}{p}}$$
$$\times\{\int_1^\infty x^{q-1}(\ln x)^{q(1-\frac{\lambda}{2})-1}g^q(x)dx\}^{\frac{1}{q}}, \qquad (6.2.65)$$

$$\int_1^\infty \frac{(\ln y)^{(p\lambda/2)-1}}{y}[\int_1^\infty k_\lambda(1,\ln x\ln y)f(x)dx]^p dy$$
$$<k_\lambda^p\int_1^\infty x^{p-1}(\ln x)^{p(1-\frac{\lambda}{2})-1}f^p(x)dx; \qquad (6.2.66)$$

for $0<p<1$, we have the equivalent reverses of (6.2.65) and (6.2.66).

In particular, setting

$$k_\lambda(x,y)=\frac{1}{(x+y)^\lambda}(\lambda>0),$$

for $p>1$, we have the following equivalent inequalities:

$$\int_1^\infty\int_1^\infty \frac{1}{(1+\ln x\ln y)^\lambda}f(x)g(y)dxdy$$
$$<B(\tfrac{\lambda}{2},\tfrac{\lambda}{2})\ \{\int_1^\infty x^{p-1}(\ln x)^{p(1-\frac{\lambda}{2})-1}f^p(x)dx\}^{\frac{1}{p}}$$
$$\times\{\int_1^\infty x^{q-1}(\ln x)^{q(1-\frac{\lambda}{2})-1}g^q(x)dx\}^{\frac{1}{q}}, \qquad (6.2.67)$$

$$\int_1^\infty \frac{(\ln y)^{(p\lambda/2)-1}}{y}[\int_1^\infty \frac{f(x)}{(1+\ln x\ln y)^\lambda}dx]^p dy$$
$$<(B(\tfrac{\lambda}{2},\tfrac{\lambda}{2}))^p\int_0^\infty x^{p-1}(\ln x)^{p(1-\frac{\lambda}{2})-1}f^p(x)dx; \qquad (6.2.68)$$

for $0<p<1$, we have the equivalent reverses of (6.2.67) and (6.2.68).

Example 6.2.19 If $u(x)=v(x)=e^x$, $x\in(-\infty,\infty)$, then for $p>1$ in (6.2.59) and (6.2.60), we have the following equivalent inequalities:

$$\int_{-\infty}^\infty\int_{-\infty}^\infty k_\lambda(1,e^{x+y})f(x)g(y)dxdy$$
$$<k_\lambda(\int_{-\infty}^\infty e^{-\frac{p\lambda}{2}x}f^p(x)dx)^{\frac{1}{p}}(\int_{-\infty}^\infty e^{-\frac{q\lambda}{2}x}g^q(x)dx)^{\frac{1}{q}}, \qquad (6.2.69)$$

$$\int_{-\infty}^{\infty} e^{\frac{p\lambda}{2}y}[\int_{-\infty}^{\infty} k_\lambda(1,e^{x+y})f(x)dx]^p dy$$

$$< k_\lambda^p \int_{-\infty}^{\infty} e^{-\frac{p\lambda}{2}x} f^p(x)dx; \qquad (6.2.70)$$

for $0<p<1$, we have the reverses of (6.2.69) and (6.2.70).

Example 6.2.20 If $u(x)=v(x)=\tan x$ $(x\in(0,\frac{\pi}{2}))$, then for $p>1$ in (6.2.59) and (6.2.60), we have the following equivalent inequalities:

$$\int_0^{\frac{\pi}{2}}\int_0^{\frac{\pi}{2}} k_\lambda(1,\tan x\tan y)f(x)g(y)dxdy$$

$$< k_\lambda\{\int_0^{\frac{\pi}{2}} \frac{(\tan x)^{p(1-\frac{\lambda}{2})-1}}{(\sec x)^{2(p-1)}} f^p(x)dx\}^{\frac{1}{p}}$$

$$\times\{\int_0^{\frac{\pi}{2}} \frac{(\tan x)^{q(1-\frac{\lambda}{2})-1}}{(\sec x)^{2(q-1)}} g^q(x)dx\}^{\frac{1}{q}}, \qquad (6.2.71)$$

$$\int_0^{\frac{\pi}{2}} \frac{\sec^2 y}{(\tan y)^{1-(p\lambda)/2}}[\int_0^{\frac{\pi}{2}} k_\lambda(1,\tan x\tan y)f(x)dx]^p dy$$

$$< k_\lambda^p \int_0^{\frac{\pi}{2}} \frac{(\tan x)^{p(1-\frac{\lambda}{2})-1}}{(\sec x)^{2(p-1)}} f^p(x)dx; \qquad (6.2.72)$$

for $0<p<1$, we have the equivalent reverses of (6.2.71) and (6.2.72).

Example 6.2.21 If $u(x)=v(x)=\sec x-1$ $(x\in(0,\frac{\pi}{2}))$, then for $p>1$ in (6.2.59) and (6.2.60), we have the following equivalent inequalities:

$$\int_0^{\frac{\pi}{2}}\int_0^{\frac{\pi}{2}} k_\lambda(1,(\sec x-1)(\sec y-1))$$

$$\times f(x)g(y)dxdy$$

$$< k_\lambda\{\int_0^{\frac{\pi}{2}} \frac{(\sec x-1)^{p(1-\lambda/2)-1}}{(\sec x\tan x)^{p-1}} f^p(x)dx\}^{\frac{1}{p}}$$

$$\times\{\int_0^{\frac{\pi}{2}} \frac{(\sec x-1)^{q(1-\lambda/2)-1}}{(\sec x\tan x)^{q-1}} g^q(x)dx\}^{\frac{1}{q}}, \qquad (6.2.73)$$

$$\int_0^{\frac{\pi}{2}} \frac{\sec y\tan y}{(\sec y-1)^{1-(p\lambda)/2}}[\int_0^{\frac{\pi}{2}} k_\lambda(1,(\sec x-1)(\sec y-1))$$

$$\times f(x)dx]^p dy < k_\lambda^p \int_0^{\frac{\pi}{2}} \frac{(\sec x-1)^{p(1-\lambda/2)-1}}{(\sec x\tan x)^{p-1}} f^p(x)dx; \qquad (6.2.74)$$

for $0<p<1$, we have the equivalent reverses of (6.2.73) and (6.2.74).

6.3. SOME MULTIVARIABLE HARDY-TYPE INTEGRAL INEQUALITIES WITH THE NON-HOMOGENEOUS KERNEL

6.3.1. A COROLLARY

By Theorem 6.2.5, we have

Corollary 6.3.1 Suppose that (p,q) is a pair of conjugate exponents with $p>0(p\neq 1)$, $\lambda\in\mathbf{R}$, $\tilde{k}_\lambda(x,y)$ is a homogeneous function $-\lambda$ -degree,

$$k_\lambda(x,y)=\begin{cases}\tilde{k}_\lambda(x,y), & 0<y\le x,\\ 0, & y>x,\end{cases}$$

and

$$\tilde{H}_\lambda(x) := \int_0^1 \tilde{k}_\lambda(1,u)u^{x-1}du$$

is a positive number in a neighborhood I_λ of $\frac{\lambda}{2}$,

$$\tilde{k}_\lambda := \int_0^1 \tilde{k}_\lambda(1,u)u^{\frac{\lambda}{2}-1}du.$$

If $\phi(x)=\|x\|_\alpha^{p(n-\frac{\lambda}{2})-n}$,

$$\psi(x)=\|x\|_\alpha^{q(n-\frac{\lambda}{2})-n}, x\in R_+^n,$$

$f(\ge 0)\in L_\phi^p(R_+^n)$, $g(\ge 0)\in L_\psi^q(R_+^n)$, such that

$$\|f\|_{p,\phi}=\{\int_{R_+^n}\|x\|_\alpha^{p(n-\frac{\lambda}{2})-n} f^p(x)dx\}^{\frac{1}{p}}>0,$$

$$\|g\|_{q,\psi}=\{\int_{R_+^n}\|x\|_\alpha^{q(n-\frac{\lambda}{2})-n} g^q(x)dx\}^{\frac{1}{q}}>0,$$

then (1) for $p>1$, we have the following equivalent inequalities:

$$\int_{R_+^n}[\int_{\{y\in R_+^n;\|y\|_\alpha\le\|x\|_\alpha\}} \tilde{k}_\lambda(1,\|x\|_\alpha\|y\|_\alpha)$$

$$\times g(y)dy]f(x)dx$$

$$=\int_{R_+^n}[\int_{\{x\in R_+^n;\|x\|_\alpha\ge\|y\|_\alpha\}} \tilde{k}_\lambda(1,\|x\|_\alpha\|y\|_\alpha)$$

$$\times f(x)dx]g(y)dy$$

$$<\frac{\Gamma^n(\frac{1}{\alpha})\tilde{k}_\lambda}{\alpha^{n-1}\Gamma(\frac{n}{\alpha})}\|f\|_{p,\phi}\|g\|_{q,\psi}, \qquad (6.3.1)$$

$$\int_{R_+^n}\|y\|_\alpha^{\frac{p\lambda}{2}-n}[\int_{\{x\in R_+^n;\|x\|_\alpha\ge\|y\|_\alpha\}} k_\lambda(1,\|x\|_\alpha\|y\|_\alpha)$$

$$\times f(x)dx]^p dy < [\frac{\Gamma^n(\frac{1}{\alpha})\tilde{k}_\lambda}{\alpha^{n-1}\Gamma(\frac{n}{\alpha})}]^p\|f\|_{p,\phi}^p, \qquad (6.3.2)$$

$$\int_{R_+^n}\|x\|_\alpha^{\frac{q\lambda}{2}-n}[\int_{\{y\in R_+^n;\|y\|_\alpha\le\|x\|_\alpha\}} \tilde{k}_\lambda(1,\|x\|_\alpha\|y\|_\alpha)$$

$$\times g(y)dy]^q dx < [\frac{\Gamma^n(\frac{1}{\alpha})\tilde{k}_\lambda}{\alpha^{n-1}\Gamma(\frac{n}{\alpha})}]^q\|g\|_{q,\psi}^q, \qquad (6.3.3)$$

where the constant factors are the best possible; (2) for $0<p<1$, we have the equivalent reverses of (6.3.1), (6.3.2) and (6.3.3) with the best constant factors (but the reverse of (6.3.3) keeps the same form).

In particular, for $n=1$, if $p>1$, then we have the following equivalent inequalities:

$$\int_0^\infty[\int_0^x \tilde{k}_\lambda(1,xy)g(y)dy]f(x)dx$$

$$= \int_0^\infty [\int_y^\infty \tilde{k}_\lambda(1,xy) f(x)dx] g(y)dy$$

$$< \tilde{k}_\lambda \| f \|_{p,\phi_1} \| g \|_{q,\psi_1}, \qquad (6.3.4)$$

$$\int_0^\infty y^{\frac{p\lambda}{2}-1} [\int_y^\infty k_\lambda(1,xy) f(x)dx]^p dy$$

$$< \tilde{k}_\lambda^{\,p} \| f \|_{p,\phi_1}^p, \qquad (6.3.5)$$

$$\int_0^\infty x^{\frac{q\lambda}{2}-1} [\int_0^x \tilde{k}_\lambda(1,xy) g(y)dy]^q dx$$

$$< \tilde{k}_\lambda^{\,q} \| g \|_{q,\psi_1}^q, \qquad (6.3.6)$$

where the constant factors are the best possible; if $0 < p < 1$, then we have the equivalent reverses of (6.3.4), (6.3.5) and (6.3.6) with the best constant factors (but the reverse of (6.3.6) keeps the same form).

6.3.2. SOME EXAMPLES

In the following examples, the conditions

$$\phi(x) = \| x \|_\alpha^{p(n-\frac{\lambda}{2})-n},$$

$$\psi(x) = \| x \|_\alpha^{q(n-\frac{\lambda}{2})-n}, x \in R_+^n,$$

$$f(\geq 0) \in L_\phi^p(R_+^n),\ g(\geq 0) \in L_\psi^q(R_+^n),$$

$$\| f \|_{p,\phi} = \{ \int_{R_+^n} \| x \|_\alpha^{p(n-\frac{\lambda}{2})-n} f^p(x)dx \}^{\frac{1}{p}} > 0,$$

$$\| g \|_{q,\psi} = \{ \int_{R_+^n} \| x \|_\alpha^{q(n-\frac{\lambda}{2})-n} g^q(x)dx \}^{\frac{1}{q}} > 0$$

(for $n = 1$, $\phi_1(x) = \| x \|_\alpha^{p(1-\frac{\lambda}{2})-1}$, $\psi_1(x) = \| x \|_\alpha^{q(1-\frac{\lambda}{2})-1}$, $x \in R_+$), and the conclusions that the constant factors are the best possible are omitted.

Example 6.3.2 If $\beta > -1, \lambda > 0,$

$$\tilde{k}_\lambda(x,y) = \frac{|\ln(x/y)|^\beta}{(\max\{x,y\})^\lambda},$$

then by Example 2.2.25, we obtain

$$\tilde{k}_\lambda = \Gamma(\beta+1)(\tfrac{2}{\lambda})^{\beta+1}.$$

By Corollary 6.3.1, (1) for $p > 1$, we have the following equivalent inequalities:

$$\int_{R_+^n} [\int_{\{y \in R_+^n; \|y\|_\alpha \leq \|x\|_\alpha\}} \frac{|\ln(\|x\|_\alpha \|y\|_\alpha)|^\beta}{(\max\{1, \|x\|_\alpha \|y\|_\alpha\})^\lambda}$$

$$\times g(y)dy] f(x)dx$$

$$= \int_{R_+^n} [\int_{\{x \in R_+^n; \|x\|_\alpha \geq \|y\|_\alpha\}} \frac{|\ln(\|x\|_\alpha \|y\|_\alpha)|^\beta}{(\max\{1, \|x\|_\alpha \|y\|_\alpha\})^\lambda}$$

$$\times f(x)dx] g(y)dy$$

$$< \frac{\Gamma^n(\frac{1}{\alpha})}{\alpha^{n-1}\Gamma(\frac{n}{\alpha})} \Gamma(\beta+1)(\tfrac{2}{\lambda})^{\beta+1} \| f \|_{p,\phi} \| g \|_{q,\psi}, \qquad (6.3.7)$$

$$\int_{R_+^n} \| y \|_\alpha^{\frac{p\lambda}{2}-n} [\int_{\{x \in R_+^n; \|x\|_\alpha \geq \|y\|_\alpha\}} \frac{|\ln(\|x\|_\alpha \|y\|_\alpha)|^\beta}{(\max\{1, \|x\|_\alpha \|y\|_\alpha\})^\lambda}$$

$$\times f(x)dx]^p dy$$

$$< [\frac{\Gamma^n(\frac{1}{\alpha})}{\alpha^{n-1}\Gamma(\frac{n}{\alpha})} \Gamma(\beta+1)(\tfrac{2}{\lambda})^{\beta+1}]^p \| f \|_{p,\phi}^p, \qquad (6.3.8)$$

$$\int_{R_+^n} \| x \|_\alpha^{\frac{q\lambda}{2}-n} [\int_{\{y \in R_+^n; \|y\|_\alpha \leq \|x\|_\alpha\}} \frac{|\ln(\|x\|_\alpha \|y\|_\alpha)|^\beta}{(\max\{1, \|x\|_\alpha \|y\|_\alpha\})^\lambda}$$

$$\times g(y)dy]^q dx$$

$$< [\frac{\Gamma^n(\frac{1}{\alpha})}{\alpha^{n-1}\Gamma(\frac{n}{\alpha})} \Gamma(\beta+1)(\tfrac{2}{\lambda})^{\beta+1}]^q \| g \|_{q,\psi}^q; \qquad (6.3.9)$$

(2) for $0 < p < 1$, we have the equivalent reverses of (6.3.7), (6.3.8) and (6.3.9) (but the reverse of (6.3.9) keeps the same form).

In particular, for $n = 1$, if $p > 1$, then we have the following equivalent inequalities:

$$\int_{R_+} [\int_{\{y \in R_+; 0 < y \leq x\}} \frac{|\ln(xy)|^\beta}{(\max\{1,xy\})^\lambda} g(y)dy] f(x)dx$$

$$= \int_{R_+} [\int_{\{x \in R_+; x \geq y\}} \frac{|\ln(xy)|^\beta}{(\max\{1,xy\})^\lambda} f(x)dx] g(y)dy$$

$$< \Gamma(\beta+1)(\tfrac{2}{\lambda})^{\beta+1} \| f \|_{p,\phi_1} \| g \|_{q,\psi_1}, \qquad (6.3.10)$$

$$\int_{R_+} y^{\frac{p\lambda}{2}-1} [\int_{\{x \in R_+; x \geq y\}} \frac{|\ln(xy)|^\beta}{(\max\{1,xy\})^\lambda} f(x)dx]^p dy$$

$$< [\Gamma(\beta+1)(\tfrac{2}{\lambda})^{\beta+1}]^p \| f \|_{p,\phi_1}^p, \qquad (6.3.11)$$

$$\int_{R_+} x^{\frac{q\lambda}{2}-1} [\int_{\{y \in R_+; 0 < y \leq x\}} \frac{|\ln(xy)|^\beta}{(\max\{1,xy\})^\lambda} g(y)dy]^q dx$$

$$< [\Gamma(\beta+1)(\tfrac{2}{\lambda})^{\beta+1}]^q \| g \|_{q,\psi_1}^q; \qquad (6.3.12)$$

if $0 < p < 1$, then we have the equivalent reverses of (6.3.10), (6.3.11) and (6.3.12) (but the reverse of (6.3.12) keeps the same form).

Example 6.3.3 If $\lambda \in \mathbf{R}, \frac{\lambda}{2} < \beta < 1,$

$$\tilde{k}_\lambda(x,y) = \frac{(\min\{x,y\})^{\beta-\lambda}}{|x-y|^\beta},$$

then by Example 2.2.26, we obtain

$$\tilde{k}_\lambda = B(1-\beta, \beta - \tfrac{\lambda}{2}).$$

By Corollary 6.3.1, (1) for $p > 1$, we have the following equivalent inequalities:

$$\int_{R_+^n} [\int_{\{y \in R_+^n; \|y\|_\alpha \leq \|x\|_\alpha\}} \frac{(\min\{1, \|x\|_\alpha \|y\|_\alpha\})^{\beta-\lambda}}{|1-\|x\|_\alpha \|y\|_\alpha|^\beta}$$

$$\times g(y)dy] f(x)dx$$

$$= \int_{R_+^n} [\int_{\{x \in R_+^n; \|x\|_\alpha \geq \|y\|_\alpha\}} \frac{(\min\{1, \|x\|_\alpha \|y\|_\alpha\})^{\beta-\lambda}}{|1-\|x\|_\alpha \|y\|_\alpha|^\beta}$$

$$\times f(x)dx] g(y)dy$$

$$< \frac{\Gamma^n(\frac{1}{\alpha})}{\alpha^{n-1}\Gamma(\frac{n}{\alpha})} B(1-\beta, \beta-\tfrac{\lambda}{2}) \| f \|_{p,\phi} \| g \|_{q,\psi}, \qquad (6.3.13)$$

$$\int_{R_+^n} \| y \|_\alpha^{\frac{p\lambda}{2}-n} [\int_{\{x\in R_+^n;\|x\|_\alpha \ge \|y\|_\alpha\}} \frac{(\min\{1, \|x\|_\alpha \|y\|_\alpha\})^{\beta-\lambda}}{|1-\|x\|_\alpha \|y\|_\alpha|^\beta} \times f(x)dx]^p dy$$
$$< [\frac{\Gamma^n(\frac{1}{\alpha})}{\alpha^{n-1}\Gamma(\frac{n}{\alpha})} B(1-\beta, \beta-\tfrac{\lambda}{2})]^p \| f \|_{p,\phi}^p, \quad (6.3.14)$$

$$\int_{R_+^n} \| x \|^{\frac{q\lambda}{2}-n} [\int_{\{y\in R_+^n;\|y\|_\alpha \le \|x\|_\alpha\}} \frac{(\min\{1, \|x\|_\alpha \|y\|_\alpha\})^{\beta-\lambda}}{|1-\|x\|_\alpha \|y\|_\alpha|^\beta} \times g(y)dy]^q dx$$
$$< [\frac{\Gamma^n(\frac{1}{\alpha})}{\alpha^{n-1}\Gamma(\frac{n}{\alpha})} B(1-\beta, \beta-\tfrac{\lambda}{2})]^q \| g \|_{q,\psi}^q; \quad (6.3.15)$$

(2) for $0 < p < 1$, we have the equivalent reverses of (6.3.13), (6.3.14) and (6.3.15) (but the reverse form of (6.3.15) keeps the same).

In particular, for $n = 1$, if $p > 1$, then we have the following equivalent inequalities:

$$\int_0^\infty [\int_0^x \frac{(\min\{1,xy\})^{\beta-\lambda}}{|1-xy|^\beta} g(y)dy] f(x)dx = \int_0^\infty [\int_y^\infty \frac{(\min\{1,xy\})^{\beta-\lambda}}{|1-xy|^\beta} f(x)dx] g(y)dy$$
$$< \frac{\Gamma^n(\frac{1}{\alpha})}{\alpha^{n-1}\Gamma(\frac{n}{\alpha})} B(1-\beta, \beta-\tfrac{\lambda}{2}) \| f \|_{p,\phi_1} \| g \|_{q,\psi_1}, \quad (6.3.16)$$

$$\int_0^\infty y^{\frac{p\lambda}{2}-1} [\int_y^\infty \frac{(\min\{1,xy\})^{\beta-\lambda}}{|1-xy|^\beta} f(x)dx]^p dy < [\frac{\Gamma^n(\frac{1}{\alpha})}{\alpha^{n-1}\Gamma(\frac{n}{\alpha})} B(1-\beta, \beta-\tfrac{\lambda}{2})]^p \| f \|_{p,\phi_1}^p, \quad (6.3.17)$$

$$\int_0^\infty x^{\frac{q\lambda}{2}-1} [\int_0^x \frac{(\min\{1,xy\})^{\beta-\lambda}}{|1-xy|^\beta} g(y)dy]^q dx < [\frac{\Gamma^n(\frac{1}{\alpha})}{\alpha^{n-1}\Gamma(\frac{n}{\alpha})} B(1-\beta, \beta-\tfrac{\lambda}{2})]^q \| g \|_{q,\psi_1}^q; \quad (6.3.18)$$

if $0 < p < 1$, then we have the equivalent reverses of (6.3.16), (6.3.17) and (6.3.18) (but the reverse form of (6.3.18) keeps the same).

Example 6.3.4 If $\lambda \in \mathbf{R}$, $\beta > \frac{\lambda}{2}$,

$$\tilde{k}_\lambda(x,y) = \frac{(\min\{x,y\})^{\beta-\lambda}}{(x+y)^\beta},$$

then we obtain

$$\tilde{k}_\lambda = \sum_{k=0}^\infty \binom{-\beta}{k} \frac{1}{k+\beta-\frac{\lambda}{2}}.$$

By Corollary 6.3.1, (1) for $p > 1$, we have the following equivalent inequalities:

$$\int_{R_+^n} [\int_{\{y\in R_+^n;\|y\|_\alpha \le \|x\|_\alpha\}} \frac{(\min\{1, \|x\|_\alpha \|y\|_\alpha\})^{\beta-\lambda}}{(1+\|x\|_\alpha \|y\|_\alpha)^\beta} \times g(y)dy] f(x)dx$$
$$= \int_{R_+^n} [\int_{\{x\in R_+^n;\|x\|_\alpha \ge \|y\|_\alpha\}} \frac{(\min\{1, \|x\|_\alpha \|y\|_\alpha\})^{\beta-\lambda}}{(1+\|x\|_\alpha \|y\|_\alpha)^\beta} \times f(x)dx] g(y)dy$$
$$< \frac{\Gamma^n(\frac{1}{\alpha})}{\alpha^{n-1}\Gamma(\frac{n}{\alpha})} \tilde{k}_\lambda \| f \|_{p,\phi} \| g \|_{q,\psi}, \quad (6.3.19)$$

$$\int_{R_+^n} \| y \|_\alpha^{\frac{p\lambda}{2}-n} [\int_{\{x\in R_+^n;\|x\|_\alpha \ge \|y\|_\alpha\}} \frac{(\min\{1, \|x\|_\alpha \|y\|_\alpha\})^{\beta-\lambda}}{(1+\|x\|_\alpha \|y\|_\alpha)^\beta} \times f(x)dx]^p dy$$
$$< [\frac{\Gamma^n(\frac{1}{\alpha})}{\alpha^{n-1}\Gamma(\frac{n}{\alpha})} \tilde{k}_\lambda]^p \| f \|_{p,\phi}^p, \quad (6.3.20)$$

$$\int_{R_+^n} \| x \|^{\frac{q\lambda}{2}-n} [\int_{\{y\in R_+^n;\|y\|_\alpha \le \|x\|_\alpha\}} \frac{(\min\{1, \|x\|_\alpha \|y\|_\alpha\})^{\beta-\lambda}}{(1+\|x\|_\alpha \|y\|_\alpha)^\beta} \times g(y)dy]^q dx$$
$$< [\frac{\Gamma^n(\frac{1}{\alpha})}{\alpha^{n-1}\Gamma(\frac{n}{\alpha})} \tilde{k}_\lambda]^q \| g \|_{q,\psi}^q; \quad (6.3.21)$$

(2) for $0 < p < 1$, we have the equivalent reverses of (6.3.19), (6.3.20) and (6.3.21) (but the reverse of (6.3.21) keeps the same form).

In particular, for $n = 1$, if $p > 1$, then we have the following equivalent inequalities:

$$\int_0^\infty [\int_0^x \frac{(\min\{1,xy\})^{\beta-\lambda}}{(1+xy)^\beta} g(y)dy] f(x)dx = \int_0^\infty [\int_y^\infty \frac{(\min\{1,xy\})^{\beta-\lambda}}{(1+xy)^\beta} f(x)dx] g(y)dy$$
$$< \frac{\Gamma^n(\frac{1}{\alpha})}{\alpha^{n-1}\Gamma(\frac{n}{\alpha})} \tilde{k}_\lambda \| f \|_{p,\phi_1} \| g \|_{q,\psi_1}, \quad (6.3.22)$$

$$\int_0^\infty y^{\frac{p\lambda}{2}-1} [\int_y^\infty \frac{(\min\{1,xy\})^{\beta-\lambda}}{(1+xy)^\beta} f(x)dx]^p dy < [\frac{\Gamma^n(\frac{1}{\alpha})}{\alpha^{n-1}\Gamma(\frac{n}{\alpha})} \tilde{k}_\lambda]^p \| f \|_{p,\phi_1}^p, \quad (6.3.23)$$

$$\int_0^\infty x^{\frac{q\lambda}{2}-1} [\int_0^x \frac{(\min\{1,xy\})^{\beta-\lambda}}{(1+xy)^\beta} g(y)dy]^q dx < [\frac{\Gamma^n(\frac{1}{\alpha})}{\alpha^{n-1}\Gamma(\frac{n}{\alpha})} \tilde{k}_\lambda]^q \| g \|_{q,\psi_1}^q; \quad (6.3.24)$$

if $0 < p < 1$, then we have the equivalent reverses of (6.3.22), (6.3.23) and (6.3.24) (but the reverse of (6.3.24) keeps the same form).

6.4. REFERENCES

1. Hong Y. On multiple Hardy-Hilbert integral inequalities with some parameters. Journal of Inequalities and Applications, Vol. 2006, Art. ID 94960: 1-11.
2. Zhong WY, Yang BC. On a multiple Hilbert-type integral inequality with the symmetric kernel. Journal of Inequalities and Applications, Vol. 2007, Art.ID 27962, 1-17.
3. Yang BC. A new Hilbert's type integral inequality. Soochow Journal of Mathematics, 2007;33(4):849-859.
4. Yang BC. On the norm of an integral operator and applications. J. Math. Anal. Appl., 2006, 321: 182-192.

Index

www.ingramcontent.com/pod-product-compliance
Lightning Source LLC
Chambersburg PA
CBHW041716210326
41598CB00007B/677
* 9 7 8 1 6 0 8 0 5 0 3 3 8 *